INTERACTIONS OF BONE WITH ORTHOPEDIC IMPLANTS AND POSSIBLE FAILURES

Orthopedic Implants and Biomaterials Series

The Orthopedic Implants and Biomaterials series offers a distinctive, pragmatic approach to the development and application of orthopedic implants and biomaterials. Volumes in the series are written by developers, engineers, and academic researchers in the fields of orthopedic implants and biomaterials, providing practical and research-led content for industry and academic professionals alike. This series covers current techniques and methods in all design and development areas of orthopedic implants and biomaterials, from design modeling and manufacturing to biological testing and clinical translation. The Orthopedic Implants and Biomaterials series provides a combined practical and research perspective, and thus is relevant to engineers, academics, students, research and development scientists and researchers, implant manufacturers, and biomaterials developers.

About the Series Editor

Amirhossein Goharian is a senior engineer working in the orthopedic implant industry. He completed his masters in mechanical engineering in biomechanics at the University Technology Malaysia in 2012. He joined the Research and Development Department of LEONIX Sdn. Bhd., Penang, Malaysia, in 2012, rising to senior research and development leader in Jan. 2015. In 2017, he joined Isfahan Orthopedic Implant Development Co. as the product development leader until Sep. 2020. Since then, he has been developing innovations in the field of orthopedic implants, based on ideas set out in his published works.

Orthopedic Implants and Biomaterials

INTERACTIONS OF BONE WITH ORTHOPEDIC IMPLANTS AND POSSIBLE FAILURES

Amirhossein Goharian

Independent Engineering Consultant in Orthopaedic Implant and Biomaterial Industry, Isfahan, Iran

Ehsan Golkar

Orthopedic Surgeon and Specialist, Ashrafi Hospital, Isfahan, Iran

ELSEVIER

ACADEMIC PRESS

An imprint of Elsevier

Academic Press is an imprint of Elsevier
125 London Wall, London EC2Y 5AS, United Kingdom
525 B Street, Suite 1650, San Diego, CA 92101, United States
50 Hampshire Street, 5th Floor, Cambridge, MA 02139, United States
The Boulevard, Langford Lane, Kidlington, Oxford OX5 1GB, United Kingdom

ISBN: 978-0-323-95410-5

For information on all Academic Press publications visit our website at https://www.elsevier.com/books-and-journals

Publisher: Mara Conner
Acquisitions Editor: Carrie L. Bolger
Editorial Project Manager: Sara Greco
Production Project Manager: Prem Kumar Kaliamoorthi
Cover Designer: Miles Hitchen

Typeset by TNQ Technologies

Contents

Preface

Bone tissue may be injured through loss, defects, resorption, and so on in interaction with orthopedic implants. Most orthopedic implants are manufactured from metallic materials (e.g., titanium, stainless steel, cobalt-chromium alloy). Considerable mechanical mismatch between bone tissue and metallic implants causes an inadequate action and reaction force between these two parts with different results of hardness and stiffness. Alternatively, the nonbiophysiologic design of implants increases the stiffness of the implant and decreases effective load transference between the bone and implant. Because of the bioactive nature of bone tissue, its morphologic structure actively changes as a result of the pattern of interaction or mechanism with orthopedic implants. Under severe conditions, bone tissue cannot restructure effectively to counteract the inadequate load interaction with implants (even over short-term or long-term implantation), and the gradual or accelerated resorption of bone tissue results. Through this mechanical and biological phenomenon, a bone defect initiates and propagates to a greater volume of bone stock near the bone–implant interface. Through the growth of bone failure interacting with orthopedic implants, the implant may be displaced, subsided to the bone, or even fracture. This necessitates revision surgery to remove the implant, reconstruct the bone loss or defect, and reimplant the new implant. Normally, failed clinical cases are reported in clinical studies, in which no further discussion and exploration of engineering or scientific aspects are provided. On the other hand, biomedical researchers do not incorporate failures and associated or possible reasons in real cases and mostly focus on the fundamentals and scientific aspects.

This book explores mechanical parameters that are effective in the interaction of bone tissue with orthopedic implants, including stress shielding, loosening, stiffness, the design geometry, subsidence, mechanical integrity, mechanical interlocking, load transference, localized stress zones, and design optimization. Based on the mechanical interaction between the bone and implant, some biological effects occur at the interface of the bone and implant, such as bone resorption, bone loss, bone ingrowth, bone ongrowth, osteoconductivity, osteoinductivity, and osseointegration. Each chapter focuses on a specific category of orthopedic implant uses these highlighted technical concepts to explore interactions between the bone and implant and possible failures that might occur based on reported cases in the clinical literature. The intention is to provide a review discussion—based reference in the engineering and clinical fields of orthopedic implants to explore the relations of bone—implant interaction and failures involving the design geometry, surface condition, structural form, material, surface porosity, surface bioactivity, and so forth. Such discussions are provided for multiple types of orthopedic implants in each category (total knee replacement [TKR], total ankle replacement [TAR], total shoulder replacement [TSR], total hip replacement [THR], trauma implants, and spinal implants); therefore, it could be useful for development engineers to modify the design, surface conditions, or structural form of the implant or prosthesis. Also, it would be useful for surgeons to select implants better according to the injury and patient conditions, including the patient's body mass index, the severity of periprosthetic bone loss, the level of muscle deficiency, postoperative planning, the patient's willingness and ability to follow instructions regarding postoperative treatment, and so on.

The book is not a fully fundamental or practical reference, but it is greatly practicable and constructive for many engineers, researchers, and clinicians who are dealing with orthopedic implants, as a reference to evaluate their knowledge and be familiar with other aspects of the matters of their profession. Today, the number of orthopedic surgeons who are designing their concepts into new products is increasing. Moreover, the number of engineers and researchers who intend to become familiar with the practical consequences of what they are designing or developing is also increasing. Thus, many startups involve surgeons, researchers, and engineers to develop novel products in the orthopedic implant industry. The aim of this book (and overall, the aim of the series *Orthopedic Implants and Biomaterials*) is to have a knowledge- and literature-based role for the new and fast-growing generation of orthopedic implant developers with the involvement and engagement of groups from various disciplines of engineering and clinical and scientific audiences.

In this book, Chapter 1 is a fundamental chapter that explores the required concepts and meanings that are reviewed and discussed in the next chapters. Chapters 2—4 focus on bone loss, defects, resorption at the bone and implant interface for joint replacement prostheses, TKR, TAR, TSR, and THR, according to published clinical studies. Interactions of bone with trauma and spinal implants and possible failures are reviewed and discussed in Chapters 5 and 6, respectively. Overall, *Interactions of Bone with Orthopedic Implants and Possible Failures* is an in-depth review discussion book on practical aspects of the orthopedic implant industry. The authors aim to extend the book's contents in the next editions to include more types of prostheses and implants and possible type of failure. This book is recommended for engineers, developers, researchers, and orthopedic and neurospine surgeons to become familiar with technical reasons for the failure of bone tissue interacting with orthopedic implants. All relevant and interested audiences and readers are welcomed to share their notes and comments with the authors via orthoimplant.develop@gmail.com to improve the book in the next editions.

Amirhossein Goharian
Ehsan Golkar
March 2022

Dedication and acknowledgments

Dedicated to the spirit of *my father*

for his efforts to give love, aid, and cure to those who were needy,

and special thanks to him

for teaching me to be kind and supportive to others, to be diligent and creative to achieve what I wish, and to have a strong spirit, high patience, and big hopes.

Dedicated to my merciful and lovely *mother*, and special thanks

for her love and selflessness in my whole life,

for her patience and encouragement,

for teaching me to be righteous, moral, hopeful, and self-confidant.

Dedicated to my lovely and elegant *wife* and special thanks

for her love and encouragement,

for her patience during the writing of this book,

for her great, soothing, and merciful companionship.

Dedicated to the spirit of my lovely *wife's mother* and special thanks

for her love and encouragement

Dedicated to my cute and elegant *daughter*

for her lovely presence in my world and making my life full of hope, passion, and energy.

Special thanks to *Dr. Ehsan Golkar*

for his efforts to review the book manuscript and valuable comments to modify the clinical aspects of the book.

Special thanks to *Prof. Mohamed Ruslan Abdullah*

for his kind and valuable comments to modify the biomechanical engineering aspects of the book.

Special thanks to *Prof. Mohammed Rafiq Abdul Kadir*

for his kind and valuable comments to modify the biomedical engineering aspects of the book.

Special thanks to *Elsevier Publisher*

for all supportive and professional editorial management and the publication of this book.

CHAPTER ONE

General concepts of interactions of bone with orthopedic implants and possible failures

1.1 Introduction

The interaction of the bone and the implant is the main factor in the long-term effectiveness of orthopedic implants in treating bone injuries. With adequate and careful consideration of the clinical aspects of surgical implantation (e.g., an adequate preoperative study of the bone injury, adequate selection of the implant based on the type of and severity of injury, patient conditions, careful surgical planning, the absence of infection at the implant site, adequate closure of the incision, effective postoperative plans, the surgeon's skills), short-term clinical success is achievable in most cases. However, after long-term implantation, bone tissue remodeling at the interface of the bone and implant or near the zone of the interface begins owing to various types of inadequacy in mechanical and biological interactions with the implant. As a bioactive structure, bone does not match the implant, which is a nonactive structure. The mechanism of load transference to bone tissue is significantly affected, which may cause the biological activity of bone formation to fail, and from which the gradual and long-term segregation of the implant and bone results. The lifetime and range of functionality of the implanted zone are increasing whereas the mean age of affected patients who need surgical intervention is decreasing, even to less than age 50 years (many patients have damaged patinas at age 20—40 years). Thus, the advanced design and development of orthopedic implants need to be investigated to increase the mechanical and biological activeness of the implant in interaction with the bone tissue to balance failures in bone remodeling after long-term implantation. This chapter reviews the main mechanical and biological interactions of the bone and implant and discusses their possible failures in general. In addition, interactions of bone with orthopedic implants are elaborated on in general terms to make this chapter become a prerequisite for Chapters two to seven.

Interactions of Bone with Orthopedic Implants and Possible Failures
ISBN: 978-0-323-95410-5
https://doi.org/10.1016/B978-0-323-95410-5.00002-5

1.2 Mechanical treatment of severe bone injuries

Bone is a mechanical structure. It is categorized as a hard tissue in the human body. Its mechanical properties are higher than many materials in nature, even more than synthesized materials such as polymers. Therefore, this tissue has been traditionally thought of as a hard material that needs to be treated with artificial devices in mechanical terms. Bone injuries may be severe and not healed by the physiologic, chemical, or biological activity of associated cells and biological factors at the injured zone. It may require using artificial devices as salvage to maintain the physical structure of the injured bone at a stable position to allow biological activity and healing of the bone. Such a condition is observed in the fixation of bone fractures to stabilize bone fragments in anatomic position and to facilitate bone fusion at fracture gaps. Likewise, to treat spinal cord compression, it is necessary to restabilize spinal vertebrae in an optimum physiologic condition at which decompression of the spinal cord may be achieved. In another type of bone injury treatment, bone cartilage at the joint has degenerated, and thus replacement surgery (arthroplasty) is required to facilitate physiologic motion conditions at the joint. Replacement of bone cartilage is managed through artificial components that are implanted through adjacent bones of the injured joint. Bone injury treatment is managed in the form of extensive bone tissue formation between two healthy bone segments, such as may be observed in the fusion treatment of degenerated intervertebral discs. In some cases, it is necessary to implant and replace the artificial device for a large volume of the bone, in which bone tissue has been extensively resorbed and the new generation of bone matrix needs to be structured through the implant (such a condition might occur as a result of bone cancer, severe infection, or another biological disorder that might progress in a portion of bone tissue). Using artificial devices or components to treat severe bone injuries is inevitable. This practice has been carried out for more than 100 years. Internal implantation of artificial components, implants, and prostheses through bone tissue or in direct contact with the bone matrix is highly challenging for long-term function and performance in interaction with bone.

1.2.1 Mechanical strength of implants

Physiologic loading conditions that are introduced or transferred to bone tissue in all parts of the skeletal body are considerable. In some application, they may increase to several thousands of newtons (several tonnage). In addition, when the geometrical and molecular integrity of bone tissue is affected owing to injury, mechanical equilibrium at the injured zone is significantly compromised and the level or extent of loading conditions sharply increases. Therefore, the device that is implanted in the bone needs to have high mechanical strength to bear such increased loading conditions. The incorporation of metallic materials such as stainless steel, titanium alloys, and cobalt-chromium alloys in the manufacture of orthopedic implants to treat bone injuries may be discussed from this aspect. Transferred loading conditions to the bone—implant construct are normally generated in the form of compressive force, bending moment, torsion torque, and shear force. Compressive force is applied as the result of body organs or tissue weight in various parts. Bending moment is generated owing to the existence of the effective arm or the distance from the applied point or area of the compressive force relative to the effective point or area at the interface of the bone and implant (Fig. 1.1). The bending moment might be generated as the result of static force or dynamic motion.

Because of the physiologic motion of human limbs or organs around the axial axis of the body (craniocaudal axis), torsion torque is created, which has an effect on the bone—implant construct as a main type of loading condition that may cause failure. The final effective physiologic load is shear force

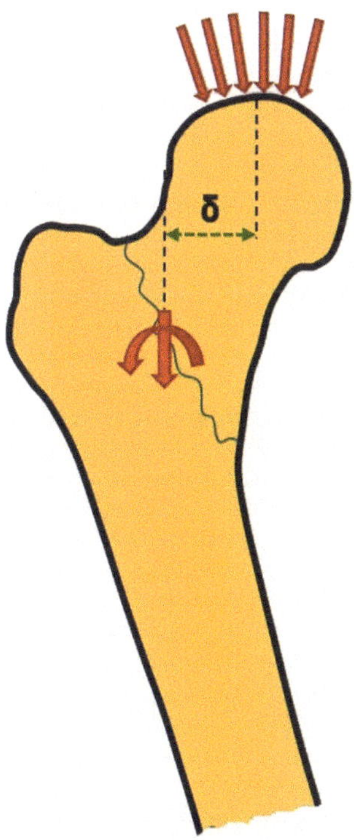

Figure 1.1 Compressive force at the hip joint (top arrows) and its influence in creating bending moment at the femoral neck (bottom arrows). δ is the arm of bending moment, which is multiplied by the force and quantifies the moment value.

generated at the bone–implant interface. The presence of relative motion to a small or large extent causes shear force to the interlayer between the bone and implant. Compressive, bending, or torsion stress in conjunction with motion of the body organ at the implanted zone is converted to shear stress, which affects the bonding or adhesion strength between the bone and implant. The major portion of these loading conditions at the injured site or zone is transferred to the implant with different patterns of distribution and localized zones. Higher mechanical properties of implant materials and the stiffness of the implant were believed to be the traditional strategy to secure the function and performance of the implant, even at localized zones at which the stress value sharply increased. Evaluation of effective stress to localized or critical zones of the implant under physiologic conditions is greatly complicated. Even a determination of concentrated stress zones is not easily predictable. Thus, high safety factors are considered in the design and manufacture of orthopedic implants, in which the stiffness of the implant is higher than what is ideally required.

1.2.2 Stiffness of implant

As highlighted in Section 1.2.1, the stiffness of the implant is increased to bear a large portion of transferred loading conditions to the implant because of the high value of safety factors in the design and manufacture of the implant. The stiffness of the implant is a parameter that needs to be optimized during the design and development of the orthopedic implants. Many geometric, material, and engineering parameters influence stiffness as the main parameter for mechanical interaction between the bone and implant. A consideration of the geometry, material, and the manufacturing processes in combination determines the stiffness of the implant. The mechanical properties of the implant's raw material is the first parameter that is effective in implant stiffness. The ductility and brittleness of materials and their physical bonding at the molecular scale form the inherent mechanical properties of materials against various types of mechanical loading conditions (e.g., compressive, tensile, bending, torsion, shear). The manufacturing process that is used influences molecular bonding strength and increases or decreases the material's inherent mechanical modulus while increasing the ductility and brittleness. The chemical and physical bonding of materials at the atomic scale are greatly effective in enhancing the mechanical properties. Metallic materials have a high bonding strength at the atomic scale that is enhanced by alloying or incorporating other elements through atomic structuring. They have a high level of ductility that is increased through sequential rolling, and which is decreased through casting with no further hardening. Ceramic materials are inherently brittle owing to the development of these materials from metal-oxide molecules. These types of materials may have a higher level of mechanical modulus, but this is lower for metallic materials. The other category of materials is polymeric materials, which are structured based on atomic bonding between carbon atoms with other carbon atoms or gas elements. With alterations in the processing parameters, conditions, and types of elements, the type of polymer and the mechanical modulus are altered. Normally, the mechanical modulus of most polymers is much lower than that of metallic and ceramic materials, whereas it has a high extent of ductility characteristics. Metallic, ceramic, or polymeric materials are composed in the various forms of powder, granules, or melts to achieve new materials with combined mechanical properties.

The other main factor in implant stiffness is the geometry of the implant. The implant geometry may be simple to complex. The implant may include holes, internal empty space, low-thickness zones, grooves, a curved profile in various planes, and other geometric features for an anatomic shape or profile of the bone at the intended zone. The implant design has a significant effect on the flow of stress through the implant under various types of physiologic conditions. In some sections, the stress density increases, and thus stress is concentrated or localized at that section. Effective and reliable design and development are carried out based on computational and experimental biomechanical evaluation methods in which physiologic loading and boundary conditions transferred to the bone–implant construct are simulated in vitro to study the effective strength of the implant. Evaluation techniques are carried out with simplifications for effective loading and boundary conditions to perform comparative studies between newly designed implants and those that were previously developed. However, judge the safety and performance of implants reliably for treating bone injuries, biomechanical evaluation condition need to be as close as possible to physiologic conditions in the intended zone. It is advantageous to consider the effect of local soft tissues such as ligaments, tendons, and muscles to establish the reliable loading equilibrium for a biomechanical evaluation of the bone–implant construct. The use of cadaver bones and high-efficient testing machines and computational software are other crucial factors that have a significant effect on the reliability of results. A consideration of chemical corrosion of the implant at the localized zone is beneficial to enhance the accuracy of results in the human body environment. When preclinical evaluation analyses and testing are managed effectively, the design optimization of the implant geometry is enhanced. As a result, implant stiffness is not insufficiently high or low.

The other influential factor for implant stiffness is the manufacturing process of the implant. Implants are normally fabricated by a main manufacturing process such as machining, casting, forging, molding, or three-dimensional printing, followed by several complementary postprocesses (eg, heat treatment, stress relief, anodizing, coating) to enhance the various types of physical, chemical, and biologic properties of the implant in the human body environment. The manufacturing process of the implant has a minor or significant influence on the inherent mechanical properties of the implant. For instance, hot forging increases the mechanical properties of the implant at the molecular scale and increases the implant's stiffness. Likewise, postprocessing methods have a significant effect on enhancing molecular bonding at bulk or surface sections of the implant that enhance or reduce stiffness. In some cases, a pos-processing method such as annealing is used to reduce the stiffness of the fabricated implant, such as annealing of an implant manufactured by hot forging.

1.2.3 Mechanical integrity of bone and implant

Mechanical integrity of the implant and bone is a key factor that is a main strategy for implanting artificial devices or components in bone. The bone—implant interface has types. One is the implant (e.g., screw) purchased through the bone, from which the implant has internal contact surfaces with both cortical and trabecular bone. The other type of implant is positioned on the bone (e.g., trauma plate or intervertebral fusion cage), which may have external contact surfaces with either cortical or trabecular bone. Normally, the combination of these types of bone—implant interfaces is required for the effective integrity of the implant and bone (Fig. 1.2). However, internal integrity is better mechanically than external integrity (external mechanical integrity is achievable by creating specific geometric features on the implant surface, such as grooves, dents, or teeth with sharp edges to allow relative subsidence of the implant through the bone; however, these might not be as good as internal integrity). Because of the high compactness of cortical bone with a mechanical modulus of 8—20 GPa, purchase of screw-type components was observed with a high level of stability and mechanical integration. Cortical bone as a semisolid bone

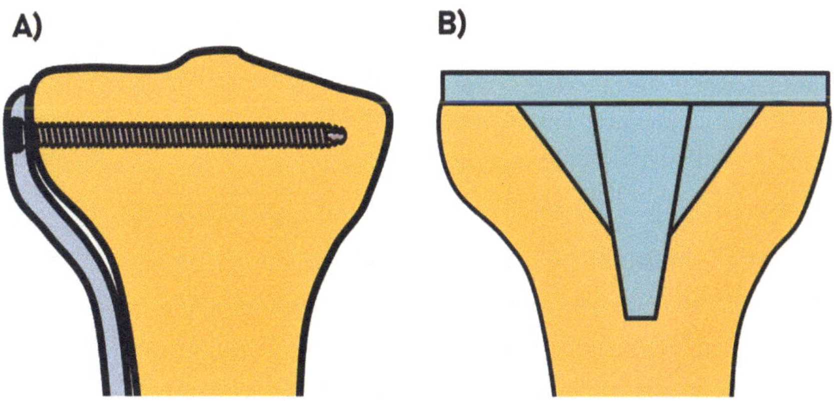

Figure 1.2 Illustration of external and internal interface of implant with bone tissue. (A) The screw has a large internal interface with bone tissue, whereas the plate has an external interface. (B) The tibial tray component has an external interface with the bone at the tray portion and an internal interface at the stem portion.

(with a porosity less than 20%) facilitates the secure fixation of various types of implants with threads or grooves on the contact surfaces with cortical bone.

The implant might be designed with an extended screw-shaped feature from its main body. Likewise, the implant might be fixed to the bone by screws with long threads through screw holes on the implant or prosthesis. The integrity of the implant with cortical bone even for a short length is crucial, particularly for early stabilization. However, many orthopedic implants are in contact with trabecular or cancellous bone, which are not as compact as cortical bone (with a porosity of greater than 70% and a mechanical modulus of less than 2 GPa). The optimal threaded characteristics of the implant at the interface with trabecular bone are different from those cortical bone. A higher contact surface of the implant at the threaded portion with trabecular bone is required. Thus, the threaded depth and pitch might be specifically designed to enhance bone–implant integrity at the trabecular zone. The level of loading conditions and bone tissue compactness have a direct influence on the effectiveness of bone–implant mechanical integrity. The implant might be designed for various types of bone quality and loading conditions. The mechanical integrity of the bone and implant at the contact interface is affected by the generated shear stress. The mechanical integrity acts as a mechanical barrier to prevent sliding or dissociation of the implant from the bone, and vice versa. This mechanical interaction between the bone and implant is the main factor that quantifies the stability of the bone–implant construct as a whole. When the mechanical integrity of the bone–implant at the interfaces is stable, the whole construct has better performance and safety under various physiologic loading conditions. Most failures occur at the interface of the bone and implant after the deterioration of mechanical integrity. This will be reviewed in general concepts in Section 1.6 and discussed in detail for various types of orthopedic implants in Chapters two to seven.

1.2.4 Load transferring between implant and bone

Load transference is an interesting and challenging concept in the design and development of orthopedic implants. Physiologic loading conditions are transferred through the bone–implant construct as the result of the effective weight and movement of body organs at the implanted zone. Loading conditions are transferred from the bone to the implant and from the implant to other parts of bone. The implant system acts as a mechanical bridge between bone fragments in one limb or bone tissues in two different limbs. The load transferring mechanism is influenced by the extent of mechanical mismatch between components. An excessive difference in the mechanical stiffness causes ineffective load transference between hard and soft components. Bone tissue normally interacts with the softer material, with respect to metallic implants as the harder material. Transferred energy to the metallic component does not have a significant amount of elastic strain owing to the large extent of mechanical modulus. If the geometry of the implant is not optimized to reduce implant stiffness, the implant maintains high rigidity against transferred loading conditions from bone. Therefore, the reaction force to bone tissue from the implant is extremely high because of the lack of absorption (or negligible energy absorption) of energy by the implant. What occurs in the bone as the composite material with a higher level of flexibility compared with the metallic implant is the absorption of energy from the point or area at which the load is applied until the interface of the bone and metallic implant. The flow of energy absorbance through the bone structure abounds at its interface with the implant. Thus, the reaction energy applies an increased level of energy to the peri-implant bone tissue at a close distance to the implant surface (Fig. 1.3). By the cyclic action and reaction of such a mechanism, bone tissue is overloaded.

Load transference between the implant and bone might be different at the internal or external interfaces. At the internal interface, the risk for osteolysis and bone remodeling is greater than when the implant and bone are in contact through their external surfaces. When the action and reaction load are transferred through the implant from external surfaces, overall bone tissue as a solid component interacts with the implant and thus has a high level of stiffness against loading conditions. Conversely, when loading conditions are transferred from the internal zone of bone, local bone tissue around the implant interface are mechanically engaged

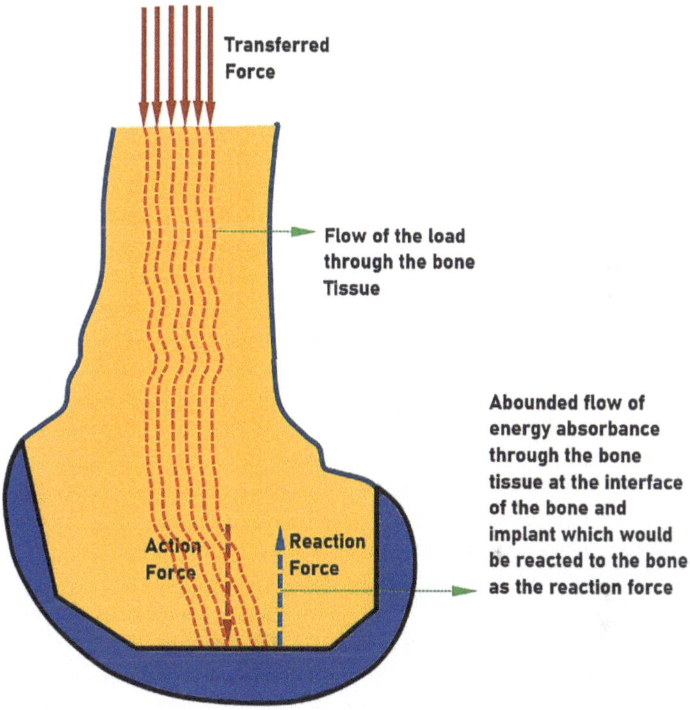

Figure 1.3 Schematic illustration of the load transference between bone and implant. The transferred load flows through the composite and porous structure of the bone tissue in either the cortical or trabecular bone sections. The transferred energy is absorbed at various levels of the bone tissue. When the load flow is reached to the implant at the bone–implant surface, because of the high mechanical modulus of the metallic implants, the transferred energy does not generate elastic strain in an effective range that can absorb energy. Therefore, a large of amount of energy reacts to bone tissue, which affects bone remodeling at the bone–implant interface and increases bone resorption at this zone, which is detected as a radiolucent line or zone on radiographic scans. The anatomy of the bone and design of the implant are schematics and do not show actual internal bone morphology or architecture at the distal portion of the femur. This figure is meant to deliver the concept of the generation of a reaction force at the bone–implant interface.

to bear action or reaction loads from the implant. This load transference causes different failures, which are reviewed in Section 1.3. Generation of shear stress at bone—implant interfaces, either internal or external interfaces, might be destructive to bone. Such a transverse nature of loading conditions greatly affects the bone—implant interface. To a large extent, the presence of this type of mechanical stress at the bone—implant interface introduces an larger load than is required to stimulate bone formation and causes overloading conditions to the bone matrix. Gradually, through cyclic loading of shear stress, peri-implant bone tissue is resorbed and bone remodeling progresses to a greater distance of the bone matrix from the bone—implant interface (Fig. 1.4).

1.2.5 Stress shielding of implant

Stress shielding has been used extensively in the literature to express the reason for bone—implant failure at the contact interfaces or severe bone

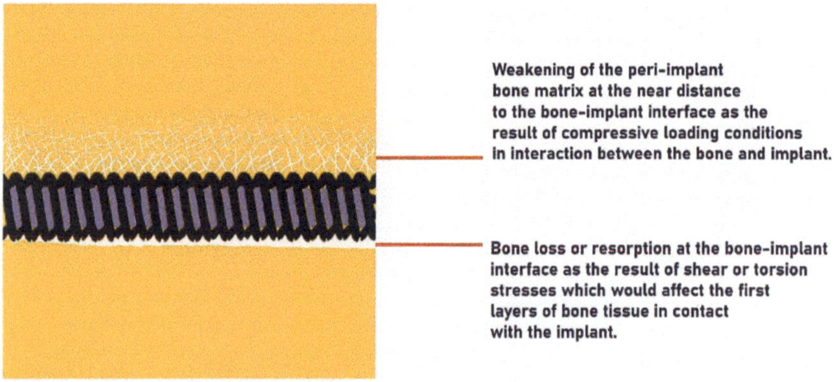

Figure 1.4 Illustration of bone tissue failure in interaction with the implant. Various types of loading conditions might be transferred at the bone—implant interface, mainly as compressive, torsion, and shear loads. The compressive load may affect the bone matrix near the bone—implant interface, whereas the shear or torsion load would cause bone resorption at the first layers of bone tissue on contact surfaces with bone. The shear load is generated as a result of the tendency toward relative transverse motion between the implant and bone. When shear stress reaches shear strength at the bone tissue at the interface, overloading conditions occur and bone starts to be remodeled by resorption. Ultimately, implant loosening takes place. Overloading conditions as a result of the compressive load do not affect the first layers of the interface. However, they weaken the bone matrix gradually. The progression of such failure in bone tissue result in bone fracture or excessive bone loss from the interface through the distance layers of bone tissue.

loss at bone tissue distant from the implant. This effect is a main mechanical interaction between the bone and implant. Stress shielding occurs when sufficient loading conditions are not transferred through bone tissue at the bone—implant interface. Under healthy or normal physiologic conditions of bone tissue, whole portions of bone as an integrated component are engaged to bear applied loading conditions. However, when the bone injury is treated with an artificial component such as an implant, bone is no longer integrated in nature and mechanically depends on the implant. Therefore, the implant does not transfer the load to all portions of the bone tissue, and in some bone sections or zones, stress is shielded. If the load direction is parallel to the normal axis of the bone—implant contact surface (Fig. 1.5), bone tissue bears the transferred load from the implant. The implant has contact surfaces different from those of the implant, and all surface normal axes are not parallel to those of the load direction, which causes a stress-shielding effect or phenomenon at the interfaces. In some cases, the implant may cause a lack of transference of the load to a large zone of the bone tissue (e.g., trochanter zone at the lateral aspect of the proximal femur in total hip replacement through the proximal third zone of the femur). This causes severe stress shielding at a large zone that might progress in the form of catastrophic failure (reviewed in Section 1.3). Finite element analysis of the bone—implant construct under feasible and reliable physiologic and boundary loading conditions demonstrates areas with a lower amount of transferred load compared with other parts of bone tissue. The stress shielding effect needs to be studied at the bone—implant interface and in areas included in unloaded or less-loaded areas after implantation at the injured zone.

1.2.6 Ideal mechanical interaction between bone and implant

The bone and implant might as one integrated structure or construct with similar mechanical functions and performance against physiologic loading conditions. In fact, this might not be achievable in the current design concept of orthopedic implants on the market. Current implants have mechanical behavior significantly different from that of bone tissue as a smart and active mechanical structure. Bone tissue can remodel itself structurally for better performance with respect to applied physiologic loading conditions. Bone morphology, structure, and compactness are different among the human body skeletal systems. On the other hand, implants are manufactured to be rigid and cannot remodel themselves at all to be strengthened or

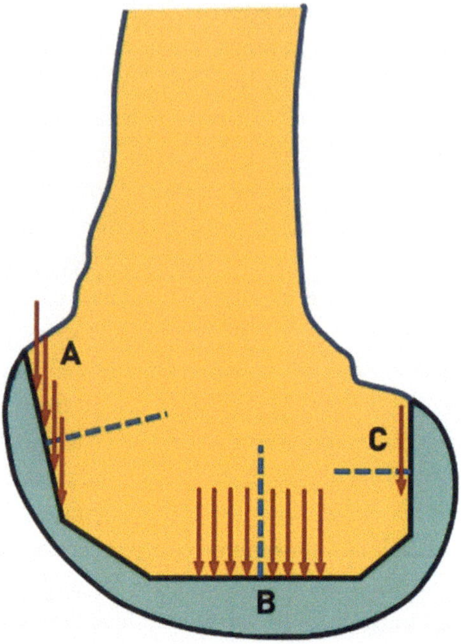

Figure 1.5 Illustration of contact surface normal direction (*blue dashed line*) compared with loading direction (*red arrows*). Parallel directions of surface normal and loading direction enable the presence of load to the bone tissue to generate mechanical signals as the physical factor for inducing bone formation. Increased loading conditions at the interface cause overloading conditions, from which bone resorption and the generation of fibrous tissue at the bone–implant interface result. Conversely, a nonparallel direction reduces transfer loading to the bone matrix and stress shielding occurs at the bone–implant interface, from which the biological activity of osteoclast cells increases and bone resorption accelerates.

weakened smartly with respect to applied loading conditions. Overall, ideal mechanical interactions of bone and implant are achievable if the mechanical performance of the implant is close to or similar to that of bone tissue. The ideal implant is remodeled at localized stress zones to be strengthened to reduce the concentration and level of stress at that zone. In some cases, weakening of the implant at the localized stress zone is helpful to reduce the rigidity of the implant at that zone and thus increase of the elastic strain and mechanical stimulation signals to bone tissue. The implant has the ability to heal itself when undesirable loading conditions are transferred to the implant in static or cyclic form and generate microcracks internally or externally at surface areas. The concept of self-healing materials was introduced in the literature for use in various industries such as aerospace, construction, and medical devices. They would be helpful to enhance implant self-activity in interacting with bone tissue. This concept of the ideal mechanical interaction of bone and implant might not be achievable in the near future; however, it could be investigated for future revolutions in design and development of orthopedic implants. The ideal biological integration of bones and implants is hypothesized in Section 1.4.5, and the overall conclusion on this topic is presented in Section 1.6.

1.3 Mechanical failure of bone–implant construct

In Section 1.2, the mechanical interaction of the bone and implant is expressed in terms of mechanical strength, stiffness, load transference, mechanical integrity, stress shielding, and ideal conditions. Ineffectively achieving the optimal conditions of bone–implant interaction under physiologic loading conditions initiates or propagates failure at the interface of the bone and implant and the bone–implant construct as a whole. Various types of failure occur and decrease bone–implant effectiveness, function, or performance in short- or long-term implantation. Because of this, clinical outcomes become unsuccessful and revision surgeries become necessary once failure conditions reach catastrophic conditions. Failure affects the bone tissue, implant, or both. The implant might be cracked or fractured at the failure section, whereas the bone tissue might be remodeled or even fractured. Failure of bone tissue is observed in different modes of thickening, increased compactness, decreased compactness, severe bone loss, cracking, or fracture. Traditionally, orthopedic implants are designed and developed with respect to the surgical plan, which might not be mechanically adequate in short- or long-term implantation at the injured zone. Orthopedic or neurospine surgeons may provide design concepts for orthopedic implants owing to their prominent knowledge regarding the whole anatomy of hard and soft tissues in conjunction with arteries and neural tissues at the affected zone. They plan how to implant the artificial device into bone tissue, although they may not have a detailed understanding of mechanical issues that might result from the design. On the other hand, when engineers receive the design concepts from surgeons, they design the mechanism or general idea of the implant in computer-aided design modeling software, and then the drafted model is completed and verified by its implantation through artificial or cadaver human bones. Engineers may not have a sufficient understanding of the clinical concepts of the design concepts. Thus, they may not be able to judge the effective mechanical response of the implant adequately after implantation. This is why time after time, the design concepts of orthopedic implants are altered and modified as a result of clinical complications that might have been achieved in long-term clinical studies. Various methods of biomechanical evaluation were carried out for a better examination of the mechanical behavior of the implant after implantation or replacement. The following subsections discuss the failure of the bone—implant construct in view of mechanical factors.

1.3.1 Implant fracture or cracking

The implant might crack or fracture under increased static or long-term cyclic loading conditions. Such a failure might occur on the implant from various types of metallic, ceramic, and polymeric materials. Increased localized stress at the section or zone causes an increase in the risk for fracture, particularly in rigid fixation of the implant to the bone with a high level of stiffness. A bone—implant construct with localized stress areas may not effectively manage a bone injury using artificial devices. Owing to surgical limitations in implanting the implant to treat the bone injury, the implant might be designed with several inevitable mechanical localized stress which would be converted to a failure point in some circumstances. The fracture might occur as a result of the increased level of the transferred static load to the implant, from which localized stress at the failure point reaches above the inherent mechanical strength of the implant material. In addition, the fracture might progressed at the implant as a result of increased localized stress greater than the fatigue strength of the inherent implant material. The fracture may be seen to be a catastrophic mechanical failure that necessitates removing the implant and managing revision surgery to implant a new implant, probably with a different design, function, and implant conditions compared with the primary failed implant. Postoperative planning after implantation are crucial in some cases to prevent full bearing of physiologic loading conditions to the bone—implant construct, particularly when the bone needs time to be remodeled or formed to complete bone—implant integrity (such a condition could be observed in the fixation of bone fractures or in fusion treatment of a degenerative spinal intervertebral disc).

1.3.2 Undesired rigidity or flexibility of implant

Ineffective rigidity or flexibility of the implant occurs as mechanical failure after implantation of the implant to the bone. As highlighted in Section 1.3.1, excessive rigidity of the implant causes increased stress at maximum or localized zones, and thus fracture of the implant. In addition, excessive rigidity has a significant effect on reducing bone—implant construct elastic strain. Thus, sufficient mechanical stimulation signals are not generated at areas where bone formation is required for healing of the injury (such as bone healing at fracture gaps in the fixation of bone fractures or bony fusion in treating degenerative intervertebral discs). Rigidity of the bone—implant construct results from excessive implant stiffness or the use of unnecessary components in the fixation of the implant structure. For instance, the unnecessary use of locking screws to fix the plating system increases the rigidity of the bone—implant construct, and fracture healing is not achieved. Excessive rigidity of the implant structure increases the risk for implant loosening from bone tissue with compactness that is much lower than under normal conditions such as what are seen in osteoporotic bone. Poor-quality bone needs to be mechanically engaged with the implant structure much more so than healthy bone, and increased rigidity of the implantation structure causes weaker mechanical integrity to the implant with poor bone matrix. Therefore, the rigidity of the implant structure is beneficial for lower interaction with low-quality bone compared with healthy bone with normal bone compactness under similar fixation or stabilization conditions (for instance, for fixing a similar bone fracture pattern, a more flexible plate might be advantageous for poorly compact bone compared with healthy bone). Compared with excessive rigidity, excessive flexibility occurs when treating bone injuries using orthopedic implants. Flexibility is observed owing to the low stiffness of the implant or the use of a lower amount of fixation components in constructing the implant structure (e.g., the use of a low number of locking or interfragmentary screws in bone fracture fixation or a low number of pedicle screws in spinal stabilization in treating spinal injuries). Excessive flexibility generates mechanical strain greater than an effective physiologic range, which may deteriorate the biological activity of bone formation or increase the rate of undesirable bone formation.

1.3.3 Dissociation between implant and bone

Dissociation between the implant and bone is a bone—implant construct failure, which is observed in some cases. Dissociation occurs at the interface of the bone and implant when the implant is externally in contact with bone tissue. In such cases, the implant and bone integrity provide compressive force, and thus no other mechanical fixator fixes the implant to the bone.

For instance, an intervertebral fusion cage is stabilized through a preload in the form of a compressive force at the superior and inferior aspects of the cage during fusion. In replacing the joint prostheses, particularly in the lower extremities, normally bone cement is used to stabilize the implant through the bone while presence of body weight loading conditions in the form of compressive force to keep the bone and implant together. Without introducing a compressive force, the implant is dissociated from the bone and is loosened at its initial implantation position, with failure of bone–implant integrity. This type of failure occurs early after implantation, when the bone and implant do not have effective mechanical interaction to be matched together. Under such conditions, the bone has not been remodeled significantly at the bone–implant interface. The risk for early dissociation between the implant and bone is why patients are advised to have physiotherapy activities with controlled programs for a couple of weeks after implantation to apply partial loading conditions to the bone–implant construct. Through such postoperative planning, the bone and implant maintain engagement until effective bone remodeling is gradually better matched with the implant. The other issue that increases the risk for implant and bone dissociation is the ineffective selection of the implant size in accordance with implantation conditions at the affected zone. A smaller or larger implant initially causes inadequate engagement of the bone and implant at the interfaces. For instance, the height of the selected intervertebral fusion cage may not be matched with the fixed position of the adjacent vertebrae, and thus the preload conditions do not stabilize the cage between the vertebrae effectively, and early cage migration or dislocation occur. As another example, the size of a femoral knee prosthesis might be selected to be larger than the created cutting sections at the distal condyles of the femur, from which early dissociation of the implant and bone might be observed. There may be other reasons for early dissociation of the bone and implant; however, they might not be related to the interaction of the bone with orthopedic implants (such as the ineffective use of surgical instruments, or the use of worn surgical instruments to prepare the implant replacement).

1.3.4 Loosening at bone–implant interface

Loosening between the implant and bone is an important type of failure observed in most bone–implant constructs. Loosening progresses as a result of mechanical mismatch between the bone and implant, from which mechanical integration or engagement of the bone and implant is affected.

As highlighted in Sections 1.3.1 and 1.3.2, the implant fails in the form of cracking, fracture, or plastic deformation and undesirable rigidity or flexibility. These influence the integrity of the bone–implant interface and loosen the implant from the bone. Bone remodeling at bone–implant interfaces is another reason for loosening of the bone from the implant. Ineffective mechanical interaction between the bone and implant as the result of stress shielding or overloading conditions accelerates the biological activity of osteoclast cells to resorb the bone matrix to a weaker or lower compactness structure and reduces bone–implant contact surfaces. Loosening of the bone and implant is detected as radiolucent lines on x-ray or computed tomography scan images at the follow-up visit of patients after implantation. Regular follow-up sessions allow surgeons to see the initiation of failed areas of bone–implant interfaces before they progress to catastrophic conditions, and to manage a proper solution if possible. Therefore, loosening is a process with an initial point that progresses until the complete loosening of the implant and bone. In some implantations, bone–implant contact surfaces are large with different load transferring mechanisms at the interfaces. For instance, the contact surfaces of the knee tibial tray have both external and internal interfaces with the bone tissue, where external interfaces mainly bear a compressive force and internal interfaces mainly bear shear and torsion stresses. In external interfaces, overloading conditions occur and in internal interfaces, a stress shielding effect progresses. Therefore, bone remodeling and radiolucent areas are observed in either external or internal interfaces. If loosening of the implant occurs in either external or internal interfaces, the overall implant may not loosen and may maintain its initial position; however, when the loosening zones occur in one part of the implant, loosening of other parts is accelerated and ultimately catastrophic dislocation or migration of the implant occurs from its initial position. In other cases, the implant may have interfaces with both cortical and trabecular bone tissue, and thus loosening might occur in either cortical or trabecular bone. This scenario is considerably observed in the implantation of pedicle screws through vertebra bone. The bone resorption zone at the interface of the bone and screw occurs at the vertebra body, where the screw shaft is in contact with trabecular bone or at the pedicle canal of the vertebra, where the screw shaft is in contact with the cortical bone. In such cases, loosening may not cause overall loosening of the screw, but it could accelerate the progression of bone resorption through the whole bone–screw interface and dislocation or migration of the screw from its initial stabilized position. It is vital to consider loosening of the implant in terms of postoperative failure in the short or long term periods. It could reduce the success rate of clinical outcomes over the years. Other factors, such as infection at the bone–implant interface, accelerate bone and implant loosening. These are not within the scope of this book and thus are not discussed here. The provided examples of failure in this section and other many cases of failure of orthopedic implants are discussed and illustrated in x-ray scans taken from the real clinical cases in Chapters two to six for various types of trauma, joint, and spinal implants.

1.3.5 Bone suppression at bone–implant interface

Bone suppression is the mechanical failure of the bone–implant construct where the bone is suppressed at internal interfaces of the bone–implant. When the implant is purchased or replaced inside bone tissue, the bone might be damaged at interfaces with the implant and dislocate the bone fragment relative to its initial position to the implant. Such a type of failure is mostly created in the interaction of the bone and screws, particularly in long-length contact surfaces. As the result of stress shielding or overloading conditions at the bone–screw interface and the progression of bone resorption, the suppression of bone tissue occurs before screw loosening, particularly in osteoporotic bone. Overall stiffness of the bone matrix at the trabecular portion of long bones such as proximal femur or vertebra bone, might not be sufficiently high to bear the reacted loading conditions from metallic screws, and severe resorption of the bone matrix could occur; thus, the whole bone tissue or fragment may collapse. Bone suppression failure does not occur in most surgical treatments of bone injuries. However, it is observed at a high rate in the fracture management of the femoral head and neck, fracture management of the humeral head, fracture management of the proximal tibia, and so on. In such cases, bone suppression is also recognized as the screw cutout phenomenon in the clinical literature (Fig. 1.6).

1.3.6 Bone fracture at bone–implant interface

Severe bone resorption at the bone–implant interface and its progression through the larger zone of bone tissue causes bone fracture at the affected zone. Under such conditions, overall stiffness of the bone structure at the affected zone is significantly reduced and an increase in mechanical strain accelerates bone resorption in a greater zone until the bone matrix does not have sufficient mechanical strength against physiologic loading conditions and fractures. The fracture of peri-implant bone occurs in cases with

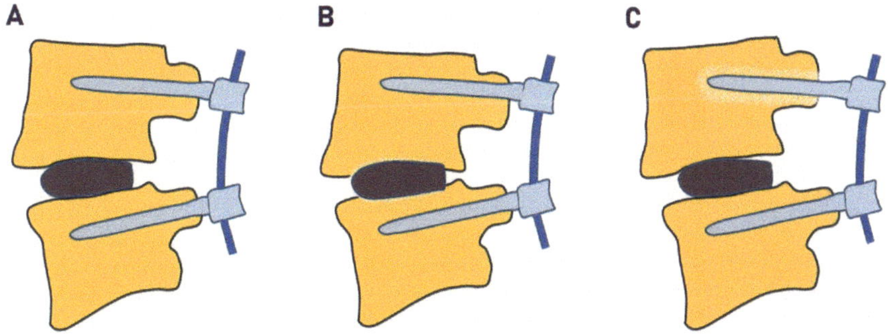

Figure 1.6 Illustration of bone suppression and subsidence in interaction of bone with implant. (A) Successful bone–implant construct in spinal stabilization of lumbar vertebrae using intervertebral fusion cages between vertebrae. (B) Occurrence of bone resorption at external interface of bone–implant between fusion cage and vertebral body, which may cause subsidence of the cage through adjacent vertebrae at the inferior and superior surfaces. (C) Occurrence of bone suppression at internal interface of bone–implant between the pedicle screw and vertebral body, which may cause dislocation of the vertebral body over the screw. The figures are illustrated schematically; the bone anatomy and illustration of the failures may not show actual conditions.

a small volume of the bone width near the bone–implant interface, such as fracture of the acetabular wall after replacement of the acetabular cup to the acetabulum, or fracture of the femur after replacement of the femoral stem to the proximal third of the femur.

1.3.7 Subsidence at bone–implant interface

The implant might penetrate the bone tissue at external interfaces with the bone, which is known as subsidence failure. Similar mechanical interaction conditions in bone suppression at internal interfaces progress during subsidence. Subsidence results from increased compressive loading between the implant and bone or a gradual reduction in bone stiffness near the implant (Fig. 1.5). (Bone stiffness is not only bone resorption at the bone–implant interface; it includes a reduction in bone compactness at the distance of the implant.) Excessive compressive force between the bone and implant results in subsidence failure owing to the mismatch of mechanical stiffness. Because of bone remodeling under excessive loading conditions, the implant subsides through the bone when the implant is made of metallic materials (e.g., titanium alloy) or polymeric materials (e.g., polyetheretherketone [PEEK]). If the implant withstands increased compressive loading between the bone and implant, the risk for implant subsidence increases (Fig. 1.6); otherwise, implant failure occurs. Subsidence progresses gradually or accidentally. Gradual subsidence is reported in patients with a high body mass index (BMI), which may be altered from a couple of weeks to a couple of years in accordance with the BMI range for obesity or overweight. Thus, subsidence has a higher rate in patients with an affected zone in the lower extremities, spinal column, cervical column, or any other zones that might be affected by the weight of body organs at the bone—implant construct.

1.4 Biological interaction of bone and implant

Biological interactions of the implant and bone have not been documented and investigated as much as have mechanical interactions. Biological interactions of an artificial component in the human body environment have been extensively evaluated as the concept of biocompatibility to explore the effect of implanted components on the activity of biological factors at the implanted zone. However, this section focuses on the effect of implants on bone tissue in view of biological aspects. Normally, orthopedic implants are made of bioinert materials with no bioactivity in the body environment. If the material has chemical interactions in ionized body fluid or induces activity in biological factors involved in bone remodeling, it is known as a material with biological activity in the human body environment (e.g., magnesium [Mg]). The induction and conduction of bone formation at the bone—implant interface are strategies for treating bone injuries to enhance the bioactivity of traditional bioinert materials to develop orthopedic implants. After implanting the artificial component or device through bone tissue internally or replacing bone tissue externally, it is ideal to establish bone growth through the implant at the interfaces to achieve bonding between the bone and implant. Greater adhesion of the implant to the bone has advantages in enhancing the clinical outcomes after surgery. In the following subsections, various aspects of the biological interaction between the bone and implant are discussed, which are prerequisite knowledge for exploring the interactions and failures of bone with certain orthopedic implants, which are reviewed and discussed in Chapters two to seven.

1.4.1 Bioactivity of implant surface

The first part of the implant that interacts with bone tissue is the contact surface. In biochemistry, there are many chemical molecules inside and outside cells (e.g., proteins, genes, growth factors) to be activated or deactivated to coordinate the remodeling of the bone matrix. Induction of the osteogenic cells and the presence of these molecules facilitate the required chemical reactions for adhesion, proliferation, differentiation, and maturation of new bone cells on the walls of the bone matrix. Conversely, the induction of bone resorption governs through the activity of other types of molecules to control and prevent excessive growth of the bone matrix. This is a chemical process involving various types of physical signaling as the result of mechanical, electrical, and thermal alterations that increase or decrease chemical reactions at the molecular scale. These complicated processes at the molecular scale establish remodeling to form new bone matured cells or resorb the dead or inactive part of the bone matrix. Establishment of bone remodeling on the surface or through the deep layers of the implant are an important key factor for enhancing the bioactivity of the implant on contact surfaces with bone. Bone remodeling is achievable only inside the bone matrix and does not be progress in contact with other materials or structures. Therefore, bone remodeling initiates and progresses under specific conditions of implant contact surfaces with bone tissue. At the bone—implant interface, when the bone matrix physically touches or impinges upon the implant, there is an amount of compressive preload conditions. The presence of such a mechanical interaction generates mechanical signals to the bone matrix and induces osteogenic or osteoclast cells to activate. Such activation may cause resorption of the bone tissue at the initial stage gradually to eliminate excessive applied mechanical loading to the bone matrix until the optimal range for induction of osteogenic cells is reached. The implantation effect is the remodeling of the bone matrix at the contact surfaces and does not affect the implant surface. It is not expected to have bone growth on the implant surface because there are no molecules or osteogenic cells on the implant surface. However, by increasing the surface roughness and/or coating certain molecules in the form of powder particles, bone remodeling is induced and conducted on the implant surface for a certain period (conductivity and inductivity of bone formation are reviewed in the next sections). Effective processing of bone remodeling, which involves adhesion, proliferation, differentiation, maturation, and resorption, is challenging to achieve in long-term implantation of the implant through bone tissue.

1.4.2 Osteoconductivity of implant

To establish bone growth on the surface of the implant, it is necessary to achieve the physical adhesion of osteogenic cells and other biological factors or molecules to the implant surface. If this adhesion process occurs early after implantation, the progression of bone remodeling is achieved. An increase in implant surface energy provides unstable physical conditions that enhance the adhesion of various types of cells at the interface of the bone and implant. Because the implant is in direct contact with the bone tissue, it has osteogenic and osteoclast cells at the surface of the bone matrix. Thus, they adhere to the implant surface to release existing energy in the micro- or nanostructure of the implant material at the contact surfaces. However, the adhesion strength of cells might not be sufficient to proliferate and be converted to osteoblast or osteocyte cells. With greater surface energy of the implant, higher adhesion strength of the cells can be achieved. Osteogenic cells are part of bone tissue, and once adherence to the implant surface is achieved, bone tissue attachment to the implant can progress. The main factor for adherence and the presence of osteogenic cells and other molecules is the flow of body fluid and vascularization through the implant surface, which is achieved by an increase in the surface energy of the implant. By converting osteogenic cells to an osteocyte cell and creating collagen fibers, mineralization of the new bone cell is initiated. Maintaining of the physical adherence of the new bone cell to the implant surface facilitates further mineralization and maturation of the new bone cell until the integrity of the matured cell to the main bone matrix occurs. By the maturation of bone cells, many osteogenic cells are created. This process is be repeated, and thus, the long-term adherence of the bone and implant is expected. However, this is an ideal condition that is not easily achieved at the bone—implant interface, and many factors are involved. Various surface strategies have been engineered to enhance osteogenic cell activities on the implant surface. These have been completely reviewed and discussed in the book *Osseoconductivity Surface Engineering for Orthopedic Implants*, written by the lead author of this book. In brief, porous structuring and bioactive coating are the main categories of surface engineering methods for creating the osteoconductive layer on the implant surface. The presence of a porous structure on the implant surface enhances the longevity and stability of bone remodeling in long term after implantation, whereas bioactive materials (e.g., calcium phosphate, growth factor, magnesium, strontium, collagen) accelerate the biological activity of osteogenic cells to be mature into new bone cells.

1.4.3 Osteoinductivity of implant

The concept of osteoinductivity was introduced as a biological interaction of the implant and bone at the contact interface. Osteoinductivity is part of bone remodeling on the implant surface. Induction of the osteogenic cells to be activated is governed through various physical and chemical factors. The chemical induction of bone remodeling is strongly influenced by the presence of a sufficient amount of biological factors and minerals in the blood flowing through the implant surface. The release of bioactive molecules or materials enhances the chemical inductivity of the new bone formation on the implant surface. Physical induction highly depends on the localized, overall mechanical interaction of the bone and implant, which has a direct influence on bone remodeling. Chemical inductivity of bone formation might focus more on the concept of osteoinductivity, which is the fundamental use of pharmaceutical drugs to enhance bone formation or reduce osteoclast cell activity to reduce bone resorption. The concept of drug delivery is a biotechnology conception of localized pharmaceutical treatment at the bone—implant interface to enhance bone remodeling. Through this, immobilization of antibacterial agents is also investigated for incorporation in the bioactive coating layer on the implant surface to enable local bacterial inactivation.

1.4.4 Peri-implant bone remodeling

Ineffective bone remodeling of the peri-implant was highlighted as a form of failure in Section 1.3, which occurs owing to inadequate mechanical interaction between the bone and implant. In biological terms, bone remodeling is highly affected by the ineffective activity of biological factors at the bone—implant interface. In some cases, there is a large area of bone loss at the implanted zone that needs to be treated by placing or injecting bone graft materials to induce bone formation at the affected zones. The specific enhancement of bone remodeling is required in sections where bone has been resorbed. Affected bone tissue at a distance from the bone—implant interface needs to be treated locally by further injection of biological factors after surgery to induce the bone remodeling at poor-quality bone zones.

1.4.5 Ideal biological interaction of implant and bone

The ideal biological interaction of the bone and implant might be achieved if the implant could allow penetration of the bone matrix formation through the internal zone of the implant. This concept is not new. Many attempts have been investigated to achieve this ideal condition. By creating a thick porous structure as the first contact layer of the implant with bone tissue, bone in-growth is conducted through the internal zone of the implant.

Bone tissue can gradually penetrate pores of the porous structure by internal vascularization and proliferation of osteogenic cells on wall surfaces. The concept of ideal conditions for the biological interaction of the bone and implant could be further developed by localizing biological activities in the affected zones after short- or long-term postimplantation. As the result of failure in mechanical interaction of the bone–implant construct, resorption of the bone matrix might be affected at the bone–implant interface. Increased biological activity at the resorbed zone might be governed by releasing bioactive molecules or particles to induce bone remodeling and prevent or reduce the rate of failure at the affected zone. Likewise, it is beneficial to enhance biological activity at a certain zone in treating bone injuries such as fracture gaps in fixing the bone fracture. Such a local increase of biological activity would need to be initiated by activating the local delivery of bioactive agents to the intended zone. This ideal concept is further discussed in Section 1.6.

1.5 Biological failures of bone–implant construct

Biological failures of the bone–implant construct correlate with mechanical failures. The achieved adherence of osteogenic cells and maturation of new bone cells on the surface or inside the implant fail after long-term implantation. At the failure point, the complete degradation or release of bioactive agents that were coated on the implant surface occurs, and thus osteoconduction and osteoinduction factors do not exist on the implant surface. Likewise, mechanical failure at the implant surface, such as cracking or significant deformation of pores in the engineered layer, generate undesirable mechanical loading conditions that affect active bone remodeling at the contact zone of the implant and bone tissue. Through ineffective osteoconductivity and osteoinductivity at the bone–implant interface, biological bonding between the implant and bone is gradually weakened. This may be observed as a radiolucent line on radiographic scans. Biological failure of bone tissue progresses to the more inner layers of bone tissue far from the bone–implant interface. Weakening of the bone tissue is governed through the inner depth of the bone matrix, from which thinning of the bone porous structure results. Such biological interactions cause a reduction in bone compactness at the affected zone. Deterioration of bone matrix stiffness at the zone near the bone–implant interface causes increased loading conditions to the bone matrix, which accelerates osteolysis and bone loss from the section at the bone–implant interface through the inner zone of the bone, which are detectable as a radiolucent zone on radiographic scans (Fig. 1.4).

1.6 Interactions of bone with orthopedic implants and possible failures

Mechanical and biological interactions and failures of the bone–implant construct are discussed in Sections 1.2–1.5. The simultaneous evaluation of these interactions and failures was performed in mechanical-biological terms to express and discuss bone–implant interactions and failures more effectively. They are greatly related to the function and performance of the bone–implant construct as one system. The biological activity of the bone tissue is highly affected by transferred loading conditions to the bone–implant construct, from which various types of failures occur. Some mechanical and biological failures do not result through ineffective interactions of the bone and implant; they occur by inherent mechanical or biological deterioration of the implant or bone. For instance, an implant material with low mechanical properties might not have sufficient strength to counteract physiologic loading conditions that would be transferred to the bone–implant construct. On the other hand, if the overall body condition of the patient is not be in good conditions (e.g., lack of a sufficient amount of minerals and biological factors in the blood or the severe activity of osteoclast cells, such as a preoperative disease), bone remodeling at the bone tissue might be significantly affected, and thus the integrity and stability of the bone–implant construct might fail. However, those failures in the bone or implant as the main portion of the bone–implant construct are crucial for evaluating bone–implant construct failure after the operation. The area of interaction would be the interactions and failures that resulted owing to simultaneous and ineffective mechanical and biological compatibility or the match of the bone and implant at the contact surfaces, which progressed far from the bone tissue to the interface line or area. Considering the acceptable (imperfect) conditions of the implant and the patient's body before implantation, matching the mechanical and biological characteristics of the bone and implant is the main challenge to the integrity and functionality of the bone–implant construct in long-term postimplantation. The mechanical modulus of the implant material; implant stiffness; surface conditions of the implant; severity of the bone injury; postoperative treatment and physiotherapy planning; anatomic design of the implant; enhanced manufacturing processes; enhanced biomechanical, biological, and clinical evaluation methods; level of bone ingrowth or on-growth through the implant; variations in implantation solution for similar types of bone injury; consideration of soft tissue deficiency; and careful implant selection are some strategies to be considered and developed for the more effective design and development of orthopedic implants with enhanced performance for treating bone injuries. The design of the implant is a main challenge in design and development. For a specific implant, various types of design models are on the market with similar basic concepts but different features and benefits. The implant design is correlated to the interaction of the bone and implant. Normally, to design an implant, various types of conception, such as greater bone stock preservation at the injured zone, anatomic or morphologic shaping of the bone, a low profile, a sufficient or optimal contact surface with the bone tissue, a stress-distributed profile, less localized stress zones, and so on, are considered for effective mechanical matching with bone tissue. Many design parameters and factors are considered in designing each type of implant. These are obtained from the intended function and performance of the implant to treat the intended bone injury. An effective implant design enables adequate load transference between the implant and bone in the physiologic direction, from which the risk for creating undesired load components is reduced. This allows adequate physiologic action and reaction loading conditions between the implant and bone, and thus effective stimulation of adhered osteogenic cells to the implant to be activated and formed as new bone tissue grows on the implant surface. Chapters two to six reviews and discusses in depth interactions for each type of orthopedic implant and reasons why failures occur.

1.7 Summary and discussion

To evaluate better the interactions of bone and orthopedic implants, it is beneficial to discuss the mechanical and biological concepts that are involved. In this chapter, the influences of implant mechanical strength, mechanical stiffness, mechanical integrity, load transference, stress shielding, and overloading on the effectiveness of the bone—implant construct function and performance were discussed in general terms. The optimal conditions for these mechanical concepts are crucial to prevent failure from occurring as the result of inadequacy in these technical aspects. The material, design geometry, and structure of the implant are vital to achieving optimal conditions with an adequate range of strength, stiffness, integrity, and load transference at the bone—implant construct while eliminating stress shielding and overloading of severe effects between the bone and implant. The current orthopedic implants with an extensive range of use in treating trauma, osteoarthritis, and spinal injuries are manufactured from metallic materials (mainly titanium alloy, cobalt-chromium alloy, and stainless steel), polymeric materials (PEEK, ultrahigh molecular weight polyethylene), and ceramic materials (mainly alumina and zirconia). These materials are all bioinert with the risk for material particle release in specific chemical or physical failures. Metallic materials have an inherent high mechanical modulus that increases the risk for mechanical failure of the bone–implant construct for various reasons. In one way, they may cause very high rigidity and stiffness of the implant in contact with the bone and the occurrence of stress shielding and overloading. In another way, they may concentrate mechanical stresses at localized zones of the implant, and thus fatigue failure may occur after a short or long period of implantation. On the other hand, metallic materials increase the mechanical safety of the implant against increased loading conditions to the implant as the result of obesity, a high range of motion, heavy object handling, falling, or accidents. Therefore, a low profile and optimized design of the implant to decrease stiffness have been the main strategies of orthopedic implant development. With an increase in the use of polymers such as PEEK in the development of orthopedic implants with a low level of load bearing, mechanical interactions of the bone and implant have significantly improved. PEEK composites such as carbon fiber-reinforced PEEK and hydroxyapatite PEEK were introduced with enhanced mechanical properties close to those of cortical bone to enhance load bearing, which is advantageous to orthopedic implants. For load transference, optimized stiffness and an anatomic design of the implant were addressed as main technical requirements to facilitate effective load sharing and transference between the implant and bone. Ineffective load transference causes stress shielding on some contact surfaces and overloading on other contact surfaces of one implant. It may cause bone loss at the distant zone of bone tissue near the bone–implant interface (e.g., the occurrence of bone loss at the trochanteric zone of the proximal femur near the femoral stem prosthesis in total hip arthroplasty). Reported mechanical failures of the bone–implant construct were a variant discussed as an implant fracture, undesired rigidity, or flexibility of the construct, bone–implant dissociation, implant loosening, bone suppression, implant subsidence, bone fracture, and extensive bone loss. These failures were all completely elaborated upon and examined in this chapter. Further exploration of the failure details for each category of orthopedic implants is extensively provided in Chapters two to six. For the biological interaction of bone and implant, the concepts of implant surface bioactivity, osteoconductivity, osteoinductivity, and peri-implant bone remodeling were expressed in this chapter. The biological activity of bone remodeling as bone formation and bone resorption need to be balanced and effectively achieved at the bone—implant interface. This is manageable by enhancing implant surface conditions for increased conduction and induction of the bone on-growth and ingrowth through the inner layers of implant contact areas with the bone tissue. In early implantation of orthopedic implants, loading conditions were lower than those of normal physiologic conditions, which enabled the effective achievement of biological adherence of osteogenic cell vascularization and flow of blood through the implant interface with the bone. With the increased presence of involved biological molecules or factors and minerals at the bone—implant interface, the proliferation, differentiation, mineralization, and maturation of new bone cells and their ultimate integrity with the bone matrix are accelerated. Achievement of effective mechanical and biological interactions between the bone and implant is achieved mainly because of mechanical advantages and characteristics of the implant and bone or biological activity of bone remodeling in the bone—implant interface or the distant zone to the interface early and in the short term period after implantation. However, in long-term implantation, simultaneous mechanical and biological interactions between the bone and implant gradually increase, which influences the effectiveness of the bone—implant construction in the treatment of bone injuries and enhances clinical outcomes. In the next chapters, interactions of the bone with orthopedic implants and possible failures are discussed in depth based on reported clinical complications in published clinical studies. The failures are illustrated radiographically to enhance the expression of incidents that occurred. The lead author of this book will attempt to provide a rigorous discussion on the reasons for the failures based on the fundamentals and concepts.

1.8 Remind and learn

In this section, some questions are provided as effective reminders for review of the content discussed in this chapter. Answering these questions enhances learning about the interactions of the bone and orthopedic implants.

Q 1.1: Express the various bone injuries that are treated by internal implantation of the artificial components, implants, or prostheses through the bone tissue.

Q 1.2: Why have metallic materials traditionally and conventionally been used to develop orthopedic implants? What types of mechanical loads might be transferred to the bone–implant construct? Express how these loading conditions might be generated at the bone–implant construct.
Q 1.3: Discuss why the stiffness value of the implant might be higher than what would ideally be required. Why would the safety factor for the design and manufacture of the implant increase?
Q 1.4: What are three main factors that would influence implant stiffness? Discuss the overall mechanical properties of various types of materials.
Q 1.5: Discuss how the design geometry of the implant influences implant stiffness. What kinds of evaluation methods might be used to achieve effective and reliable implant design geometry in mechanical interactions with the one tissue?
Q 1.6: What types of manufacturing processes might be used to fabricate implants? Discuss how the manufacturing process influences the stiffness of the implant.
Q 1.7: What types of mechanical integrity could be considered in structuring the bone–implant construct? Discuss how these types of mechanical integrity are achieved at the interface of the bone and implant.
Q 1.8: Discuss bone–implant integrity in cortical and trabecular bone tissues. What mechanical integrity at the interface of the bone and implant is crucial for the long-term effectiveness of the implant? What type of loading condition affects the mechanical integrity of the bone and implant at the interfaces?
Q 1.9: What is the overall mechanism of load transference between the bone and implant? Discuss why the level of reaction force or energy to bone tissue is in a high range at the bone–implant interface.
Q 1.10: Discuss the mechanical strength of bone against action or reaction load transference from the internal and external interface with the implant. What loading condition might be more destructive to bone tissue at the bone–implant interface? What failures might occur at the bone–implant interface or far from the bone matrix from the bone–implant interface? (Consider Fig. 1.4 in discussing your answer.)
Q 1.11: Discuss the stress shielding effect and why it may cause undesirable mechanical interactions to occur between the bone and implant. What bone areas might be affected by stress shielding? Discuss how the loading direction and contact surface normal axis may influence the occurrence of stress shielding at the bone–implant interface.

Q 1.12: Describe the ideal mechanical interaction of the implant and bone. Why might the deterioration of implant stiffness at concentrated stress zones be beneficial?

Q 1.13: Discuss why the design concepts of developed orthopedic implants might not be carried out effectively. (Consider the limitations of surgeons and engineers for effectively generating the design concepts for the enhanced mechanical interaction of the implant and bone.)

Q 1.14: Why might localized stress zones be generated at the implant? How might localized stress reach the failure point? What postoperative planning is helpful to reduce implant failure?

Q 1.15: What are the concepts of rigidity and flexibility of the implant in its construction with bone tissue? What is the relation between implant rigidity and localized stress? Discuss the inadequate mechanical interaction of poor-quality bone and highly rigid implants. What are the possible failures of an excessively flexible implant structure constructed with bone tissue?

Q 1.16: Discuss the concept of bone and implant dissociation. What loading conditions provide an effectively stabilized implant in contact with bone in cases with a risk for dissociation? Under what conditions do such failures occur? Discuss why early dissociation of the implant from bone might occur and what postoperative planning might be effective in preventing the occurrence of this failure.

Q 1.17: Discuss the concept of implant loosening from bone tissue. What are the main reasons for implant loosening? Discuss whether implant loosening is catastrophic and under what conditions revision surgery is required as a result of implant loosening.

Q 1.18: What is bone suppression at the bone–implant interface? What type of internal or external interfaces between the bone and implant are more suppressed? What are the main reasons for bone suppression? Discuss why the interface of the bone and screw might be at a high risk for such failure. Provide some examples of the most critical zones at which bone suppression or screw cutout might occur.

Q 1.19: What is the main reason for peri-implant bone fracture implantation of the implant to the bone? What cases might be riskier for the occurrence of peri-implant fracture?

Q 1.20: What is subsidence in mechanical interaction between the bone and implant? Discuss the main reasons for subsidence failure. What is the reaction between the implant material and subsidence failure?

Q 1.21: Describe the overall aspects of the biological interaction between the implant and bone. Express the overall aspects of bone remodeling in bone tissue. Discuss the challenges of bone remodeling on the implant surface.

Q 1.22: Discuss the adherence conditions of osteogenic cells and other biological factors through the implant surface. How is bone formation conducted on the implant surface at its interfaces with bone tissue?

Q 1.23: What is the chemical and physical inductivity of bone formation at the bone–implant interface? Discuss how the concept of chemical osteoinductivity is used for localized pharmaceutical treatment at the bone–implant interface.

Q 1.24: Describe how peri-implant bone resorption or loss could be treated biologically.

Q 1.25: Discuss the mechanism of biological failure of bone tissue at the bone–implant interface and far from the interface.

Q 1.26: Discuss the mechanical and biological failures that are not associated with interactions between the bone and implant. What types of technical parameters are considered to achieve an effective interaction between the bone and implant?

CHAPTER TWO

Interaction of bone with knee and ankle joint replacement prostheses and possible failures

2.1 Introduction

The knee and ankle joints are positioned at the lower limbs of the body musculoskeletal system. These joints are articulated to be convex and concave with multiple degrees of freedom that allow six translation and rotation motion in body movement for various activities. The range of motion (ROM) at the knee joint is higher than that at the ankle joint. The ankle joint is highly restricted by available ligaments for higher stability of the musculoskeletal system during gait, stair climbing, running, and other activities. With greater ROM at the knee joint, it is essential to have a lower ROM at the ankle joint for overall stability of skeletal system motion while bearing body weight loading conditions. Excessive loading conditions at the knee and ankle joints causes gradual degeneration of the cartilage and meniscus tissues where the tissue at two articulated ends of the bones has direct contact and generates inflammation and severe pain at the affected joint. Such severe conditions are treated by replacing the artificial components at the joints to prevent direct contact between bones while providing physiologic ROM without pain. Interactions of the replaced prostheses with the bone in terms of mechanical and biological parameters is complex. Thus, it is necessary to create appropriate design concepts in accordance with physiologic loading conditions and the anatomic shaping of bones at the joint. This process has been challenging for clinicians and engineers. Through the growth of technology in various aspects of engineering, testing, analysis, image processing, bone cellular biomechanics, bone biochemistry, surface engineering, and so on, prosthesis developers attempted to challenge old concepts and create new hypotheses to enhance the development of prosthesis designs and performance at the operated joint. Ineffective compatibility and interactions of bone with the prosthesis design and mechanical properties increase the risk for prosthesis loosening, migration, subsidence, and fracture at the implanted site. Therefore, the overall treatment goal of

Interactions of Bone with Orthopedic Implants and Possible Failures
ISBN: 978-0-323-95410-5
https://doi.org/10.1016/B978-0-323-95410-5.00001-3

joint replacement fails, necessitating revision surgery. In this chapter, failure by loosening of the prosthesis, owing to ineffective interaction between the prosthesis and bone, is reviewed and discussed for prostheses of the knee and ankle replacements. Prostheses such as the tibial tray, femoral knee, hinge knee prosthesis, ankle tibial tray, and talar component are studied in this chapter. In accordance with reported loosening cases for each prosthesis, data provided for each prosthesis have different extents or technical issues. The reviewed and discussed contents in this chapter are not complete for all designs of prostheses, particularly for those developed in very recently years, which have not been studied and reported clinically. This chapter attempted to address the most critical incidents of loosening. In next edition, the content will be extended to more design variations of knee and ankle joint prostheses.

2.2 Interactions of bone with knee joint prostheses and possible failures

2.2.1 Loosening of knee joint prostheses

Knee replacements are used to treat the severe loss of cartilage or multifragmentary fractures at the most distal area of femur or proximal area of tibia near the knee joint. In such cases or indications, the cartilage and a portion of bone are removed and resected to replace artificial components. The replaced components or prostheses permanently act as the knee joint and facilitate the functionality of the joint close to that of a healthy joint. The replaced prostheses transfers the body weight load from the femur through the tibia and from the tibia to the femur. The prostheses of femoral knee, patella, and tibial tray are inserted through the bone and the tibial insert is positioned between the femoral knee and tibial tray prostheses (Fig. 2.1). The inserted prostheses to the bone might be loosened from its initial position, which was fixed during surgery. Loosening of the prosthesis results in revision surgery. The number of revision surgeries is increasing, as reported in 2013 [1] and 2014 [2] and highlighted in a published clinical study at the Department of Trauma and Hand Surgery, Heinrich Heine University Hospital, Duesseldorf, Germany [3]. Loosening of prostheses occurs as a result of the cytokine level at the interface of the bone and implant, from which the activity of osteoclast cells is exceeded [4]. Reasons for an undesirable increase in osteoclast activity and excessive level of bone resorption are extensive. They include various mechanical and biological interactions between the bone and implant and will be discussed in this chapter. Another

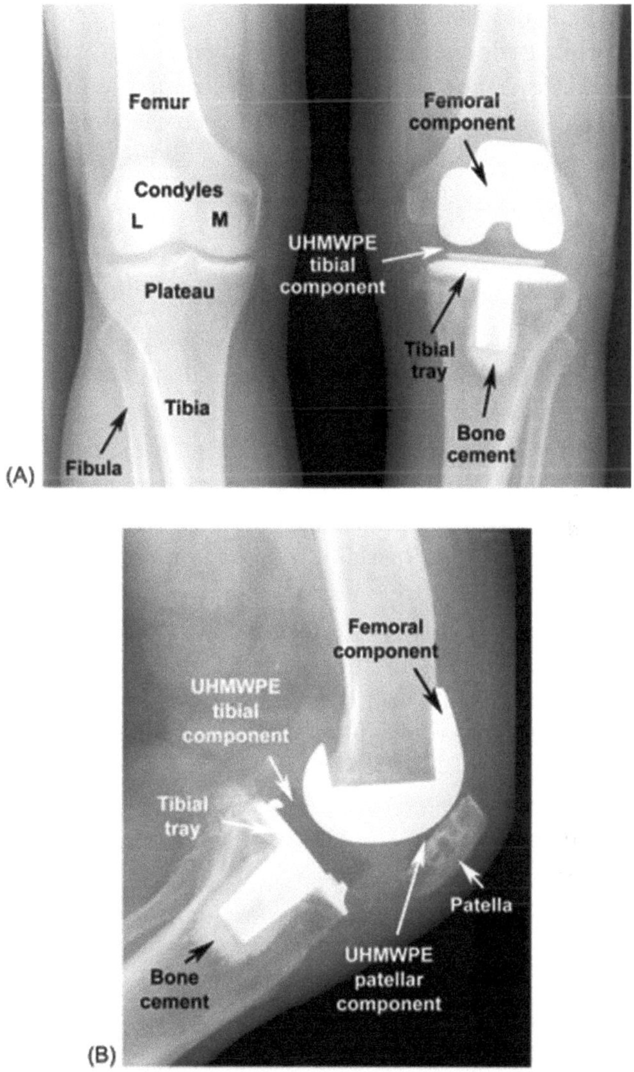

Figure 2.1 Illustration of knee joint and total knee replacement (TKR) prostheses. TKR consists of four prostheses, three of which are inserted to the femur, tibia, and patella. The fourth is stabilized between two prostheses of the femoral and tibial tray to act as bearing prostheses and constrain the femoral condyles in six degrees of freedom. *UHMWPE*, ultrahigh molecular weight polyethylene. *Through the courtesy of S.K.-U.B. Handbook, undefined 2009, The Origins and Adaptations of UHMWPE for Knee Replacements, Elsevier, n.d.. https://www.sciencedirect.com/science/article/pii/B9780123747211000079. (Accessed 6 October 2020)and get permission from Elsevier.*

reason for implant loosening is the presence of low virulent bacteria [5]. The diagnosis of prosthesis loosening is carried out by measuring the synovial cell count, joint fluid aspiration, or C-reactive protein [3]. Normally, loosening of the prosthesis as a result of interactions between the bone and prosthesis occurs long after implantation (up to 16 years, as observed in the investigation by Schiffner et al. [3], which was carried out on 149 patients).

2.2.2 History of knee joint design concept

In the 18th century, knee replacement was developed with a hinged joint to relieve arthritis pain and other complications at the knee joint [7]. In that type of knee implantation treatment, the knee joint was converted to a hinged joint that provided only flexion-extension rotation. Replacing the hinged knee replacement system required cutting a large portion of the femur and tibia. Gradually, with the growth of image processing techniques, manufacturing processes, contact loading measurement sensors, kinematic and wear analysis and testing machines, and design modeling software, the concept of bone resection with a maximal bone preservation volume was introduced as a key development strategy. Such a concept was developed to substitute the multiaxis knee replacement system with the one-axis hinged knee system. The designed knee replacement system was revolutionary in dedicating the condylar profile of the distal femur and tibia plateau. This design concept is marketed in various types. A multiaxis design of knee replacement is constrained at the articulation surface between the femur and tibia. The constraint is implemented geometrically through a convex—concave pattern under the compressive force of body weight and physiologic constraints of collateral ligaments and muscle—tendon systems between the femur and tibia. Such constraint conditions allow relative movement of the prostheses in 6 axes of rotation and translation: (1) flexion-extension rotation, (2) varus-valgus rotation, (3) internal-external (IE) rotation, (4) anteroposterior translation, (5) medical-lateral translation, and (6) superior-inferior translation. The multiaxis knee system is recognized as a condylar knee design system in which femoral and tibial insert prostheses are shaped similar to the femoral and tibial condyles. (The geometry of the condyles is not exactly similar to the anatomic condyles of the femur and tibia. The design is altered to provide effective constraint to facilitate physiologic motion of the knee joint while the anterior or/and posterior cruciate ligaments are sacrificed.) The condylar knee system is used for knee osteoarthritis in which cartilage and meniscal tissue at the knee joint has degenerated severely and the patient reports severe pain and a significant reduction of ROM. There are other cases for which a large portion of bone needs to be resected or removed from either the distal portion of the femur or the proximal portion of the tibia (a large resection or cutting of the bone might be managed owing to the abnormal growth of bone tissue that formed as neoplasms). Likewise, collateral ligament functionality might have significantly deteriorated and ligament deficiency might have occurred. Severe injuries to bone tissue and collateral ligaments at the knee joint do not allow the use of conventional condylar knee systems. Therefore, the product developers have used the concept of the hinged knee system that was introduced in the middle of the 18th century. However, the fixed hinged design was converted to a rotating hinged design with enhanced functional and performance. Based on research carried out by the author, the loosening of knee prostheses occurs in both condylar [8–13] and rotating hinged designs [14–20]. In the following sections, loosening of knee replacement prostheses is reviewed and discussed.

2.2.3 Loosening of tibial tray prostheses

The tibial tray prosthesis is a component of the tibial knee replacement (TKR) system. It is implanted through the proximal portion of the tibia in contact with trabecular tissue at the metaphyseal portion of the proximal tibia. This prosthesis bears a large amount of compressive, torsional, and shear stresses transferred at the knee joint as a result of body weight and other physiologic motion in daily activities. Because of the large interface of this prosthesis with trabecular bone, its interactions with this type of porous and spongy bone matrix has a significant effect on its long-term effectiveness. Loosening of the tibial tray was reported from various areas of this prosthesis for various reasons, which make its design challenging. Thus, different tibial trays have been designed and commercialized, which makes it challenging for surgeons to select the appropriate prosthesis based on the patient's physiologic condition. In the next sections, the relation of design concepts to loosening is reviewed and discussed for tibial tray prostheses. An in-depth discussion elaborates upon why loosening occurs based on the patient's condition and the design features of the prosthesis.

2.2.3.1 Asymmetric design of tibia tray at the plate section

Loosening of knee prostheses has been observed in tibial tray prostheses in more cases than in femoral components. Tibial tray prostheses are designed in a *T* shape composed of two sections: a plate shape and a stem shape

(Fig. 2.2). The plate section is designed symmetrically or anatomically (Fig. 2.3). In an anatomic design, medial and lateral sections are asymmetric and dedicated to the proximal tibial surface after resection or cutting to

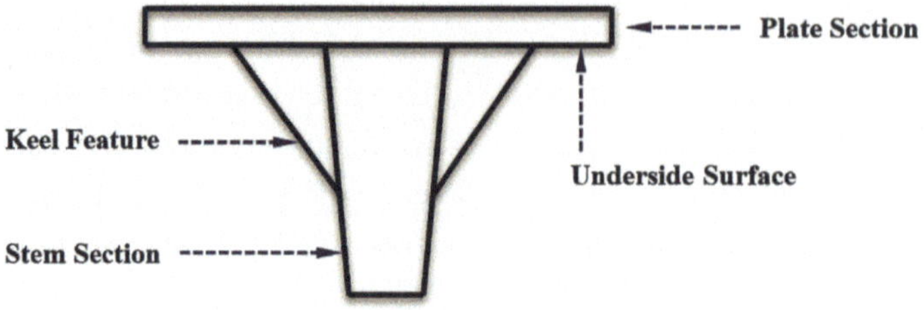

Figure 2.2 Illustration of the tibial tray in frontal view.

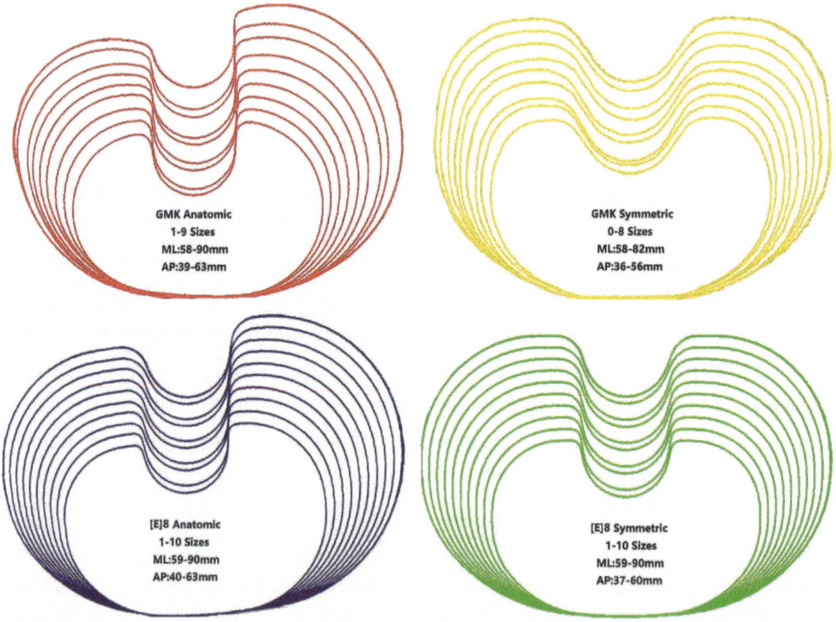

Figure 2.3 Anatomic (left side) and nonanatomic (right side) or asymmetric and symmetric designs of the plate section of a tibial tray prosthesis. An asymmetric design enhances the interaction between the bone and knee prostheses. Because of a greater amount of compressive stress at the medial condyles, the higher area of the tibial tray at the medial side reduces stress and thus better load transference between the femoral knee and tibial tray (GMK: *Gemini Mobile Knee*; ML: *Medial-Lateral*; AP: *Anterior-Posterior*). *Through the courtesy of L. Shao, X. Wu, T. Wang, X. Liu, W. Xu, W.H.-S. Reports, undefined 2020, Approximating the Maximum Tibial Coverage in Total Knee Arthroplasty Does Not Necessarily Result in Implant Malrotation, Nature.Com, n.d.. https://www.nature.com/articles/s41598-020-67613-2. (Accessed 22 September 2020) and Springer Nature.*

replace the tibial tray. The aim of the asymmetric design of the tibial tray at the plate section is to establish more physiologic load transfer from the femoral prosthesis through the tibia[21]. This physiologic or anatomic design reduces the generation of excessive stress at the interface of the tibial insert and tibial tray prostheses, where the risk for wear of the polymeric tibial insert at this interface is significantly reduced. Furthermore, the subjected compressive stress from the medical condyle of the femoral prosthesis to the medial aspect of the tibial tray is reduced, from which concentrated stress at the medial aspect of the tibial tray is reduced. This reduction lessens the risk for tibial tray fracture from the medial part, as reported by Callaghan et al. [8] and Rytter et al. [22] (Fig. 2.4). About 60%–70% of the compressive load during the gait cycle is loaded through the medial part of the tibia. Hence, because of the symmetric design of the femoral and tibial insert prostheses in lateral and medical condyles, similar load distribution is expected at the articulation of the tibial insert and tibial tray, whereas the amount of stress is increased on the medial side. The anatomic design of the tibial tray is greatly advantageous for a fixed bearing tibial assembly in which the polymeric tibial insert is interlocked to the tibial tray at its plate portion. Almost all well-established manufacturers of total knee replacement (TKR) system have developed versions with an anatomic design of the tibial tray at the medial aspect. This design necessitates a correlated design of the tibial insert matched with the anatomic tibial tray from which greater contact with the articulation surface features a concave profile of the tibial insert at the medial aspect. This distributes the contact force and reduces contact stress, ultimately decreasing the wear rate in accordance with Archard's law. On the other hand, the femoral component and tibial insert are designed to reduce the effect of increased compressive stress at the medial aspect of the tibial tray.

2.2.3.2 Effectiveness of deep bone ingrowth in the bone prosthesis

As illustrated in Fig. 2.5, the tibial tray may have rotational migration owing to prosthesis subsidence through trabecular bone at the lateral aspect of the proximal tibia. Reducing transferred energy from the metallic tibial tray prosthesis to bone reduces the catastrophic effect of bone overloading and the risk for prosthesis subsidence is reduced. The creation of a transition layer at the interface of the tibial tray prosthesis (particularly at the underside surface) and bone could act as an energy-absorbing layer, which might be achieved by establishing of a thick porous structure on the underside surface as an osteoconductive surface engineered layer. The effectiveness of an

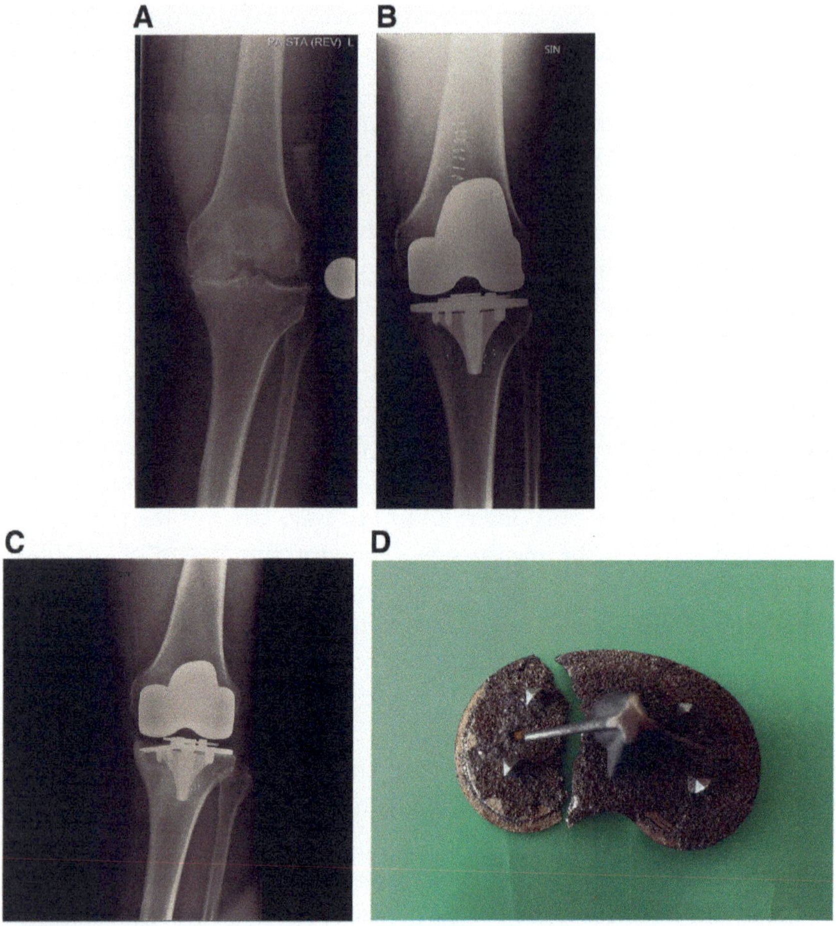

Figure 2.4 Illustration of breakage in a modular tibial tray with fine stem and keel. The radiographs (A–C) belong to a patient with a body mass index of 34 kg/m^2. In this type of modular tibial tray, the stem is completely modular. It is installed during surgery at the plate section of the tibial tray. The installation mechanism causes a gap between the stem and plate section of the prosthesis. This gap eliminates the load-bearing advantage of the stem keel against the compressive force, from which the plate section bears the transferred compressive load on both the medial and lateral sides. Thus, this prosthesis is not secure for use for obese or overweight patients. Another issue of this retrieved prosthesis is removing the porous coated layer at the underside surface. This may show evidence of a large amount of shear stress in the mediolateral and IE axes. Using a fined keel may not give an effective load-bearing advantage against torsional torque in IE rotation, where the edge of the tibial tray at the medial edge is affected. *Through the courtesy of S. Rytter, F. Madsen, C. Jepsen, M.S.-T. Knee, undefined 2019, Implant Fracture of the Regenerex® Modular Metal Tibial Component: A Report of Three Cases, Elsevier, n.d.. https://www.sciencedirect.com/science/article/pii/S0968016019301255. (Accessed 3 October 2020) and get permission from Elsevier.*

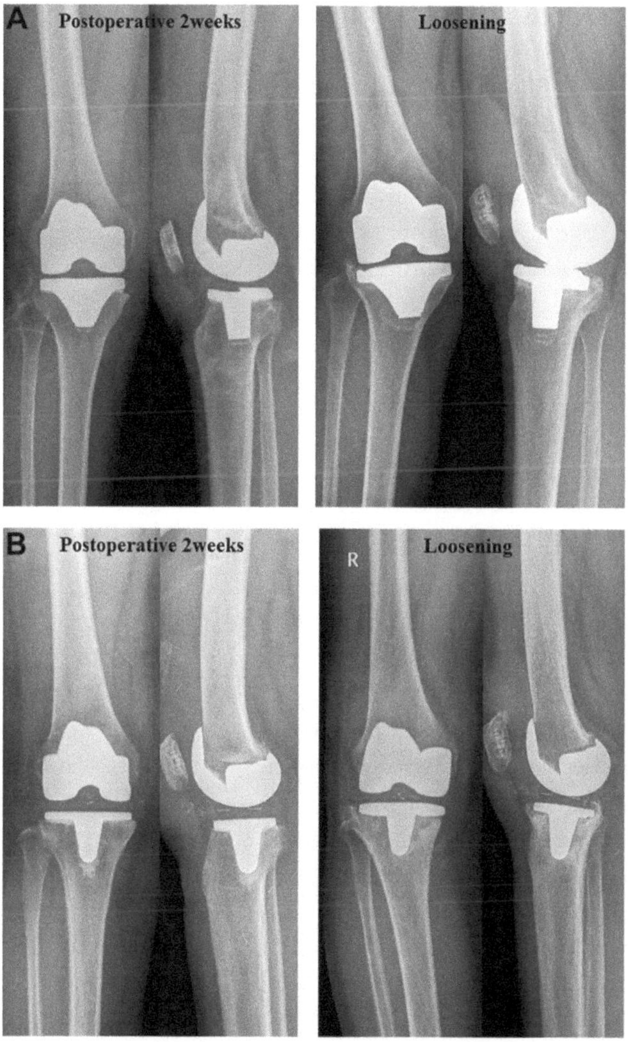

Figure 2.5 Radiographic illustration of early tibial tray subsidence and loosening in just 2 weeks after surgery in (A) fixed bearing, and (B) mobile bearing tibial knee replacement systems. These failures were reported in a patient with preoperative severe varus deformity of the lower limb with a mechanical axis of greater than 8°. *Through the courtesy of S. Song, H. Lee, D. Bae, C.P.-T.J. of arthroplasty, undefined 2020, High Incidence of Tibial Component Loosening After Total Knee Arthroplasty Using Ceramic Titanium-Nitride-Coated Mobile Bearing Prosthesis in Moderate to Severe Varus, Elsevier, n.d. https://www.sciencedirect.com/science/article/pii/S0883540319311064. (Accessed 20 September 2020) and get permission from Elsevier.*

osteoconductive engineered surface (which might be processed through titanium plasma spraying or direct selective laser porous structuring of the tibial tray) is increased through the greater formation of bone matrix at the interface of the tibial tray prosthesis and trabecular bone. In addition, with the higher stability of the tibial tray prosthesis, the formed bone matrix at the interface of the bone—prosthesis is more calcified and the strength of the interface layer against compressive stress is promoted; hence, the effect of overloading is reduced and the risk for prosthesis subsidence is reduced. Such enhanced biological integration reduces the risk for malrotation or migration of the tibial tray.

2.2.3.3 Interaction of bone with tibial tray stem

The other section of the tibial tray is designed in a stem shape. (In some designs, the stem is modeled as multiple pegs used in minimally invasive surgical procedures.) The various designs of stem sections of tibial tray prostheses are displayed in Fig. 2.6 [23]. Normally, the stem portion is designed in two parts: the middle part, which is in cylindrical or conical, and the keel part. The stem portion replaces the tibial tray through the proximal tibia, which is inserted into the metaphyseal zone of the proximal tibia up to the intramedullary canal at the diaphyseal portion of the tibia. The keel portion acts as a mechanical support to enhance the load bearing extent of the plate section under compressive loading conditions at the knee joint. Likewise, with the angular design of the keel (Fig. 2.6) relative to the coronal plane, the strength of the tibial tray against varus-valgus moment, torsional moment, and mediolateral force is significantly increased, and the level of strain introduced to the trabecular bone at the metaphyseal portion of the proximal tibia is maintained at a lower level. The stem section of the tibial tray is designed in various shapes, each with advantages and disadvantages that make it difficult to select the appropriate design to treat knee osteoarthritis. Improper selection of the tibial tray based on its stem design may induce loosening of the tibial tray in some cases.

Loosening of the tibial tray prosthesis with the minikeel design was reported by Kajetanek et al. [24] (Fig. 2.7). In their investigation, 12 tibial tray loosening cases in 212 investigated patients with the minikeel tibial tray prosthesis were observed. (A minikeel prosthesis is used for a minimally invasive subvastus approach in which tibiofemoral dislocation is not carried out. Thus, the tibial tray should have a short stem length to be implantable through this approach.)

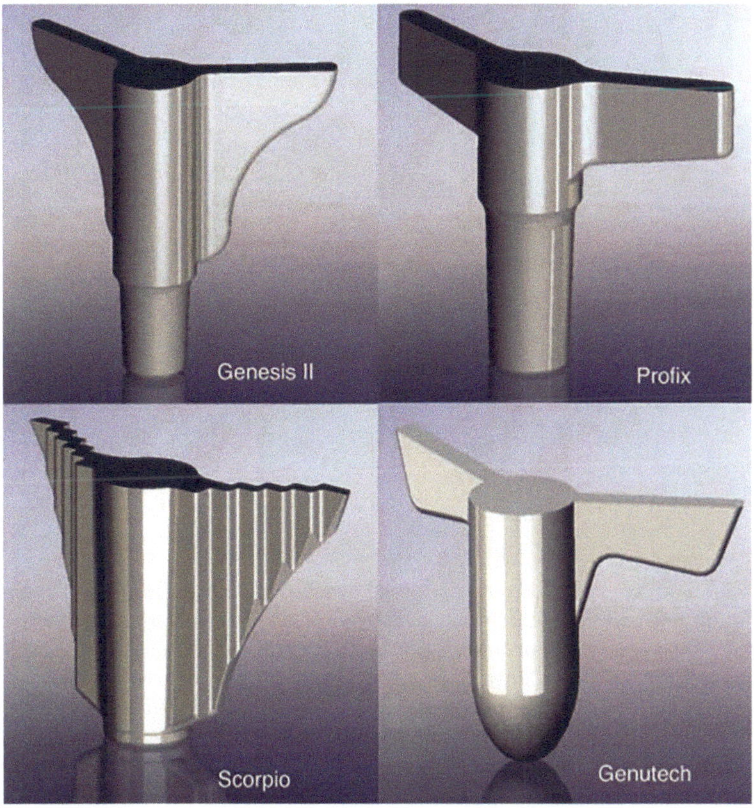

Figure 2.6 Illustration of the stem portion of the tibial tray prosthesis in various commercialized tibial knee replacement systems. Various designs of the stem keel are apparent. In general, the tibial tray stem is designed with no keel, a minikeel, a conventional keel, or a transverse-extended keel. The keel can be transversely extended in both mediolateral and posteroanterior axes to enhance the engagement of a greater area of the tibial tray prosthesis with trabecular bone at the metaphyseal portion of the proximal tibia. Likewise, it could distribute transferred torsional and compressive stress through a greater volume of periprosthetic trabecular bone and reduce concentrated stress at the interface of the prosthesis and bone. This reduces overloading of the periprosthetic bone and bone resorption at the interfaces and the risk for loosening of the prosthesis. *Through the courtesy of S. David, J. Martínez, I. Bermúdez, … F.M.-R.E., undefined 2014, The Geometry of the Keel Determines the Behaviour of the Tibial Tray Against Torsional Forces in Total Knee Replacement, Elsevier, n.d. https://www.sciencedirect.com/science/article/pii/S1988885614001011. (Accessed 10 October 2020) and get permission from Elsevier.*

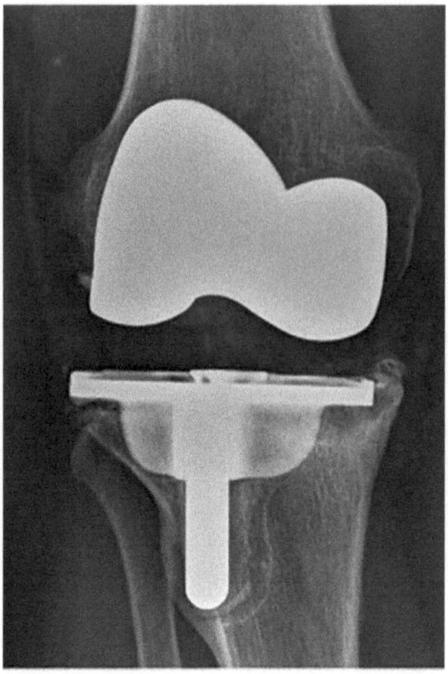

Figure 2.7 Radiographic illustration of prosthesis loosening in a tibial tray with a minikeel design. The minikeel reduces the invasive intervention of a surgical procedure but it is not a secure solution for overweight or obese patients. Because of the patient's reduced activity after total knee arthroplasty, body weight increases over time and the patient's body mass index increases. This is essential in considering implant section and the overall treatment plan. The minikeel tibial tray has a lower surface area; thus, transferred torsional torque in the internal-external axis is applied to the lower trabecular bone volume at the metaphyseal portion of the proximal tibia. Because of the greater distribution of the body weight load (in the forms of compressive force and torsional moment) at the medial aspect of the tibia, the risk for prosthesis migration or subsidence from the medial side is higher, as illustrated in this figure. *Through the courtesy of C. Kajetanek, B. Bouyer, M. Ollivier, ... P.B.-... & T.S., undefined 2016, Mid-Term Survivorship of Mini-keelTM Versus Standard Keel in Total Knee Replacements: Differences in the Rate of Revision for Aseptic Loosening, Elsevier, n.d. https://www.sciencedirect.com/science/article/pii/S187705681630055X. (Accessed 24 September 2020) and get permission from the Elsevier.*

2.2.3.4 Interaction of bone with mobile bearing tibial assembly

Normally the tibial tray prosthesis in a mobile bearing TKR is designed with a ribbed conical stem without a keel. The stem has a large hollow to facilitate inserting the stem portion of the polymeric tibial insert through the tibial tray (this allows rotation of the tibial insert in IE rotation and constrains other axes of rotation and translation). For such an advantage in a mobile bearing TKR, the stem of the tibial tray has to be designed in a tube pattern. This requires stem to be cone-shaped to enhance the stability of the tibial insert inside the tibial tray.

Tibial tray migration is observed in cases with preoperative varus deformity greater than 8° (deviation in the mechanical axis of the lower limb, which begins at the hip joint and ends at the ankle joint, shows the degree of deformity at the knee joint). Such anatomic deficiency occurs in elderly patients who postpone treatment of severe arthritis. Varus deformity might also be observed in young patients who had this issue from birth and for whom the deformity increased at a young age. Song et al. [12] observed seven cases of tibial tray loosening in 100 patients who underwent total knee arthroplasty (TKA) using a nonkeeled mobile bearing tibial tray, because they had a preoperative moderate to severe varus deformity. In their study, one case of loosening for the keeled fixed bearing tibial tray was reported, which shows the lower risk of tibial tray loosening in a fixed bearing knee design compared with that of a mobile bearing design. In mobile bearing knee replacement, loosening of the tibial tray begins with resorption of preimplant bone tissue, as highlighted by Kutzner et al. [25]. In their study, 32 retrieval tibial trays (mobile bearing type with ribbed and nonkeeled stem) for patients with loosening complications were investigated. The radiolucent line or area has been addressed as the analysis parameter to access the tibial tray loosening. It was observed that [25] there is a large radiolucent area near the stem part of the tibial tray in the series of failed cases. Radiolucency is apparent when bone mineral density is significantly reduced. This issue occurs as a result of ineffective and nonphysiologic interactions between the bone and prosthesis. In mobile bearing knee replacement, the tibial tray stem has a tubular conical shape with no keel. This design of the stem increases stress shielding owing to significant deviation between the normal axis to the stem surface and axial body weight loading, which is transferred at the knee joint through the proximal tibia. Keel featuring of the stem adds a load bearing advantage to the stem, from which the periprosthetic trabecular bone at the interface with the stem is induced mechanically to form the bone matrix. Increased roughness of the stem and keel surfaces with porous structuring and bioactive coating acts as an osteoconductive platform to conduct bone ingrowth through the deep layers of the stem surface and enhance osseointegration, preventing progressed resorption of periprosthetic bone at the stem interfaces.

Fabrication of tubular conical stem, by current manufacturing solution such precise forging and investment casting, might be challenging (these two manufacturing processes are normally utilized for fabrication of titanium alloy and cobalt-chromium alloy tibial tray followed by complementary machining processing). To overcome this issue, some manufacturers have designed tibial tray prostheses for use in a mobile bearing TKR with a design similar to that used for a fixed bearing TKR. Thus, there is no need for conical sharing of the tibial tray stem, in which the stem could be featured by the keel portion to prevent loosening of the tibial tray in a mobile bearing TKR. In such a mobile bearing design concept, to provide rotation of the tibial insert around the tibial tray, a peg is used on the bearing section of the tibial tray (Fig. 2.8D) for insertion into the correlated hole at the undersurface of the polymeric tibial insert (Fig. 2.8E) [26]. Mechanical interaction between the tibial insert and tibial tray at this peg may damage the tibial insert at the interface, and the tibial insert may dislocate or migrate, as highlighted by Kobayashi et al. [26].

2.2.3.5 Interaction of bone with cemented and cementless tibial tray

Early mechanical stability of the cementless tibial tray was investigated through micromotion analysis at the interface of the bone and implant [10] compared with the cemented tibial tray with a similar basic design [26,27]. In that study, the cemented prosthesis had a porous surface, four pegs, and a keeled stem (Fig. 2.9), whereas the cementless prosthesis had a similar design without a porous undersurface layer and pegs (Fig. 2.8). As expected, micromotion was greater in the cementless tibial tray prosthesis. However, the difference was lower than $<150\ \mu m$. As mentioned in that study, biomechanical testing experiments were managed to qualify the comparison of micromotion under cyclic axial compressive loading conditions between the cemented and cementless tibial tray prostheses. Based on what the researchers achieved [10], the level of micromotion was higher in the medial and anterior aspects of the proximal tibia, whereas there was greater micromotion on the posterior aspect for other designs of the tibial tray. This shows the design dependence of the micromotion effect under physiologic loading conditions. The design of the stem influences the distribution of the loading conditions to the plate portion of the tibial tray. The investigated tibial tray in Crook et al. [10] had a large keel that provided mechanical support from the midline through the posterior aspect of the prosthesis, from which there was a lower level of micromotion posteriorly compared with the anterior aspect.

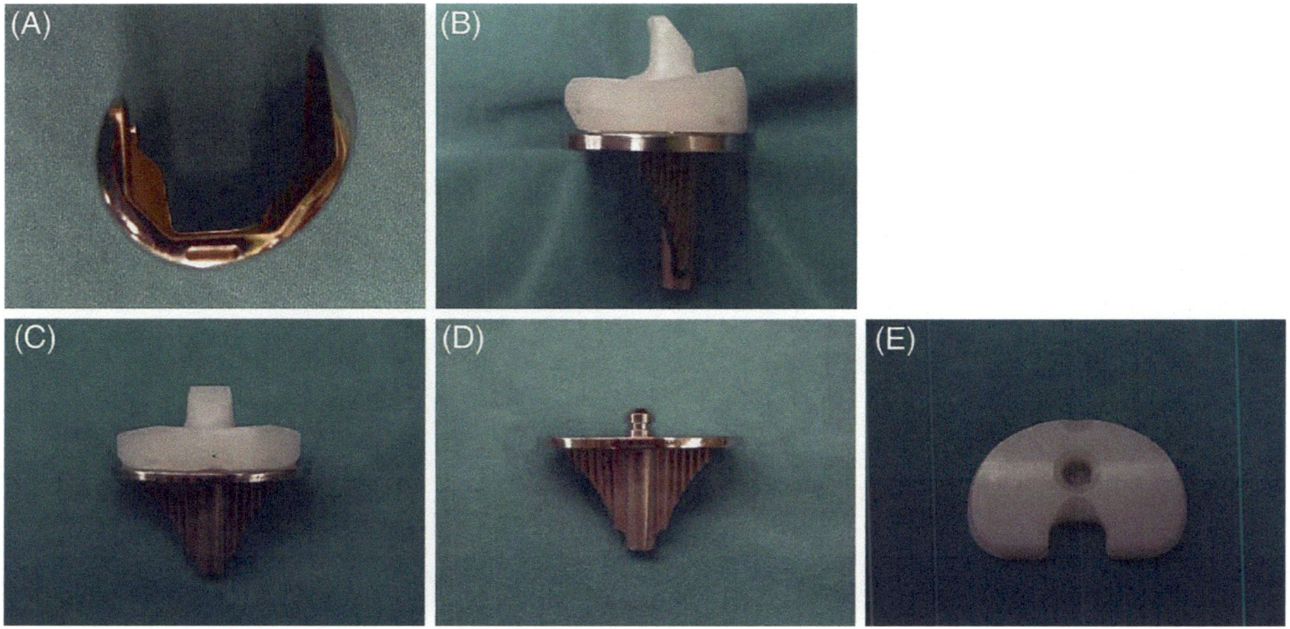

Figure 2.8 Illustration of a tibial tray with an extended keel in the mediolateral axis oblique to the posterior aspect. Such a keel design is used for both fixed and mobile bearing tibial knee replacement. In the mobile bearing type, the tibial insert features an internal hole that coincides with the external peg at the bearing surface of the tibial tray. *Through the courtesy of H. Kobayashi, N. Mitsugi, Y. Mochida, N. Taki, Y. Akamatsu, M. Aratake, H. Ota, K. Ishii, K. Harigane, T. Ideno, T. Saito, Mid- term results of stryker® scorpio plus mobile bearing total knee arthroplasty, Sports Medicine, Arthroscopy, Rehabilitation, Therapy & Technology 4 (2012). https://doi.org/10.1186/1758-2555-4-38 and Biomed Central Ltd.*

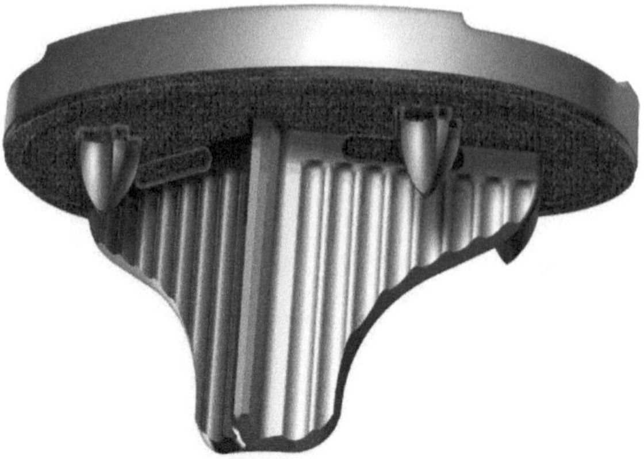

Figure 2.9 Illustration of a tibial tray prosthesis manufactured from cobalt-chromium (CoCr) alloy using laser melting. A tibia tray is normally fabricated from CoCr or Ti alloy using investment casting or forging. However, to create an effective long-term porous osteoconductive layer on the underside surface, the whole prosthesis is manufactured through additive manufacturing. This process reduces the risk for porous surface debonding at a high level of loading conditions that would be transferred at the interface of the tibia tray prosthesis and bone. Such a porous structure enhances cement interdigitation. *Through the courtesy of A. Miller, J. Stimac, L. Smith, ... A.F.-T.J. of, undefined 2018, Results of Cemented vs Cementless Primary Total Knee Arthroplasty Using the Same Implant Design, Elsevier, n.d. https://www.sciencedirect.com/science/article/pii/S0883540317310537. (Accessed 14 February 2021) and get permission from Elsevier.*

2.2.3.6 Interaction of bone with long stem in obese patients

Obesity or overweighting (body mass index [BMI] >30 kg/m^2) increases the risk for tibial tray loosening [9,28]. Obesity increases the effect of varus misalignment and generates excessive compression force at the interface of the bone—tibial tray prosthesis, in which the level of micromotion at the interface and the rate of bone resorption increase. Obesity causes loosening of the tibial tray in the form of tibial tray subsidence to the bone or fracture of tibial tray, as shown in Figs. 2.10 and 2.11.

With a shorter stem keel, a higher level of micromotion may be transferred or localized at the posterior aspect. The effect of the stem length on tibial tray loosening was studied by Garceau et al. [9]. In their research, loosening of the tibial tray was studied in two groups of tibial trays with a short stem length and an increased stem length using an extension to the short-stemmed tibial tray. Subsidence of the tibial tray was observed at the posterior aspect of the proximal tibial (Fig. 2.11) in a patient with a BMI greater than 40 kg/m^2 who had surgery with the short-stemmed tibial

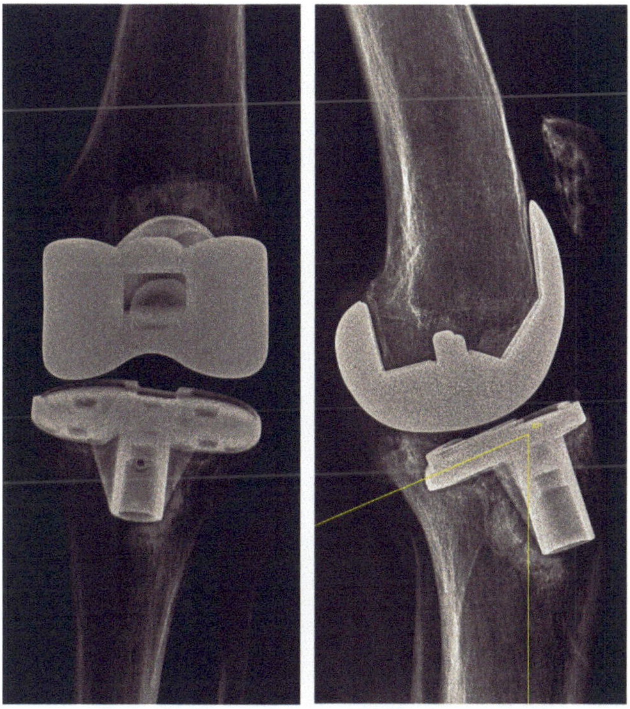

Figure 2.10 Radiographic illustration of severe tibial tray migration and subsidence through the metaphyseal portion of the proximal tibia. This radiograph belongs to a patient with a high body mass index (>35 kg/m^2). Use of a short stem tibial tray was considered by Garceau et al. to be for such patients [9]. A longer stem length for engagement with the diaphyseal portion of the proximal tibial might be advantageous for obese patients. Engagement with cortical bone at the diaphyseal portion enhances the stability of the tibial tray prosthesis under excessive compressive loading conditions at the knee joint and prevents subsidence in the sagittal plane. Another point is the use of the tibial tray with an extension keel to the posterior aspect of the tibial tray. The stem features keels in the mediolateral axis. In this radiograph, a radiolucent line is detectable at the posterior and anterior aspects of the femoral knee. *Through the courtesy of S. Garceau, S.P. Garceau, N.H. Harris, D.L. Felberbaum, G.M. Teo, A.I. Weinblatt, W.J. Long, Reduced aseptic loosening with fully cemented short-stemmed tibial components in primary cemented total knee arthroplasty primary knee, Journal of Arthroplasty 35 (2020). https://doi.org/10.1016/j.arth.2020.01.084 and get permission from Elsevier.*

tray. The stem extension was engaged with the diaphyseal portion of the proximal tibia, in which the stem was constrained by compact cortical bone. This engagement acts as a support base to keep the tibial tray in position and prevent rotation over the mediolateral axis or in the sagittal plane, whereas the short stem tibial tray was engaged with trabecular bone at the metaphyseal portion of the proximal tibia, in which the stem could not be effectively constrained as well as an extended stem length. Therefore, the use of a short stemmed tibial tray is risky in patients with a BMI in the overweight or obese zones. A high BMI in the obese zone may cause

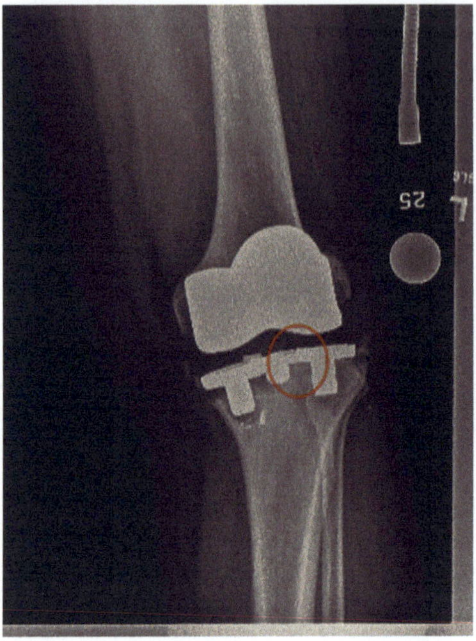

Figure 2.11 Radiographic illustration of tibial tray severe loosening. The tibial tray prosthesis has loosened first at the medial aspect. Radiolucent bone cysts are seen at the periprosthetic bone around the medial peg of the prosthesis. Because of resorption of the bone at this zone, higher loading conditions are transferred to the medial side of the prosthesis, which break the tibial tray at the highlighted area. Such a tibial tray is used in a minimally invasive surgical procedure. However, in patients with a high body mass index or with a perioperative varus deformity, the amount of load increases at the medial side of the knee joint, and thus the risk for prosthesis loosening and breakage increases. Overloading at the interface of the bone and prosthesis affects prosthesis osseointegration even in a highly porous structured implant such as tibial tray used in this patient. This shows the importance of mechanical interaction compared with biological bone ingrowth through the pores of the porous implant. *Through the courtesy of W. Scully, M. Deren, M.B.-A. today, undefined 2019, Catastrophic Tibial Baseplate Failure of a Modern Cementless Total Knee Arthroplasty Implant, Elsevier, n.d. https://www.sciencedirect.com/science/article/pii/S235234411930113X. (Accessed 27 September 2020) and get permission from Elsevier.*

loosening in the first year after implantation and a BMI in the overweight zone may result in loosening after 5 years postimplantation However, exact judgment cannot be provided to estimate possible loosening of the tibial tray after implantation. Using a long stem tibial tray establishes the position of the tibial tray through the midcanal of the tibia, in which the stem is fixed by the cortical bone. A stem that is quite rough, particularly at the distal portion, where the stem is inserted through the cortical bone, enhance the scratch-fit advantages of the tibial tray stem for better integration with the cortical bone. Such roughness is addressed as achievable by bead sintering and is reviewed in Chapter 6 of *Osseoconductive Surface Engineering for Orthopedic Implants*, written by the author.

2.2.3.7 Concept of energy shielding for the interaction of bone with tibial tray and bone

The discussion in this Section (2.2.3) shows the importance of design modeling parameters to enhance the interactions of the tibial tray prosthesis in contact with the proximal tibia. Design parameters have significant effects on the biomechanical behavior of the tibial tray against physiologic loading conditions at the knee joint that are transferred to the tibial tray component (compressive force in the axial axis, shear forces in multiple axes, and torsional moment over the axial axis).Because of the engagement of the tibial tray with the metaphyseal portion of the proximal tibia and the high level of transferred loading conditions to the tibial tray prosthesis, the level of micromotion at the interface of the prosthesis and trabecular bone might increase to over than 150 μm, at which bone resorption or osteolysis increases at the interface; thus, the prosthesis could migrate or subside through the spongy bone of the proximal tibia (Fig. 2.9). This necessitates various types of biomechanical and biological considerations to relieve the severe mechanical effect at the interface of the bone and prosthesis at the proximal portion of the tibia. The tibial tray is fabricated from CoCr or Ti alloys. The CoCr tibial tray is manufactured through highly precise forging or investment casting processes, whereas the Ti prosthesis is manufactured by conventional forging or selective laser melting (SLM). Manufacturing through forging increases the mechanical strength of the prosthesis so that it is high enough to endure a cyclic compressive load that is introduced to the tibial tray in the gait cycle, stair climbing, and other activities in a normal lifestyle. Fabricating the prosthesis by investment casting and hydroisostatic pressure (HIP) or an annealing solution supplementary process significantly reduces the fracture toughness compared with forging. As a result of melting spherical powders in SLM to fabricate solid or porous components, lower fracture toughness is expected for SLM-processed tibial tray prostheses compared with forging and investment casting processes. According to this view, the optimal rigidity or fracture toughness of the tibial tray is different for various types of manufacturing processes, which influences the absorbance of transferred energy in the tibial tray prosthesis. Lower transference of energy to trabecular bone at the proximal portion of the tibia is an advantageous material parameter for reducing applied strain to the bone matrix and the risk for bone resorption at the interface of the bone and prosthesis. Greater energy shielding or energy absorbance by the prosthesis causes the lower transference of energy from the prosthesis to bone tissue, which is beneficial at the interface of the metallic tibial tray and tibia. The porous structure of contact surfaces of the tibial tray with the bone at a high thickness allows a thickly layered bone matrix at the bone—prosthesis interface and the creation of a mechanical transition layer to reduce transferred energy gradually from the prosthesis to the bone. A developed tibial tray prosthesis with a porous structured underside surface is displayed in Fig. 2.9.

2.2.3.8 Effectiveness of biological integration in interaction of bone with tibial tray

The rate of loosening is increased in patients with a BMI in the obesity range (>30 kg/m^2) because of increased loading conditions between the implant and bone. Based on the previous discussion, biological integration is not effective as a load transferring condition at the interface of the bone and implant in obese patients. Failures that occur may be associated with the design aspects of the prosthesis and the extent of loading conditions, but the biological integration of the tibial tray prosthesis with the bone enhances the effectiveness of tibial tray stability in young patients who demand a greater level of motion and activity in normal life. Surface engineering processes such as porous structuring and bioactive coating promote the conduction of bone formation through the surface of the tibial tray prosthesis to facilitate biological integration at the bone—prosthesis interface. A greater level of micromotion in young patients is introduced to the interface of the bone and prosthesis by increasing the number of physiologic loading cycles. For instance, if an older patient walks 45 min/day (as recommended by surgeons) at about 1 Hz, young patients should walk, run, or climb stairs for more than 2 h/day at an average of 1.5 Hz. Thus, the number of cycles is increased to about three times in young patients compared with older patients. With an assumption of lower body weight in younger patients (because they are more active than elderly patients), critical loading conditions are thought of as fatigue loading conditions introduced in the form of shear stress in mediolateral and anteroposterior translations. The effect of shear stresses is significantly reduced by the enhanced biological integration of the prosthesis and bone. Various osteoconductive surface engineering methods such as porous structuring and bioactive coating (preferably in combination) establish a composite layer at the interface of the bone and implant composed of bone matrix and metallic beads or a porous structure, which act as a biological-mechanical transition layer to absorb the energy of shear stress and compressive energy at the bone—prostheses interface.

2.2.4 Loosening of femoral knee prosthesis

The femoral knee prosthesis is a component of the TKR. It is fabricated from CoCr alloy using investment casting. It is developed from zirconium alloy using forging for metal-sensitive patients. The prosthesis has a complex design among all orthopedic implants owing to the anatomic shape of the femoral condyles and patella groove at the distal portion of the femur. The femoral knee prosthesis might be loaded to a fivefold body weight compressive load, at which the prosthesis would be at risk for loosening at the periprosthetic bone interfaces. The interaction of the bone with the prosthesis is complex; thus, the bulk design and surface conditions of the prosthesis need to be optimized for effective mechanical and biological integration with the bone in long-term implantation. Figs. 2.12 and 2.13 show the schematic layout and femoral knee prosthesis, respectively.

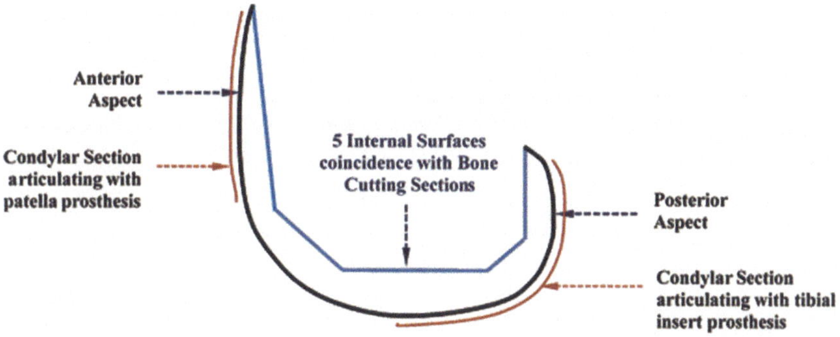

Figure 2.12 Illustration of a femoral knee prosthesis from the lateral view or anteroposterior axis. The articulation and internal sections are highlighted. The anterior and posterior aspects of the prosthesis are seen. The femoral knee is articulated with a polymeric tibial insert in contact with trabecular bone at the metaphyseal portion of the distal tibia.

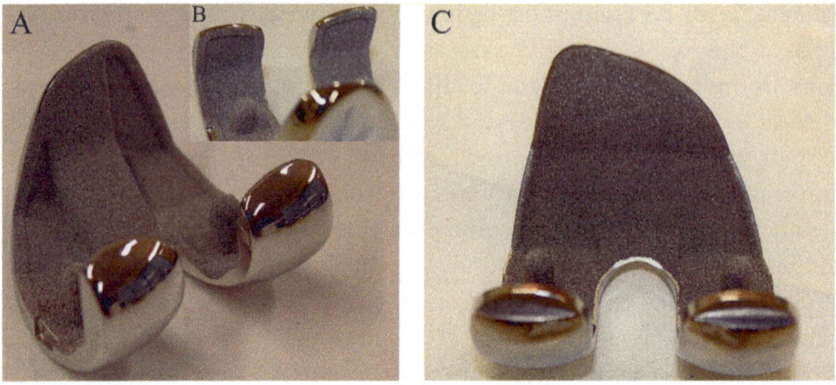

Figure 2.13 Illustration of a femoral knee prosthesis for replacement at the resected distal femur in total knee arthroplasty. This prosthesis is manufactured from CoCr alloy. In the cementless type, the internal surface of the prosthesis is coated or structured in a porous form with titanium, tantalum, or CoCr materials. The prosthesis is sintered with CoCr beads to enhance bone ingrowth at the bone–prosthesis interface. *Through the courtesy of S. Berahmani, M. Hendriks, D. Wolfson, A. Wright, D. Janssen, N. Verdonschot, Experimental pre-clinical assessment of the primary stability of two cementless femoral knee components, Journal of the Mechanical Behavior of Biomedical Materials 75 (2017) 322–329. https://doi.org/10.1016/j.jmbbm.2017.07.043 and get permission from Elsevier.*

2.2.4.1 Interaction of resected bone with femoral prosthesis

Loosening of the femoral knee prosthesis has been highlighted in the literature [28,30–38]. The anterior and posterior aspects of the femoral knee prosthesis are the critical zones in occurrence of the femoral knee loosening, as illustrated in Fig. 2.14 [39]. The femoral knee component is designed to have a condylar profile at the bearing surfaces, similar to contact surfaces created through five cut sections by the associated surgical instruments (cutting guide and saw) during replacement of the femoral knee prosthesis. Overall, these five cut sections would not determine the condylar profile. This is why load transference between the femur and femoral knee prosthesis is not physiologically distributed through the cutting sections (the femoral condyles at the medical and lateral aspects are circular in a healthy knee joint). This design-related feature of the femoral knee prosthesis in internal contact with the resected distal femoral condyles alters physiologic load transferring mechanisms or patterns at the knee joint, in which the reaction force from the knee bearing articulation interface between the femoral knee and tibial insert components is patterned, as shown in Fig. 2.15.

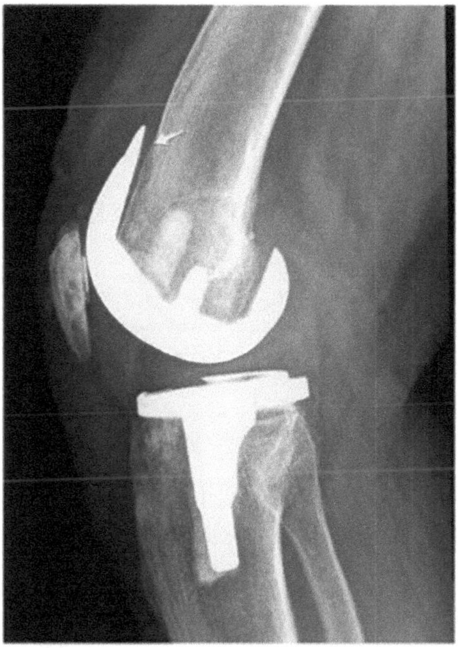

Figure 2.14 Radiographic illustration of femoral knee loosening at the anterior aspect of the prosthesis. A radiolucent line and bone cysts are seen at the most anterior internal section of the prosthesis with the cutting bone at this zone. As discussed in Section 2.2.3.6, the low range of conditions at this section increases stress shielding and progressive bone resorption in long-term implantation, ultimately loosening the prosthesis. *Through the courtesy of R. Dehdari, M.T.-B.M. Imaging, undefined 2013, Orthopedic Hardware and Complications, McGraw Hill Prof. n.d. and get permission from McGraw Hill – need permission.*

The axis of reaction force in angular rotation of the femoral knee relative to the tibial insert in the flexion-extension axis is not perpendicular to the cutting section. Therefore, homogeneous load distribution is not obtained at the whole surface of the cutting sections. This cause an inhomogeneous reaction of bone tissue to counteract the effect of loading conditions at the interface of the bone and femoral knee prosthesis. In the tibial tray prosthesis, the axis of transferred force through the bone—prosthesis interface is perpendicular to the cutting section, in which bone tissue is effectively oriented in an axis similar to the loading axis. This advantage is not expected at the bone—femoral prosthesis to be as effective as that which occurs at the interface of the bone and tibial tray prosthesis. Both the femoral knee and tibial tray are in contact with trabecular bone, in which reorientation or

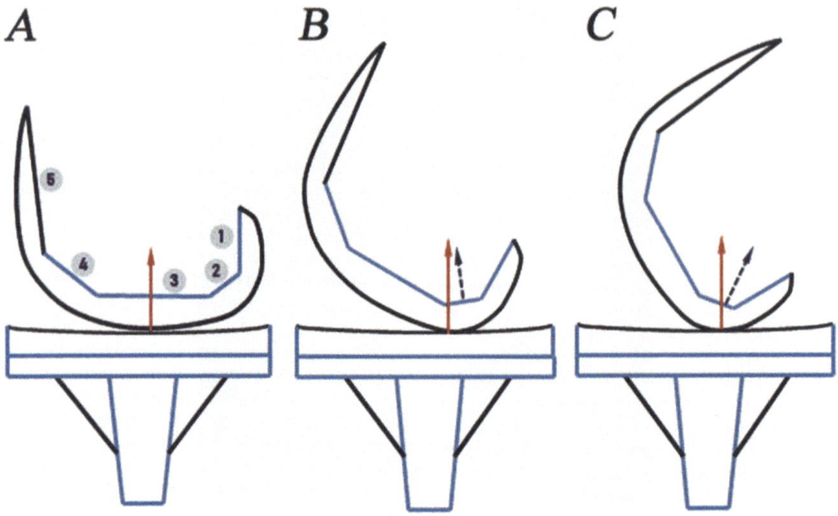

Figure 2.15 Illustration of deviation of reaction force axis with perpendicular axis to femoral knee prosthesis at the interface surface with bone at different flexion angles: (A) 0 degrees; (B) 30 degrees; and (C) 60 degrees. The reaction force is generated at the articulation of the femoral knee and tibial insert prosthesis, and its axis is considered axial or coincident with the femur mechanical axis at all flexion angles. Thus, deviation of the reaction force and interface axes at the interfaces of the femoral knee prosthesis and cutting bone sections causes ineffective transference of physiologic loading through the periprosthetic bone, particularly in most posterior and anterior aspects of the femoral knee prosthesis. This increases the risk for prosthesis loosening, as illustrated in Figs. 2.16, 2.18, and 2.19.

rearchitecture of the bone matrix is actively in process owing to the high porosity of trabecular bone compared with cortical bone matrix.

2.2.4.2 Interaction of bone with femoral prosthesis at various flexion-extension angles

A sufficient level of loading conditions is transferred to bone tissue at the interface of the femoral knee prosthesis and bone at a range of flexion angles, 0°—60°, which is more engaged in the gait cycle. For interfaces in the flexion angle of 60°—100°, the reaction load from the prosthesis to the femoral condyles is applied in specific activities. For interfaces positioned at the anterior aspect of the femoral knee prosthesis, the femoral knee prosthesis is in contact with the patella or prosthesis. In complete extension, the patella is in contact with most anterior section of the femoral knee prosthesis, and by increasing the flexion angle, it moves through the patella groove of the femoral knee prosthesis. Contact force between the patella and femoral knee prosthesis at the extension of the knee joint is in the low range and gradually increases with an increase in the flexion angle. According to this view, most anterior cutting sections (cutting Sections no. 4 and 5 in Fig. 2.15A) are not loaded to a high extent. Similar conditions occur for most posterior cutting sections (cutting Section no. 1 in Fig. 2.15A). Through this loading distribution at the interface of the resected distal femur and femoral knee prosthesis, insufficient loading conditions are transferred to periprosthetic trabecular bone tissue at these zones. This phenomenon causes a low rate of micromotion or strain required to induce osteogenic cells to activate at the interface zones, in which the bone resorption or activity of the osteoclasts gradually increases and bone–implant integration or interaction is lost. Through this failure, femoral knee loosening occurs, as highlighted in clinical studies [31,38]. Some radiographic scan images of femoral knee prosthesis loosening at these zones are illustrated in Figs. 2.14, 2.16, and 2.17.

2.2.5 Loosening of hinged knee prostheses

The rotating hinged knee system is at risk for loosening, as highlighted in clinical studies [12,20,23,28,34,40–43]. The hinged knee system is used to treat severe osteoarthritis with significant bone loss in the metaphyseal section of the distal femur or proximal tibia. Use of the hinged knee system is necessary revision surgeries for loosening prostheses that were replaced in primary TKA. Hinged knee prostheses feature a long stem for engagement with the intramedullary canal of the femur and tibia for effective mechanical stability in long-term implantation. The risk for reloosening of the hinged knee system needs to be low to prevent revision surgery. Revision surgeries incur severe clinical complications to patients and significantly reduce the satisfactory quality of life. However, the risk for the hinged knee prostheses reloosening remains high (possibly above 20%). Because of the high rate of prosthesis loosening (particularly the femoral part) in a fixed-hinged knee system, the rotating hinged knee system was developed with design and features developed for current primary TKRs. The rotating knee system reduces excessive stress concentration at the bone–implant and bone–cement interfaces, in which the rate of femoral loosening is significantly decreased. Other design features such as a medial pivot design of the tibial insert, a deeper design of the patellofemoral groove, and similar cutting sections of the distal femur as that of the primary TKA, were

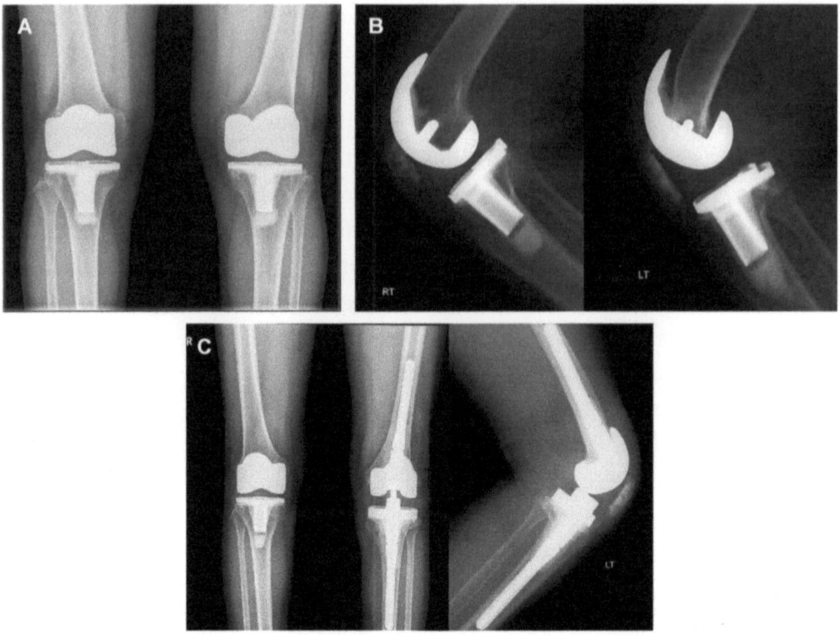

Figure 2.16 Radiographic illustration of severe resorption of periprosthetic bone at the posterior and anterior aspects of femoral knee prosthesis interfaces with the bone. This patient was operated on both knees: the right knee with conventional tibial knee replacement (TKR) and the left knee with high-flexion TKR (A). Severe periprosthetic bone loss was detected in high-flexion TKR, as seen in the right radiograph (B). The patient was then reoperated on with replacement of the hinged knee system, as seen in radiograph image (C). *Through the courtesy of Y. Kim, J. Park, J.K.-T.J. of Arthroplasty, undefined 2017, Do High-Flexion Total Knee Designs Increase the Risk of Femoral Component Loosening?, Elsevier, n.d. https://www.sciencedirect.com/science/article/pii/S088354031730058X. (Accessed 20 September 2020) and get permission from Elsevier.*

considered for newest versions of the rotating hinged knee system. Consideration of IE rotation in designing the rotating hinged knee system allows physiologic laxity in tibiofemoral articulation to prevent excessive IE torque at the articulation surface and transferring of increased torque at the interface of the femoral prosthesis and bone. As highlighted by Kouk et al. [40], midterm and long-term clinical studies report complications associated with old designs of rotating and fixed hinged knee systems. Thus, the success rates for more recent developments with the design of hinged knee systems might not be relevant. However, this chapter attempts to discuss the

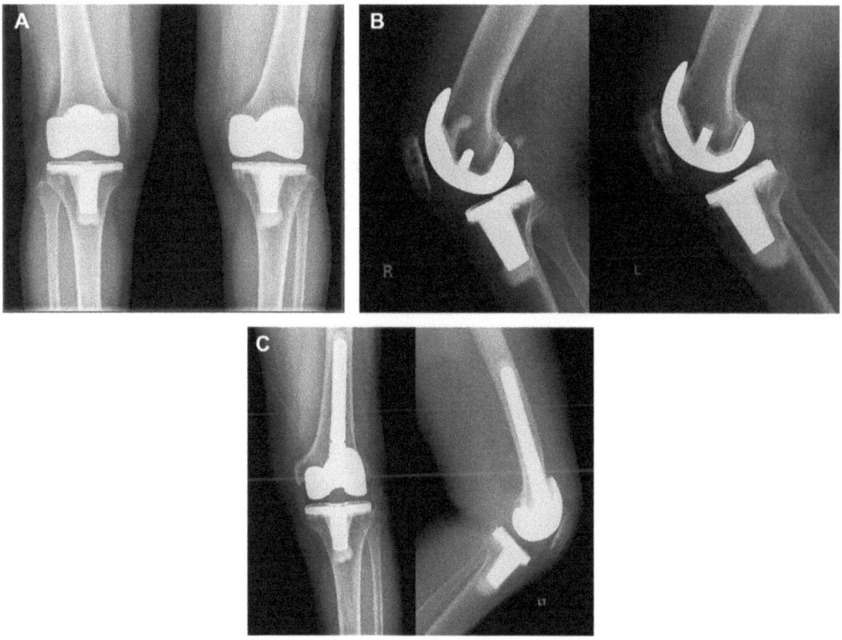

Figure 2.17 Radiographic illustration of femoral knee loosening in primary total knee arthroplasty (TKA) (A, B). The loosened femoral prosthesis was replaced with a new femoral prosthesis from the knee revision system (C). As illustrated in the right radiograph of image (B), a radiolucent line at the posterior interface and bone cysts at the anterior interface is detected, showing a loss of integration between the bone and implant at these zones. *Through the courtesy of Y. Kim, J. Park, J.K.-T.J. of Arthroplasty, undefined 2017, Do High-Flexion Total Knee Designs Increase the Risk of Femoral Component Loosening?, Elsevier, n.d. https://www.sciencedirect.com/science/article/pii/S088354031730058X. (Accessed 20 September 20, 2020) and get permission from Elsevier.*

effectiveness of the newer development concepts in reducing prosthesis loosening of hinged knee systems.

2.2.5.1 Interactions of bone with developed third generation hinged knee systems

Aside from the design considerations highlighted in the previous section, other design concepts have been contemplated in versions of rotating hinged knee systems. A reductions in the hinged joint box in the femoral prosthesis of the hinged knee system enhances the stability of the hinged knee system in the metaphyseal portion of the distal femur. This results in greater bone preservation in the femoral prosthesis in hinged knee replacement. The femoral prosthesis is completely constrained by the inserted hinged joint component, which is positioned through the tibial tray prosthesis, from which translation of the femoral prosthesis in the mediolateral and posteroanterior axes is entirely fixed. Stress shielding at the interfaces of the femoral prosthesis in a hinged knee system is increased compared with a femoral knee prosthesis in the primary knee system. In Section 2.2.4.1, the stress shielding effect at the cutting sections at the interface of the femoral knee prosthesis and bone are highlighted in conjunction with an in-depth discussion on possible reasons. In addition to those cutting sections at the femoral condyles, the interface of the hinged joint box with trabecular bone at the metaphyseal portion of the distal femur is at the risk for stress shielding because of the lack of transference of effective loading stresses through these interfaces. It needs to be clarified through long-term clinical observation whether loosening of a femoral prosthesis with a metaphyseal distal femur causes loosening of the whole femoral prosthesis with an extended stem through the diaphyseal portion of the femur. The mechanical functionality of the hinged joint at the metaphyseal portion of the distal femur depends on the integrity of the extended stem portion with the diaphyseal cortical portion of the femur and integrity with the trabecular may not be as influenced. The engaged interfaces of the prosthesis stem at the diaphyseal cortical bone can bear greater loading conditions whereas the metaphyseal portion cannot have load-bearing functionality. However, the long-term stability of the femoral prosthesis in a rotating hinged system with loosening in the metaphyseal zone and engagement with the diaphyseal bone in the distal third of the femur might is not effectively achieved because of overloading at the diaphyseal zone and gradual bone resorption at this zone and loosening of the prosthesis. Therefore, the enhanced load-bearing share of the prosthesis at the metaphyseal zone is beneficial for the long-term functional effectiveness of the femoral knee prosthesis of the rotating hinged knee system. One solution could be a transverse porous component in the screw or bar design to be assembled to the box portion of the femoral prosthesis of the hinged knee system in the mediolateral axis to enhance the engagement of the femoral prosthesis with trabecular bone while bearing a portion of physiologic loading conditions. Through the growth of the bone matrix through the pores of the assembled porous component, biological integration of trabecular bone at the metaphyseal zone to the femoral prosthesis would be effectively achieved. This is an idea that could be more explored by product developers along with adequate biomechanical, biological, and animal evaluations. Another design feature in current versions of the rotating hinged knee system is an effective hinged joint at the knee joint for enhanced ROM, particularly in the flexion-extension axis, with no dislocation of the femoral prosthesis from the hinged joint component. Likewise, the long-term durability of components, particularly at the hinged zone, is crucial for the long-term effectiveness of the hinged knee system.

2.2.5.2 Interaction of bone with hinged knee replacement using porous metaphyseal fixation

Porous structuring of the hinged prostheses at contact surfaces with the metaphyseal portion of the distal femur and proximal tibia is best to conduct bone ingrowth through the surface of the prosthesis and reduce the risk for loosening at these interfaces. Shear stress might be generated by micromotion in the caudocranial axis of the femur and tibia, where loading conditions are stimulated at the interface of the prosthesis and metaphyseal trabecular bone to induce osteogenic cells to reconstruct bone matrix through the porous structured interfaces. The porous structure could be processed with methods such as bead sintering and titanium plasma spraying. In some cases of prosthesis loosening in primary TKA, bone loss or defects at the metaphyseal portion of the distal femur or proximal tibia might occur. Thus, it is essential to remove the defective bone and replace it with an adequate prosthesis or implant to fill the gap between the resected bone and knee prosthesis at the metaphyseal zone. This is known as metaphyseal fixation of the hinged knee prosthesis in the clinical literature. This additive prosthesis is designed as a sleeve or cone. The sleeve prosthesis is thinner than the cone prosthesis and is entirely structured as a bulk form, whereas the cone is thick and is manufactured in fully porous or combined bulk-porous forms. Small bone defects in metaphyseal bone are replaced with the sleeve, which is categorized as type 1 or 2A in the Anderson Orthopedic Research Institute (AORI) classification method. Larger bone defects, categorized as AORI type 2B or 3, are treated with a cone prosthesis.

Fabrication of porous structured cones is managed with tantalum (Ta) or titanium (Ti) alloy through vapor deposition or SLM, respectively. All well-established orthopedic implant manufacturers have developed various designs for metaphyseal porous cones for insertion through the metaphyseal portion of the distal femur and proximal tibia to enhance severe bone loss treatment in knee revision surgery or primary TKA. The mechanical stiffness of the porous cone needs to be low compared with bulk titanium or tantalum bulk form. By the adequate design of the porous cone in CAD modeling software, its mechanical stiffness can be examined by finite element software to optimize the model with enhanced mechanical properties close to those of trabecular bone [44—49]. Because of the higher mechanical modulus of Ta compared with Ti alloy, a higher porosity level can be obtained for the Ta porous cone close to that of trabecular bone (up to 85%) while maintaining mechanical properties to those of trabecular bone. This allows effective mechanical interaction of the porous cone with trabecular bone at the metaphyseal portions of the distal femur and proximal tibia. Development of a fully porous cone was initiated from tantalum material by a well-established manufacturer in the orthopedic industry. A porous cone was designed in bulk-porous form and fabricated from titanium alloy. The internal surface was lasered in bulk form and the external surfaces in contact with metaphyseal bone are lasered in porous form (Fig. 2.18). The bulk portion enhances the mechanical strength of the cone while reducing the cone's thickness. Fig. 2.19 illustrates the position of the hinged knee prostheses in metaphyseal portions of the femur and tibia in conjunction with the titanium cone. To enhance the osteoconductive effectiveness of the porous cone, it can be coated with a bioactive material (calcium phosphate) to enhance bone formation through the pores of the porous cone [47].

With the use of the porous cone at the metaphyseal distal femur or proximal tibia, the rate of prosthesis loosening significantly decreases. With the progression of bone ingrowth through the porous cone, stronger integration of the prosthesis at the metaphyseal zones with trabecular bone tissue is achieved. The effective stability of the hinged knee prosthesis at the stem part with diaphyseal cortical bone enhances load transference between the trabecular bone and porous cone. Because of the lack of collateral ligaments in revision surgeries, the stability of the knee prosthesis in lateral-medial and posteroanterior axes is challenging, and replacing a long length of the prosthesis stem through the diaphyseal zone might be insufficient to hold the prostheses at its initial replacement position in long-term implantation. The porous cone acts as a mechanical-biological transition layer for the effective transfer of load in the lateral-medial and posteroanterior axes, enhancing induction of bone remodeling and preventing the severe transfer of energy to the trabecular bone and damage to the periprosthetic bone

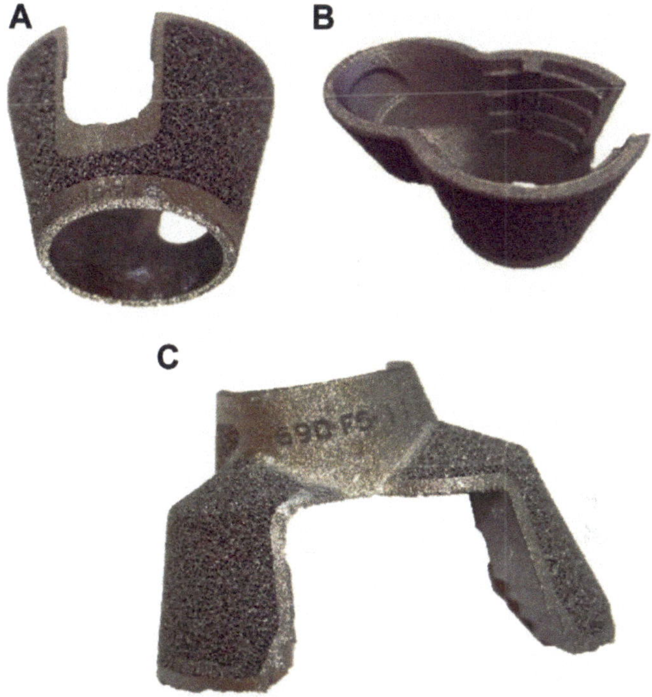

Figure 2.18 Illustration of porous cone for use with hinged or revision knee system. The porous cone is manufactured from titanium alloy using SLM or from tantalum using vapor deposition coating on a polymeric foam structure. The titanium cone is fabricated in a bulk-porous form, as illustrated in this figure, to be optimized in a size suitable for small bone loss. The tantalum cone is fabricated in a porous form entirely in a larger thickness and pore size suitable for significant bone loss at the metaphyseal portion of the distal femur and proximal tibia. *Through the courtesy of K. Denehy, S. Abhari, ... V.K.-T.J. of, undefined 2019, Metaphyseal Fixation Using Highly Porous Cones in Revision Total Knee Arthroplasty: Minimum Two Year Follow Up Study, Elsevier, n.d. https://www.sciencedirect.com/science/article/pii/S0883540319303018. (Accessed 29 September 2020) and get permission from Elsevier.*

matrix and resorption of the bone in the final stage. The use of the porous cone to reconstruct trabecular bone loss at these zones was introduced with a lower rate of hinged knee loosening in midterm and long-term follow-up studies. Figs. 2.20 and 2.21 display cases treated with hinged knee system in conjunction with porous tantalum and titanium cones, respectively.

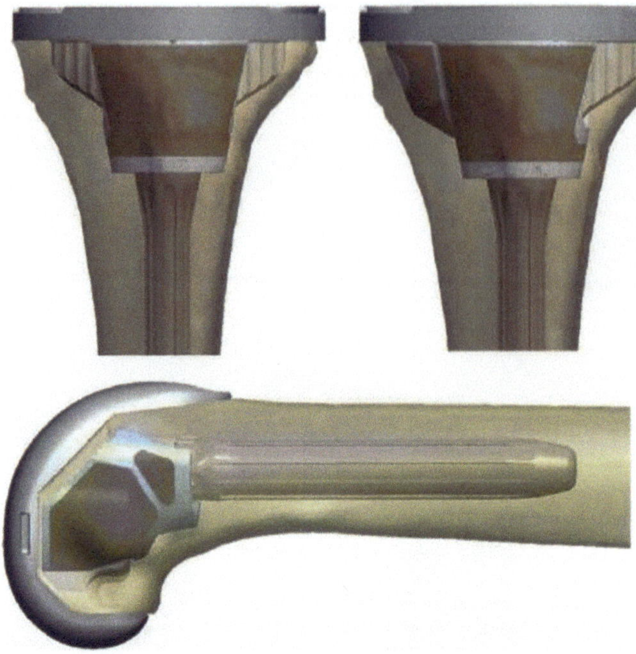

Figure 2.19 Illustration of porous cone replacement in contact with the femoral and tibial prostheses of the hinged knee system. The porous cone shields the femoral and tibial prostheses while filling bone loss zones. The porous structure of the cone enhances conduction through the pores for better integration of the overall prosthesis with trabecular bone tissue at the metaphyseal zone. Effective load transference at the interface of the cone and trabecular bone enhances calcification of the new bone matrix formed in the pores of the porous cone. As highlighted by Abdelaziz et al. [49], loosening of the hinged knee system with a porous cone was observed in some cases. *Through the courtesy of A. Faizan, M. Bhowmik-Stoker, V. Alipit, … A.K.-T.J. of, undefined 2017, Development and Verification of Novel Porous Titanium Metaphyseal Cones for Revision Total Knee Arthroplasty, Elsevier, n.d. https://www.sciencedirect.com/science/article/pii/S0883540317300153. (Accessed 29 September 2020) and get permission from Elsevier.*

2.3 Interactions of bone with ankle joint prostheses and possible failures

In this section, the interactions of ankle prostheses with the distal tibia and talar are explored and possible reasons for reported failures in the clinical literature are discussed [50–61].

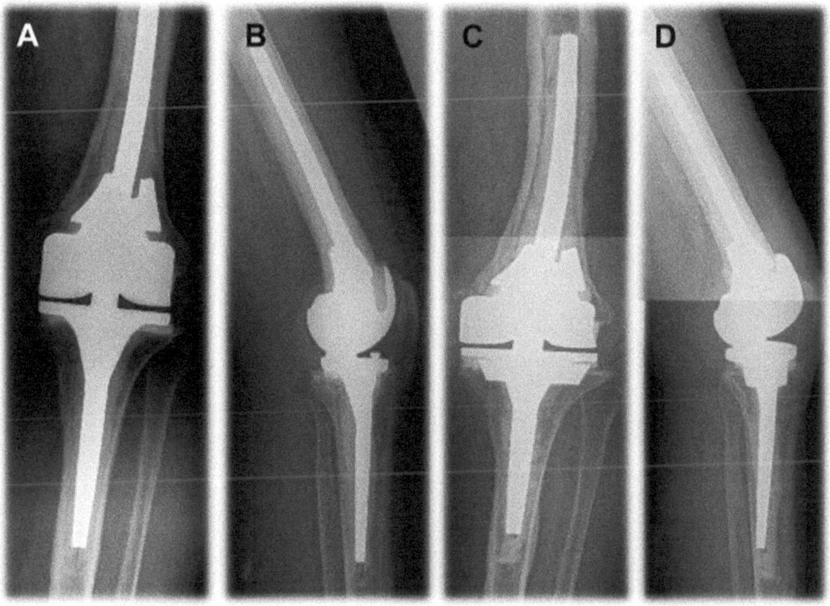

Figure 2.20 Radiographic illustration of hinged knee system and tantalum porous cone in revision of a failed primary total knee arthroplasty. In this case, bone loss was detected at the metaphyseal distal femur at the time of revision operation, which necessitated the use of a porous cone. After 5 years, the tibial tray prosthesis of the implanted hinged knee system loosened or migrated anteriorly (A, B), affecting the performance of the hinged joint. Thus, a rerevision operation used the porous cone for metaphyseal fixation of the prosthesis at the proximal tibia (C, D). *Through the courtesy of H. Abdelaziz, R. Jaramillo, T. Gehrke, … M.O.-T.J. of, undefined 2019, Clinical Survivorship of Aseptic Revision Total Knee Arthroplasty Using Hinged Knees and Tantalum Cones at Minimum 10-Year Follow-Up, Elsevier, n.d. https://www.sciencedirect.com/science/article/pii/S0883540319306485. (Accessed 27 September 2020) and get permission from Elsevier.*

2.3.1 Ankle joint complex

The ankle joint complex consists of two main joints: tibiotalar and subtalar. The ankle complex bears up to five times the body weight loading conditions in normal daily activities. The extent of loading conditions exceed the body weight when the foot is completely positioned on the ground, through which the extent of the reaction force from the ground increases multiple times compared with the body weight force. Three types of motion are observed in movements of the ankle joint complex: (1) dorsiflexion-plantar flexion, (2) eversion-inversion, and (3) abduction-adduction. The combination of eversion and abduction rotation makes protonation motion,

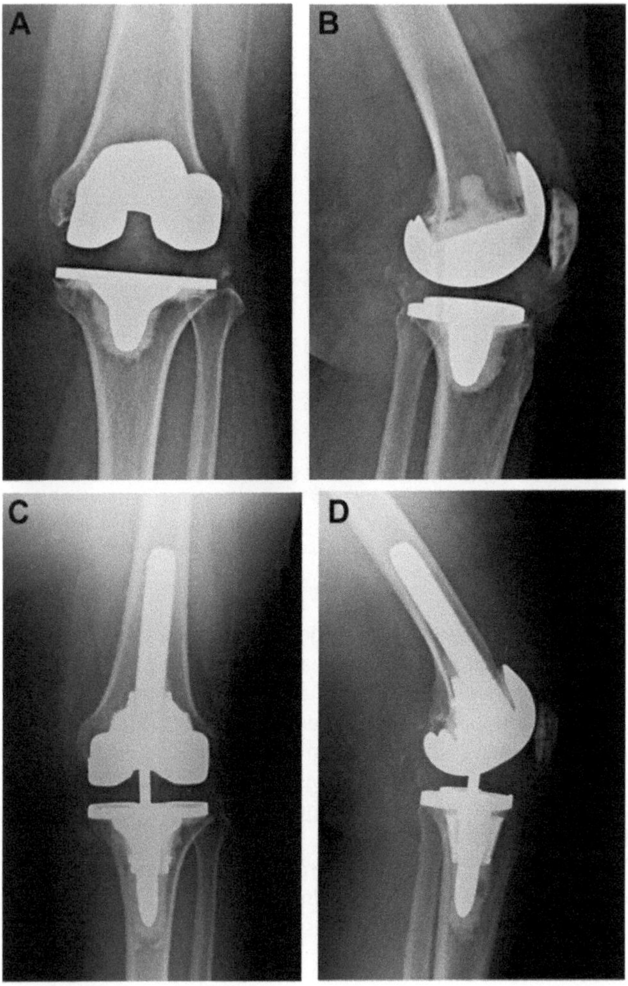

Figure 2.21 Radiographic illustration of the hinged knee system and titanium porous cone (C, D) in revision of a failed primary total knee arthroplasty (A, B). In this case, both the tibial tray and femoral prostheses loosened and their instability caused complications for the 62-year-old woman. A titanium porous cone was used to replace removed bone lost at the metaphyseal distal femur and proximal tibia. Bone loss occurred owing to bone resorption at the interface of the bone and implant or as the result of removal of the failed prosthesis and bone cement, from which a portion of trabecular bone at the metaphyseal zone might have been affected. *Through the courtesy of K. Denehy, S. Abhari, ... V.K.-T.J. of, undefined 2019, Metaphyseal Fixation Using Highly Porous Cones in Revision Total Knee Arthroplasty: Minimum Two Year Follow Up Study, Elsevier, n.d. https://www.sciencedirect.com/science/article/pii/S0883540319303018. (Accessed 29 September 2020) and get permission from Elsevier.*

whereas combined inversion and adduction rotations of the ankle joint complex create supination motion. In these motions at the ankle joint complex, both tibiotalar and subtalar joints are involved, in which the tibiotalar joint is more active in dorsiflexion-plantar flexion motion and subtalar is more involved in supination-pronation motion. The ankle joint complex is constrained by strong ligamentous soft tissues to limit ROM in protonation at 15–30°, supination at 45–60°, dorsiflexion up to 20°, and plantar flexion up to 50°. Compared with other skeletal joints such as the knee, hip, shoulder, and elbow, such ROM is significantly lower to keep the joint stiffer for better bearing of large loading conditions transferred to the ankle joint complex.

2.3.2 Tibiotalar joint injuries

Conventionally, end-stage arthritis at the tibiotalar joint is treated through arthrodesis or fusion of tibia and talus at the talocrural or tibiotalar joint. Arthrodesis of the tibiotalar joint is a temporary solution for reducing pain at this joint with no movement of the ankle in gait and other normal activities in normal life. By subjecting body weight loading conditions through the fused joint, other joints are at risk for arthrosis owing to overloading compared with physiologic conditions. Likewise, a low rate of satisfaction is with arthrodesis of the tibiotalar joint, particularly in patients who need to return to daily activities. Thus, a fused tibiotalar joint is not as effective and forces the patient to limp during walking and the gait cycle. Because of the high rate of complications in treating ankle arthrodesis in the tibiotalar joint, surgeons are motivated to invent solutions to establish movement conditions at the ankle joint by replacement of an artificial joint composed of one inserted prosthesis through the distal tibia and one through the talus. Treatment of the ankle joint is clinically controversial regarding the selection of the best implantation solution and satisfactory results. This is because of the complexity of available bones in this part of the body and various types of bone injuries that occur simultaneously. Normally, bone fracture occurs in the ankle joint as a result of high loading conditions in severe incidents such as falling from a height and sports. Such conditions occur at a younger age, particularly in athletes. In elderly patients, arthrosis of the tibiotalar joint occurs owing to long-term compressive force generated by standing on the feet for long-time when the patient has a BMI in the obese or overweight zone. Treatment of bone fractures at the distal portion of the tibia or in the talus results in arthrosis at the tibiotalar joint; when cartilage tissue at this joint is affected, it is known as posttraumatic arthritis. Arthrosis causes the direct contacting of bone to bone and generates severe inflammation and pain at the joint.

2.3.3 Interaction of bone with tibiotalar joint replacement

To treat arthritis in the tibiotalar joint, multiple generations of ankle replacements have been developed since the 1970s. The first total ankle replacement (TAR) generations had various complications and failures, in which loosening of the tibial and talar prostheses had a high rate (up to 40%). Subsidence and collapse of the talar prosthesis were clinically observed to necessitate revision surgery. The bone stock in both distal tibia and talus to replace tibia and talar prostheses is not large. Cortical bone over the total area at the cross-section of the distal tibia and proximal aspect of the talus considerably enhances load bearing. There is a small area of trabecular bone in the center of the cutting resections to replace tibia and talar prostheses. These small areas in the distal tibia and proximal talus are considered for the design of underside surface pegs for insertion and stabilization of tibia and talar prostheses through the bone. In the third generation, the peg of the talar prosthesis is small, whereas the peg for the tibia prosthesis is larger or like a stem for insertion through the distal third of the tibia. The other design point is to completely seat the tibial and talar prostheses on the cortical area of the cutting sections of the distal tibia and proximal talus for effective load transference between the bone and prostheses to prevent prosthesis subsidence through the trabecular bone. The talar underside surface is designed to be circular to coincide anatomically with the curvature profile of the talus at the tibiotalar joint. This necessitates resection of the talus in a curved or circular form for enhanced seating of the underside surface. It is advantageous to resurface the talus as low as possible to preserve the surface anatomic profile. Through this feature, physiological load transference from the tibia to the talus is managed with the highest load bearing capacity. The anatomic design of the underside surface in a talar prosthesis is achieved at the lateral areas and the center zone features a peg to fix the talar prosthesis to the talus. The anatomic design is also carried out for the topside surface of the talar prosthesis, where the prosthesis is in articulation with the polymeric insert prosthesis. The design profile of the talus at the tibiotalar joint in a sagittal plane or in the anteroposterior axis would be completely modeled as far as possible. This design concept would allow physiologic load distribution through the talar surface in dorsiflexion-plantar flexion motion, from which the contact force is maintained to have maximal contact with the surface, significantly reducing the risk for articulation wear. To replace the tibial prosthesis, up to 9 mm of distal tibia is cut to create sufficient space to place the tibial tray and tibial insert prostheses. The tibial tray prosthesis is flat in peripheral areas and it has a peg in the center area. The cutting section of the distal tibia is prepared in accordance with the underside surface of the tibial ray for the highest conformity with the prosthesis contact surfaces with the bone. The tibial tray is extended to project at the medial aspect to be seated on the medial malleolus aspect of the distal tibia. This featuring distributes physiologic loading conditions through the tibiotalar and tibiocalcaneal ligaments similar to what would be physiologically obtained in a healthy ankle joint. A design for TAR is illustrated in Fig. 2.22 and its replacement at the tibiotalar joint is radiographically shown in Fig. 2.23.

2.3.4 Interaction of bone with cementless tibiotalar joint replacement

Because of the small zone for replacing the tibial and talar prosthesis, as discussed in Section 2.3.3, fixation of the ankle prostheses is not be manageable

Figure 2.22 Illustration of a developed total ankle replacement (TAR) with the latest design concept as the fourth-generation TAR system. The tibial base or tray prosthesis (*top gray component*) is manufactured from titanium alloy implanted into the distal portion of the tibia. The tibial insert prosthesis (*middle white component*) is manufactured from ultrahigh molecular weight polyethylene polymer as the bearing prosthesis. The talar prosthesis (*bottom gray components*) is manufactured from cobalt-chromium alloy with appropriate bearing articulation for a minimal contact wear rate with a polymeric tibial insert. The contact surface of the metallic prostheses with the bone feature a porous osteoconductive engineered surface that might be titanium plasma sprayed, CoCr bead sintered, titanium bead sintered, or bonded with a tantalum three-dimensional porous structure. *Through the courtesy of C. Rushing, R. Law, C.H.-T.J. of F. and A. Surgery, undefined 2020, Early Experience with the CADENCE Total Ankle Prosthesis, Elsevier, n.d. https://www.sciencedirect.com/science/article/pii/S1067251620303252. (Accessed 7 October 2020) and get permission from Elsevier.*

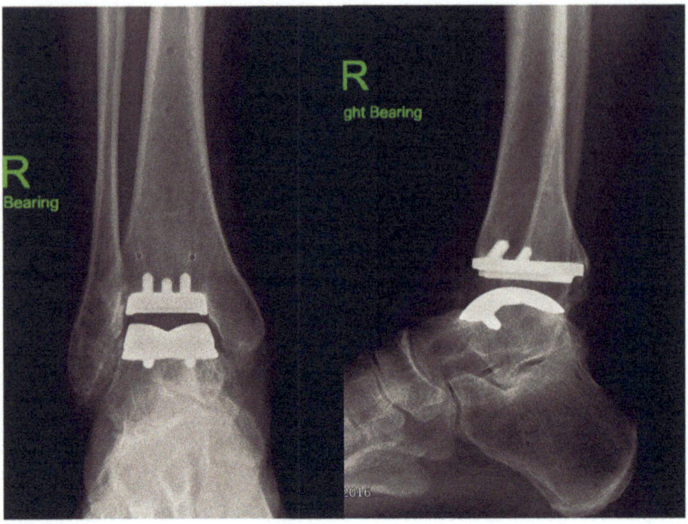

Figure 2.23 Radiographic illustration of total ankle replacement (TAR) at the implanted site. The polymeric tibial insert prosthesis is radiolucent and is not observed on radiographic imaging. The design of illustrated TAR system is the fourth generation of this type of joint replacement since the 1970s [60]. Up to 9 mm of the distal tibia is cut to place the tibial base prosthesis, whereas the talar prosthesis is implanted with maximal preservation of the talus at the tibiotalar joint. *Through the courtesy of A. King, N. Bali, A. Kassam, A. Hughes, … N.T.-F. and A., undefined 2019, Early Outcomes and Radiographic Alignment of the Infinity Total Ankle Replacement with a Minimum of Two Year Follow-Up Data, Elsevier, n.d. https://www.sciencedirect.com/science/article/pii/S1268773118300821. (Accessed 7 October 2020) and get permission from Elsevier.*

with bone cement compared with replacements in the knee, hip, or shoulder. Therefore, fourth-generation TARs were developed with cementless prostheses, where the contact surfaces of prostheses with bone tissue features an osteoconductive layer to allow bone ingrowth for enhanced biological integration of the prosthesis through bone tissue [62]. Use of a titanium porous structure through plasma spraying and porous bead sintering and coating of bioactive materials through the pores of the porous structured layer are important engineering processes for developing cementless joint prostheses that are used for ankle replacement prostheses. Porous structured pegs in tibial and talar prostheses enhance integration of the pegs with trabecular bone (pegs of talar and tibia prostheses are in contact with trabecular bone, whereas the stem type of tibial tray prosthesis has interfaces with cortical bone at the diaphyseal zone of the distal third tibia in the proximal portion and with trabecular bone at the distal section). The rod-type peg or stem does not have effective mechanical interaction with trabecular bone; thus, stress shielding occurs at the interface of these stabilization features with trabecular bone. The creation of grooves or projections at the surface of the peg or stem generates mechanical stress at the interface with trabecular bone. A tibial tray could be developed from porous tantalum similar to that of a tibial tray prosthesis in TKR. Through this, the polymeric tibial insert may be injection molded through the pores of the porous tibial tray for mechanical interlocking.

2.3.5 Loosening of tibiotalar replacement prostheses

Previous sections discussed interactions of tibiotalar prostheses with the tibia and talar. Design parameters that influence loosening of the TAR prostheses were reviewed and their effectiveness in enhancing ankle arthroplasty was explored. Prosthesis loosening was not much discussed in the literature for fourth-generation TAR tibial and talar prostheses. However, as highlighted in previous sections, there are causes of loosening, migration, and subsidence of prostheses at the implanted site. Anatomic and physiologic shaping of the tibia and talar prostheses are key factors for reducing the risk for failure of TAR prostheses. Figs. 2.24–2.26 display some incidents of TAR prostheses loosening, reported in the literature, and explore the possible reasons. Clinical studies investigate one or two types of TAR developed by a single manufacturer. Such studies report prosthesis failure without discussing or clarifying the possible reasons. Thus, an exploration of prosthesis loosening in Figs. 2.24–2.26 does not reveal the low quality of TAR systems displayed in these figures.

2.4 Summary and discussion

This chapter reviewed and discussed loosening of knee and ankle prostheses. The loosening of prostheses such as the knee tibial tray, knee femoral, ankle tibial tray, and ankle talar has a high rate for obese and overweight patients. A midterm clinical follow-up study (2–5 years) showed that more than 30% of prostheses loosened for obese patients (BMI >30 kg/m^2), whereas the rate of loosening increased in overweight patients (BMI >25 kg/m^2) in long-term clinical studies (>5–15 years). This scenario was observed in patients with preoperative varus deformity of the lower limb mechanical axis >8° as well. Long-term clinical studies are mostly for joint replacements systems from about 20 years ago to about

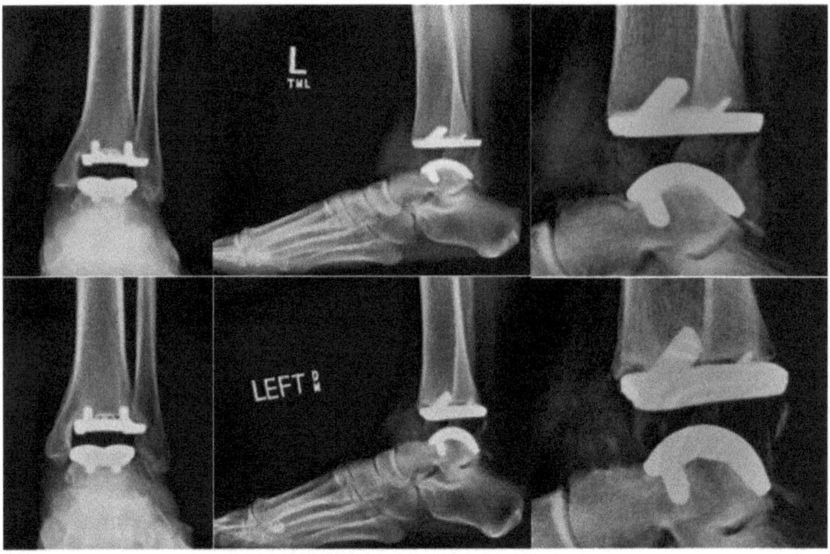

Figure 2.24 Radiographic illustration of tibial component at its interface with resected surface of distal tibia. Rushing et al. [66] chose this prosthesis along with other three TAR systems as fourth-generation prosthesis. The greater convex design of the talar prosthesis and oblique short pegs of the tibial prosthesis are key characteristics of this latest generation of TAR systems. A small radiolucency line is detectable at the posterior aspect of the tibial prosthesis in the top radiograph images, which were taken 6 weeks after the operation. In radiograph images taken at the last follow-up (more than 1 year after surgery), a progressive radiolucency line is seen from the posterior to the anterior aspects of the tibial component at the interface with the resected periprosthetic distal tibia. Rushing et al. reported that such an issue was seen in about 15% of patients; however, overall clinical outcomes (range of motion and level of pain) were good. Longer follow-up might be required to examine bone resorption at the interface with the tibial prosthesis on clinical outcomes and the level of patient satisfaction. The positive point about fourth-generation tibial prostheses is that there is no sign of a radiolucency bone line or zone in periprosthetic bone at the interface of the tibial pegs at the implanted site in midterm clinical studies. This might be the result of the load distribution between the prosthesis and trabecular bone at the operated site. However, the prosthesis would be at risk for loosening if bone resorption progresses at a deeper distance from the prosthesis, where fixation of the pegs is reduced. *Through the courtesy of C. Rushing, R. Law, C.H.-T.J. of F. and A. Surgery, undefined 2020, Early Experience with the CADENCE Total Ankle Prosthesis, Elsevier, n.d. https://www.sciencedirect.com/science/article/pii/S1067251620303252. (Accessed 7 October 2020) and get permission from Elsevier.*

10 years ago. Systems developed more recently have not passed long-term implantation, and thus no reliable evaluation can be provided for such joint replacement systems. Based on analyses and observations from clinical

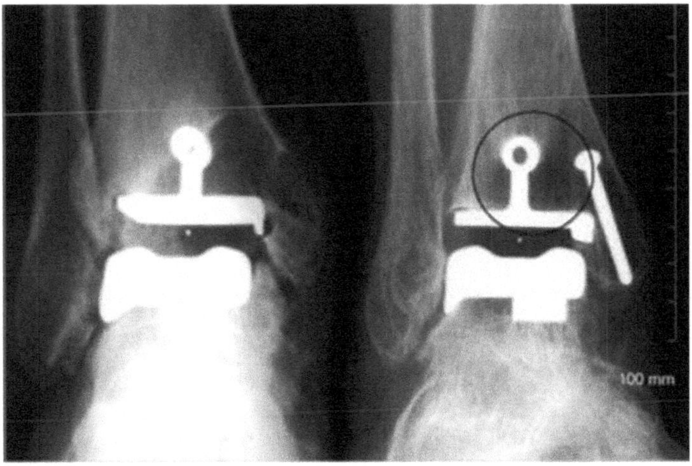

Figure 2.25 Radiographic illustration of bone cysts at the surrounding zone of the tibial prosthesis at the keel portion. In this type of keel design, the designer intended to enhance stabilization of the tibial prosthesis while it was implanted at the distal tibia compared with the short stem type. This feature of the tibial prosthesis is mechanically advantageous for preventing loosening or malalignment owing to a high reaction force and moment loaded during long-term dorsiflexion-plantar flexion motion at the tibiotalar. However, the top tube portion intervene with the stress path in the caudal-cranial axis. This causes stress shielding effect to occur to the trabecular bone matrix underneath the top tube portion, where the undesired reaction of bone is observed as the weakening of bone compactness at the affected zone. This is recognized in the radiographic images as a radiolucent zone and as highlighted in the circle in the right radiograph scan. Preyssas et al. [64] reported this issue in more than 40% of cases using a similar TAR system. In the midterm, such bone cysts may not cause loosening of the tibial prosthesis owing to engagement of the tibial keel with cortical bone at two ends of the tube portion. However, because of the growth of bone resorption at the affected zone, the cortical layer is affected, which necessities long-term follow-up studies to reveal this effect. *Through the courtesy of P. Preyssas, E. Toullec, M. Henry, … J.N.-… & T.S., undefined 2012, Total Ankle Arthroplasty—Three-Component Total Ankle Arthroplasty in Western France: A Radiographic Study, Elsevier, n.d. https://www.sciencedirect.com/science/article/pii/S1877056812000722. (Accessed 6 October 2020) and get permission from Elsevier.*

studies, particularly those from midterm and long-term investigations, concepts have been generated in which various modifications (or even revolution) in design, materials, manufacturing processes, surgical instruments, and surgical techniques enhanced clinical outcomes. This has resulted in a multitude of designs and processes for knee and ankle joint prostheses. Normally, design concept are first generated by the orthopedic surgeons who use the prosthesis and following the outcomes. They attempt to cooperate with

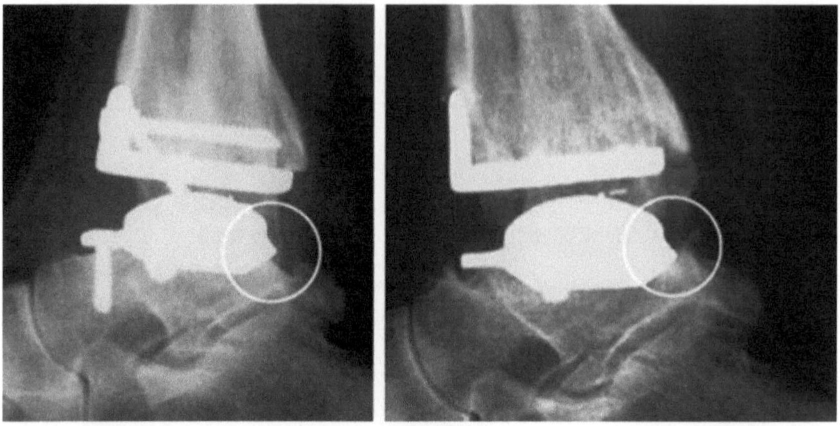

Figure 2.26 Radiographic illustration of subsidence of talar prosthesis to talus. Subsidence of the talar prosthesis occurs in each developed type of TAR in obese or osteoporotic patients. In this specific case, subsidence occurred after the fixation screw was removed at the anterior aspect of the talar prosthesis (right radiograph scan). Overlapping of the talar prosthesis on the cortical section of the resected bone lowered the risk for prosthesis subsidence owing to the greater mechanical strength of cortical bone compared with trabecular bone. Another effective factor is designing both bearing and underside surfaces circularly to dedicate the anatomy of the talus at the tibiotalar joint. This anatomic designing provides effective load transference through the talus in accordance with the physiologic orientation of the trabecular architecture, from which the interaction of the talar prosthesis and talus is significantly enhanced. *Through the courtesy of P. Preyssas, E. Toullec, M. Henry, … J.N.-… & T.S., undefined 2012, Total Ankle Arthroplasty—Three-Component Total Ankle Arthroplasty in Western France: A Radiographic Study, Elsevier, n.d. https://www.sciencedirect.com/science/article/pii/S18770568 12000722. (Accessed 6 October 2020) and get permission from Elsevier.*

engineering experts in mechanical, biomechanical, materials, and materials engineering for the appropriate development of joint replacement prostheses. However, complications of prosthesis loosening owing to ineffective interactions with bone tissue, particularly trabecular bone, are detectable in clinical studies. Loosening in knee tibial tray prostheses was reported as subsidence of the prosthesis through the trabecular bone at the metaphyseal portion of the proximal tibia. Subsidence occurs through the medial and posterior aspects of the proximal tibia owing to higher transferred loading conditions on the medial side with no extension of the stem and keel to the posterior side of the tibial tray. A radiolucent line and area are detectable on the tibial tray radiograph at the interface of the underside surface and stem with the trabecular bone. Severe load transference at the interface of the underside surface causes overloading of periprosthetic bone and acceleration of bone resorption at this zone, ultimately weakening the bone matrix. A tibial tray is fabricated from titanium or CoCr alloy; thus, energy shielding of the tibial tray is low and a large amount of energy is transferred to periprosthetic bone. First, periprosthetic bone mineral density is increased to counteract increased energy or loading conditions. However, this bioactive performance of bone tissue is gradually affected by the creation of undesired fibrous tissues at the interface of the periprosthetic bone and prosthesis to increase the load bearing of bone tissue. Because such loading conditions are cyclical and routinely subjected to periprosthetic bone, the activity of fibrous tissue gradually increases. From a biological point of view, the formation of fibrous tissue at the interface of the bone and implant affects the activity of osteogenic cells, in which osteoclast cell activity increases and the bone resorption rate ultimately is exceeded. Such a scenario is observed at the interface of the ankle talar prosthesis and periprosthetic bone. Another effect is stress shielding at the interface of the tibial tray stem, femoral knee pegs, and anterior and posterior aspects of the femoral knee. Stress shielding in these interfaces occurs when the axis of the load does not coincide with the normal axis to the interface. Overloading and stress shielding and their influence on prosthesis loosening were reviewed and discussed for various scenarios in this chapter for knee and ankle prostheses. Mechanical load transference at the interface of the bone and prosthesis has significant effects on the behavior of periprosthetic bone, and its resorption propagates through the distant bone matrix from the bone–prosthesis interface, with catastrophic migration for subsidence of the prosthesis. Osteoconductivity of the prosthesis surface by porous structuring and bioactive deposited materials is not enhanced if loading conditions are large or poor at the interface. However, if loading conditions are maintained at an appropriate range for bone formation, an osteoconductive engineered surface prevents the progression of microfailure at the bone–prosthesis interface in long-term implantation, and the effectiveness of bone–prosthesis integration is significantly improved. Further discussion in this issue is provided in Chapter 7, which discusses bone–implant interactions and failures for all types of orthopedic implants.

2.5 Remind and learn

In this section, some questions are provided as effective reminders of the contents reviewed and discussed in this chapter. Answering these questions enhances learning about the interactions and failures of the bone–implant construct for total knee and ankle replacement prostheses.

Q 2.1: What is the main reason for implant loosening at the bone–implant interface? Discuss your answer in view of the increase in the biological activity of osteoclast cells.

Q 2.2: Describe the history of the development of TKR from hinged to multiaxis design. What are the benefits of the condylar design of knee replacement prostheses? What are the main reasons for a rotating hinged knee replacement system?

Q 2.3: Describe the overall characteristics of the tibial tray prosthesis and its portions. What are the benefits of the asymmetric design of the tibial tray at the plate or tray section compared with a symmetric design? Discuss how such a design enhances interaction between the bone and knee prostheses.

Q 2.4: Describe the overall characteristics of the modular tibial tray with a fine stem and keel. What are the disadvantages of this kind of tibial tray prosthesis for obese or overweight patients? Explain possible failures that might be occur for this prosthesis.

Q 2.5: Discuss how the formation of bone tissue on the prosthesis surface enhances mechanical interaction between the tibial tray and bone interfaces. What kind of failure might be reduced through this achievement?

Q 2.6: Describe the overall characteristics of the stem portion of the tibial tray. What are the benefits of the keel part in enhancing interaction between the bone and implant? Likewise, discuss the benefits of the transverse-extended keel in both the mediolateral and anteroposterior axes. Describe the advantages and disadvantages of the tibial tray stem with a minikeel design.

Q 2.7: Describe the overall characteristics of the tibial tray prostheses in mobile bearing knee replacement. What is preoperative varus deformity in the knee joint? Compare the efficacy of tibial tray interaction with bone in mobile and fixed bearing types. What is a possible failure in a mobile bearing tibial tray prosthesis? Discuss the reasons for failure at the interface of bone and this prosthesis. What is a possible solution to enhance the tibial tray design in a mobile bearing system?

Q 2.8: What is the level of micromotion in a cemented and cementless tibial tray against compressive loading conditions?

Q 2.9: Describe the processing conditions of the illustrated tibial tray in Figs. 2.10 and 2.11 and the associated benefits. What are the advantages of this prosthesis in enhancing load transference at the interface of the tibial tray and underneath bone? Discuss your answer considering load distribution in the posterior, anterior, medial, and lateral sides.

Q 2.10: Describe failures that occur in obese or overweight patients using short-stem tibial tray prosthesis in total knee arthroplasty. What is a reasons for the progression of the tibial tray subsidence at the posterior zone? Discuss why even a highly porous structured tibial tray might fail in obese patients considering deep bone ingrowth through the prosthesis.

Q 2.11: Discuss the benefit of the long-stem tibial tray interacting with the proximal third of the femur in obese patients.

Q 2.12: Describe the concept of prosthesis or implant energy shielding and discuss the benefit of energy shielding at the interface of the tibial tray prosthesis and tibia. What is the relation between energy shielding and the type of manufacturing in the development of the tibial tray prosthesis?

Q 2.13: What is the main benefit of enhanced biological integration at the interface of the tibial tray prosthesis and tibia in young patients? Discuss the effect of biological integration compared with the design aspects of the prosthesis and loading conditions in obese patients.

Q 2.14: Describe the overall characteristics of the femoral knee prosthesis. Describe the main design features of this prosthesis. What aspects of this prosthesis might loosen in interaction with bone? Discuss the loading mechanism or pattern at the interface of the femoral knee prosthesis and resected distal femoral condyles compared with physiologic loading distribution in a healthy knee joint.

Q 2.15: What contact sections of the femoral knee prosthesis and distal femoral condyles are greatly at risk for failure? Discuss the reasons considering loading distribution in various flexion angles. What is a possible solution for treatment of failure of the femoral knee prosthesis in primary total knee arthroplasty?

Q 2.16: Describe the overall characteristics of the rotating hinged knee system. What are design features in developments of this specific knee system? What indications are considered for the use of the rotating hinged knee system?

Q 2.17: What are the design improvement strategies in the development of the third-generation rotating hinged knee system? Discuss the mechanical and biological integrity of the acetabular prosthesis of the hinged knee system at the diaphysis and metaphysis in long-term stability and functionality. Likewise, describe possible failures owing to ineffective interaction of this prosthesis with the femur. What design concepts might be useful for further development of bone—prosthesis integrity at the metaphyseal zone of the femur in replacing the rotating hinged knee system?

Q 2.18: Describe metaphyseal fixation of the hinged knee prostheses. Describe indications that such an implantation treatment might be needed. What types of components might be used for metaphyseal fixation of the hinged knee prostheses? Describe the structure of these components.

Q 2.19: What types of materials might be used to fabricate the metaphyseal cone? Discuss which material is more suitable for a complete porous cone and which would be adequate for a bulk-porous cone. Describe the reasons based on the achieved mechanical stiffness and properties of the cone compared with trabecular bone.

Q 2.20: Discuss the performance and interaction of the metaphyseal cone with trabecular bone at the metaphyseal portions of the distal femur and proximal tibia. What is the benefit of the metaphyseal cone in enhancing load transference in the transverse plane?

Q 2.21: Describe the ankle joint complex and associated physiologic motion. What is the role of the tibiotalar and subtalar joints in these physiologic motions? What is the range of rotation angle in physiologic motions of the ankle complex joint? What is the main difference of the ankle joint compared with other skeletal joints, and why?

Q 2.22: Describe arthrodesis treatment of arthritis at the tibiotalar joint. What are the disadvantages of such a treatment method? Describe the reasons for injuries at the tibiotalar joint in older or young patients. What is posttraumatic arthritis, and why might it occur in the tibiotalar joint?

Q 2.23: Describe total ankle arthroplasty (TAA) and associated prostheses. What types of failures might occur after TAA?

Q 2.24: Discuss the extent of trabecular bone stock for the implantation of the TAR prosthesis through the distal tibia and talar. What design features are considered for the effective replacement of a TAR prosthesis?

Q 2.25: What design concepts are considered to replace a talar prosthesis with enhanced physiologic load transference at its interface with the talus? What are the benefits of such a concept with greater bone preservation and a reduction in the wear rate at the articulation surface?

Q 2.26: Describe the replacement of tibial prostheses through the distal tibia. What is the benefit of the medial projection in tibial prosthesis?

Q 2.27: Describe the fourth generation of TAR as illustrated in Fig. 2.22. Discuss why the cemented TAR prostheses might not be effectively implanted through the distal tibia and talus. What kinds of surface engineering methods could be used for biological integration of cementless TAR prostheses with the bones at the tibiotalar joint?

Q 2.28: What is a possible design concept to reduce stress shielding effect at the interface of the ankle tibial tray component with a stem feature through the trabecular bone of the distal tibia?

Q 2.29: Considering Figs. 2.24–2.26, describe failures for the TAR prostheses and discuss possible reasons.

2.6 Think and challenge

In this section, some questions are provided for further thought about the reviewed and discussed contents of this chapter. The aim is to challenge the current facets of interactions and failures of the bone with TKR and TAR prostheses by answering these questions, which may enhance new design concepts for the future development of prostheses.

Q 2.30: As highlighted in this chapter, the use of the TKR and TAR systems in obese or overweight patients might result in various types of failure at the bone–implant interfaces. In most cases, to relieve severe pain at the affected joint, the surgeon may have to manage arthroplasty in the knee and ankle joints. What novel design concepts might be hypothesized to increase the effectiveness of TKA and TAA in long-term implantation?

Q 2.31: As discussed in some parts of this chapter, the biological interaction of TKR and TAR prostheses with bone tissue may have a lower effect than mechanical interaction to achieve the long-term stability and performance of prostheses. Discuss how the influence of biological interaction on the long-term effectiveness of the bone–implant construct stability could be enhanced.

Q 2.32: As discussed, implantation of TKR and TAR joint prostheses through bone is achievable using peg, stem, or keel features. Clinical studies have reported a stress shielding phenomenon, as reviewed and illustrated in this chapter. Such failures of the prosthesis at these replacement features seem inevitable. Discuss what new design concepts could be hypothesized for the mechanical stability of the prosthesis through bone while eliminating stress shielding at the bone–prosthesis interface in these replacement features.

References

[1] D. Dalury, D. Pomeroy, R. Gorab, M.A.-T.J. of Arthroplasty, undefined 2013, Why Are Total Knee Arthroplasties Being Revised?, Elsevier, n.d. https://www.

sciencedirect.com/science/article/pii/S0883540313004889 (Accessed 21 September 2020).
[2] P. Sharkey, P. Lichstein, C. Shen, … A.T.-T.J. of, undefined 2014, Why Are Total Knee Arthroplasties Failing Today—Has Anything Changed After 10 Years?, Elsevier, n.d. https://www.sciencedirect.com/science/article/pii/S0883540313005421 (Accessed 21 September 2020).
[3] E. Schiffner, D. Latz, A. Karbowski, … J.G.-J. of Clinical, undefined 2020, Loosening of Total Knee Arthroplasty—Always Aseptic?, Elsevier, n.d. https://www.sciencedirect.com/science/article/pii/S0976566219302164 (Accessed 20 September 2020).
[4] C.S.M. Hoenders, M.C. Harmsen, M.J.A. Van Luyn, The local inflammatory environment and microorganisms in "aseptic" loosening of hip prostheses, J. Biomed. Mater. Res. B Appl. Biomater. 86 (2008) 291—301, https://doi.org/10.1002/jbm.b.30992.
[5] C. Park, P. White, M. Meftah, A.R.- Orthopedics, undefined 2016, Diagnostic Algorithm for Residual Pain After Total Knee Arthroplasty, Healio.Com., n.d. https://www.healio.com/orthopedics/journals/ortho/2016-3-39-2/%7B18020ac8-17f6-4da9-9c0c-ec40ef8e1cb9%7D/diagnostic-algorithm-for-residual-pain-after-total-knee-arthroplasty (Accessed 21 September 2020).
[6] S.K.-U.B. Handbook, undefined 2009, The Origins and Adaptations of UHMWPE for Knee Replacements, Elsevier, n.d. https://www.sciencedirect.com/science/article/pii/B9780123747211000079 (Accessed 6 October 2020).
[7] C. Lavernia, J.A.-I.J.C. Rheumatol, Knee Arthroplasty: Growing Trends and Future Problems, undefined 2010, Pdfs.Semanticscholar.Org, 2010, https://doi.org/10.2217/IJR.10.49.
[8] J. Callaghan, D. DeMik, N. Bedard, A.O.-A. Today, undefined 2018, Tibial Tray Fracture in a Modern Prosthesis With Retrieval Analysis, Elsevier, n.d. https://www.sciencedirect.com/science/article/pii/S2352344117301814 (Accessed 22 September 2020).
[9] S. Garceau, S.P. Garceau, N.H. Harris, D.L. Felberbaum, G.M. Teo, A.I. Weinblatt, W.J. Long, Reduced aseptic loosening with fully cemented short-stemmed tibial components in primary cemented total knee arthroplasty primary knee, Artic. J. Arthroplast. 35 (2020), https://doi.org/10.1016/j.arth.2020.01.084.
[10] P. Crook, J. Owen, S. Hess, … S.A.-H.-T.J. of, undefined 2017, Initial Stability of Cemented vs Cementless Tibial Components Under Cyclic Load, Elsevier, n.d. https://www.sciencedirect.com/science/article/pii/S0883540317302759 (Accessed 20 September 2020).
[11] M. Abdel, G. Bonadurer III, … M.J.-T.J. of, undefined 2015, Increased Aseptic Tibial Failures in Patients with a BMI≥ 35 and Well-Aligned Total Knee Arthroplasties, Elsevier, n.d. https://www.sciencedirect.com/science/article/pii/S0883540315005914 (Accessed 20 September 2020).
[12] S. Song, H. Lee, D. Bae, C.P.-T.J. of Arthroplasty, undefined 2020, High Incidence of Tibial Component Loosening After Total Knee Arthroplasty Using Ceramic Titanium-Nitride-Coated Mobile Bearing Prosthesis in Moderate to Severe Varus, Elsevier, n.d. https://www.sciencedirect.com/science/article/pii/S0883540319311064 (Accessed 20 September 2020).
[13] C. Hampton, Z. Berliner, J. Nguyen, … L.M.-T.J. of, undefined 2020, Aseptic Loosening at the Tibia in Total Knee Arthroplasty: A Function of Cement Mantle Quality?, Elsevier, n.d. https://www.sciencedirect.com/science/article/pii/S0883540320301893 (Accessed 20 September 2020).
[14] J. Zhang, J. Wang, D. Tian, D. Jiang, J. Li, Y.H.-T. Knee, undefined 2020, Spherical Center and Rotating Platform Hinged Knee Prosthesis: Finite-Element Model

Establishment, Verification and Contact Analysis, Elsevier, n.d. https://www.sciencedirect.com/science/article/pii/S0968016020301174 (Accessed 27 September 2020).

[15] W. Ward, D. Haight, P. Ritchie, S. Gordon, J.E.- JBJS, undefined 2005, Dislocation of Rotating Hinge Knee Prostheses: A Report of Four Cases, Journals.Lww.Com., n.d. https://journals.lww.com/jbjsjournal/Fulltext/2005/05000/Dislocation_of_Rotating_Hinge_Knee_Prostheses__A.25.aspx (Accessed 27 September 2020).

[16] D. Pacha-Vicente, A. Malik, ... E.C.-F.-T.J. of, undefined 2008, Dislocation of Rotating-Hinge Knee Prostheses with Antidislocation Mechanism, Elsevier, n.d. https://www.sciencedirect.com/science/article/pii/S088354030600876X (Accessed 27 September 2020).

[17] M. Mazaleyrat, L. Le Nail, G. Auberger, ... D.B.-... & T.S., undefined 2020, Survival and Complications in Hinged Knee Reconstruction Prostheses After Distal Femoral or Proximal Tibial Tumor Resection: A Retrospective Study of 161 Cases, Elsevier, n.d. https://www.sciencedirect.com/science/article/pii/S1877056820300475 (Accessed 27 September 2020).

[18] E. Cavaignac, J. Tricoire, R. Pailhé, ... J.M.-... & T.S.., undefined 2014, Recurring Intraprosthetic Dislocation of Rotating-Hinge Total Knee Prosthesis. Effect of Implant Design on Intrinsic Stability, Elsevier, n.d. https://www.sciencedirect.com/science/article/pii/S1877056814001881 (Accessed 22 September 2020).

[19] H. Abdelaziz, R. Jaramillo, T. Gehrke, ... M.O.-T.J. of, undefined 2019, Clinical Survivorship of Aseptic Revision Total Knee Arthroplasty Using Hinged Knees and Tantalum Cones at Minimum 10-Year Follow-Up, Elsevier, n.d. https://www.sciencedirect.com/science/article/pii/S0883540319306485 (Accessed 27 September 2020).

[20] A. Abdulkarim, A. Keane, S. Hu, ... L.G.-... & T.S.., undefined 2019, Rotating-Hinge Knee Prosthesis as a Viable Option in Primary Surgery: Literature Review & Meta-Analysis, Elsevier, n.d. https://www.sciencedirect.com/science/article/pii/S1877056819302841 (Accessed 22 September 2020).

[21] L. Shao, X. Wu, T. Wang, X. Liu, W. Xu, W.H.-S. Reports, undefined 2020, Approximating the Maximum Tibial Coverage in Total Knee Arthroplasty Does Not Necessarily Result in Implant Malrotation, Nature.Com, n.d. https://www.nature.com/articles/s41598-020-67613-2 (Accessed 22 September 2020).

[22] S. Rytter, F. Madsen, C. Jepsen, M.S.-T. Knee, undefined 2019, Implant Fracture of the Regenerex® Modular Metal Tibial Component: A Report of Three Cases, Elsevier, n.d. https://www.sciencedirect.com/science/article/pii/S0968016019301255 (Accessed 3 October 2020).

[23] S. David, J. Martínez, I. Bermúdez, ... F.M.-R.E., undefined 2014, The Geometry of the Keel Determines the Behaviour of the Tibial Tray Against Torsional Forces in Total Knee Replacement, Elsevier, n.d. https://www.sciencedirect.com/science/article/pii/S1988885614001011 (Accessed 10 October 2020).

[24] C. Kajetanek, B. Bouyer, M. Ollivier, ... P.B.-... & T.S., undefined 2016, Mid-Term Survivorship of Mini-Keel TM Versus Standard Keel in Total Knee Replacements: Differences in the Rate of Revision for Aseptic Loosening, Elsevier, n.d. https://www.sciencedirect.com/science/article/pii/S187705681630055X (Accessed 24 September 2020).

[25] I. Kutzner, G. Hallan, P.J. Høl, O. Furnes, Ø. Gøthesen, W. Figved, P. Ellison, Early aseptic loosening of a mobile-bearing total knee replacement, Acta Orthop. 89 (2018) 77–83, https://doi.org/10.1080/17453674.2017.1398012.

[26] H. Kobayashi, N. Mitsugi, Y. Mochida, N. Taki, Y. Akamatsu, M. Aratake, H. Ota, K. Ishii, K. Harigane, T. Ideno, T. Saito, Mid- term results of stryker® scorpio plus

mobile bearing total knee arthroplasty, Sports Med. Arthrosc. Rehabil. Ther. Technol. 4 (2012), https://doi.org/10.1186/1758-2555-4-38.

[27] A. Miller, J. Stimac, L. Smith, … A.F.-T.J. of, undefined 2018, Results of Cemented vs Cementless Primary Total Knee Arthroplasty Using the Same Implant Design, Elsevier, n.d. https://www.sciencedirect.com/science/article/pii/S0883540317310537 (Accessed 14 February 2021).

[28] W. Scully, M. Deren, M.B.-A. Today, undefined 2019, Catastrophic Tibial Baseplate Failure of a Modern Cementless Total Knee Arthroplasty Implant, Elsevier, n.d. https://www.sciencedirect.com/science/article/pii/S235234411930113X (Accessed 27 September 2020).

[29] S. Berahmani, M. Hendriks, D. Wolfson, A. Wright, D. Janssen, N. Verdonschot, Experimental pre-clinical assessment of the primary stability of two cementless femoral knee components, J. Mech. Behav. Biomed. Mater. 75 (2017) 322–329, https://doi.org/10.1016/j.jmbbm.2017.07.043.

[30] M. Kahan, C.E.-A. Today, undefined 2020, Failure at the Femoral Stem Extension-Condylar Interface in a Rotating Hinge Knee Without Radiographic Evidence of Loosening, Elsevier, n.d. https://www.sciencedirect.com/science/article/pii/S2352344119301463 (Accessed 20 September 2020).

[31] Y. Kim, J. Park, J.K.-T.J. of Arthroplasty, undefined 2017, Do High-Flexion Total Knee Designs Increase the Risk of Femoral Component Loosening?, Elsevier, n.d. https://www.sciencedirect.com/science/article/pii/S088354031730058X (Accessed 20 September 2020).

[32] H. Han, M.L.-T. Knee, undefined 2017, Cementing Technique Affects the Rate of Femoral Component Loosening After High Flexion Total Knee Arthroplasty, Elsevier, n.d. https://www.sciencedirect.com/science/article/pii/S096801601730203X (Accessed 24 September 2020).

[33] M. Bahraminasab, B. Sahari, K.E.-M.& Design, undefined 2012, Aseptic Loosening of Femoral Components—A Review of Current and Future Trends in Materials Used, Elsevier, n.d. https://www.sciencedirect.com/science/article/pii/S0261306912003561 (Accessed 24 September 2020).

[34] T. Saari, J. Uvehammer, L. Carlsson, L.R.-T. Knee, undefined 2006, Posterior Stabilized Component Increased Femoral Bone Loss After Total Knee Replacement. 5-Year Follow-Up of 47 Knees Using Dual Energy X-ray Absorptiometry, Elsevier, n.d. https://www.sciencedirect.com/science/article/pii/S0968016006001372 (Accessed 27 September 2020).

[35] C. Higuera, J. Parvizi, 18 Causes and diagnosis of aseptic loosening after total knee replacement, in: T.H. Michael (Ed.), Unhappy Total Knee Replace, Springer International Publishing, Cham, 2015, pp. 225–237.

[36] K. Chen, J. Chu, C.L.-J. of M. Sciences, undefined 2017, Radiotherapy Inducing Total Knee Prosthetic Component Loosening for Recurrent Pigmented Villonodular Synovitis Following Joint Replacement, Jmedscindmc.Com, n.d. http://www.jmedscindmc.com/article.asp?issn=1011-4564;year=2017;volume=37;issue=4;spage=168;epage=171;aulast=Chen (Accessed 27 September 2020).

[37] C.C.-D. and Interventional Imaging, undefined 2016, Imaging of Knee Implants and Related Complications, Elsevier, n.d. https://www.sciencedirect.com/science/article/pii/S221156841630002X (Accessed 27 September 2020).

[38] H. Levine, S. Madoff, J.S. Newman, Imaging of total knee arthroplasty, in: V.B. James (Ed.), Revision Total Knee Arthroplasty, Springer International Publishing, Cham, 2017, pp. 41–60.

[39] R. Dehdari, M.T.-B.M. Imaging, undefined 2013, Orthopedic Hardware and Complications, McGraw Hill Prof, n.d.

[40] S. Kouk, P. Rathod, … A.M.-J. of Clinical, undefined 2018, Rotating Hinge Prosthesis for Complex Revision Total Knee Arthroplasty: A Review of the Literature, Elsevier, n.d. https://www.sciencedirect.com/science/article/pii/S0976566217304381 (Accessed 22 September 2020).
[41] M. Chen, J. Hung, C. Chang, S. Lee, H.S.-T. Knee, undefined 2020, Periprosthetic Knee Infection Reconstruction with a Hinged Prosthesis: Implant Survival and Risk Factors for Treatment Failure, Elsevier, n.d. https://www.sciencedirect.com/science/article/pii/S0968016020300673 (Accessed 22 September 2020).
[42] R. Grunfeld, U. Aydogan, P.J.-M. Clinics, undefined 2014, Ankle Arthritis: Review of Diagnosis and Operative Management, Medical.Theclinics.Com, n.d. https://www.medical.theclinics.com/article/S0025-7125(13)00153-3/abstract (Accessed 29 September 2020).
[43] S. Song, H. Lee, C.P.-T.J. of Arthroplasty, undefined 2020, A Current Prosthesis With a 1-mm Thickness Increment for Polyethylene Insert Could Result in Fewer Adjustments of Posterior Tibial Slope in Cruciate-Retaining Total, Elsevier, n.d. https://www.sciencedirect.com/science/article/pii/S0883540320307142 (Accessed 20 September 2020).
[44] K.L. Scott, M.P. Abdel, A.D. Hanssen, Metaphyseal sleeves and cones in revision total knee arthroplasty, in: V.B. James (Ed.), Revision Total Knee Arthroplasty, Springer International Publishing, 2017, pp. 131−141.
[45] A. De, P. Mozella, R.R. Olivero, H. Alexandre, A.B. Cobra, Use of a trabecular metal cone made of tantalum, to treat bone defects during revision knee arthroplasty. Revista Brasileira de Ortopedia (English Edition), 2014, https://doi.org/10.1016/j.rboe.2014.03.009. SciELO Bras, https://doi.org/10.1016/j.rboe.2014.03.009.
[46] K. Denehy, S. Abhari, … V.K.-T.J. of, undefined 2019, Metaphyseal Fixation Using Highly Porous Cones in Revision Total Knee Arthroplasty: Minimum Two Year Follow Up Study, Elsevier, n.d. https://www.sciencedirect.com/science/article/pii/S0883540319303018 (Accessed 29 September 2020).
[47] M. Ohlmeier, C. Lausmann, M. Wolff, … H.A.-A. of O., undefined 2020, Preliminary Clinical Results of Coated Porous Tibia Cones in Septic and Aseptic Revision Knee Arthroplasty, Springer, n.d. https://link.springer.com/content/pdf/10.1007/s00402-020-03434-2.pdf (Accessed 29 September 2020).
[48] A. Faizan, M. Bhowmik-Stoker, V. Alipit, … A.K.-T.J. of, undefined 2017, Development and Verification of Novel Porous Titanium Metaphyseal Cones for Revision Total Knee Arthroplasty, Elsevier, n.d. https://www.sciencedirect.com/science/article/pii/S0883540317300153 (Accessed 29 September 2020).
[49] M.A. Dharia, S.M. Humphrey, K.A. Roby, G.D. Stebbins, R. Zubok, D. Lewallen, T. Sekundiak, L. Kwong, S. Jit Singh Bajwa, N. Gupta, Effect of physiological loading on fretting corrosion of Zimmer® trabecular MetalTM coupled tibial cones interacting with tibial baseplates, Br. J. Med. Med. Res. 9 (2015) 1−7, https://doi.org/10.9734/BJMMR/2015/16954.
[50] A. Shane, H.S.-C. in P.M. and, undefined 2019, Total Ankle Replacement Options, Podiatric.Theclinics.Com, n.d. https://www.podiatric.theclinics.com/article/S0891-8422(19)30042-4/abstract (Accessed 29 September 2020).
[51] G. McCollum, M.M.-C. in P. Medicine, undefined 2013, Failure of the Agility Total Ankle Replacement System and the Salvage Options, Podiatric.Theclinics.Com, n.d. https://www.podiatric.theclinics.com/article/S0891-8422(12)00136-X/abstract (Accessed 29 September 2020).
[52] P. Yalamanchili, B. Donely, M. Casillas, … A.A.-O.T. in, undefined 2008, Salto Talaris Total Ankle Replacement, Elsevier, n.d. https://www.sciencedirect.com/science/article/pii/S1048666609000408 (Accessed 29 September 2020).

[53] S. Robati, A. Salih, K. Ghosh, P.V.-J. of Orthopaedics, undefined 2016, The Scandinavian Total Ankle Replacement and the Ideal Biomechanical Requirements of Ankle Replacements, Elsevier, n.d. https://www.sciencedirect.com/science/article/pii/S0972978X15001087 (Accessed 29 September 2020).

[54] B. Zhou, K.T.-A.-A.C. Review, undefined 2016, The Current Trend of Total Ankle Replacement, Intechopen.Com, n.d. https://www.intechopen.com/books/arthroplasty-a-comprehensive-review/the-current-trend-of-total-ankle-replacement (Accessed 29 September 2020).

[55] A. Henricson, S. Popelka, U.R.-F. and A. Surgery, undefined 2020, Six Year Results of the Rebalance Mobile Bearing Total Ankle Replacement, Elsevier, n.d. https://www.sciencedirect.com/science/article/pii/S1268773120300308 (Accessed 29 September 2020).

[56] R. Sopher, A. Amis, J. Calder, J.J.-M.E., undefined 2017, Total Ankle Replacement Design and Positioning Affect Implant-Bone Micromotion and Bone Strains, Elsevier, n.d. https://www.sciencedirect.com/science/article/pii/S1350453317300346 (Accessed 29 September 2020).

[57] A.S.-F. and Ankle Surgery, undefined 2011, A Review of the STAR Prosthetic System and the Biomechanical Considerations in Total Ankle Replacements, Elsevier, n.d. https://www.sciencedirect.com/science/article/pii/S1268773111000348 (Accessed 29 September 2020).

[58] P. Preyssas, E. Toullec, M. Henry, … J.N.-… & T.S., undefined 2012, Total Ankle Arthroplasty—Three-Component Total Ankle Arthroplasty in Western France: A Radiographic Study, Elsevier, n.d. https://www.sciencedirect.com/science/article/pii/S1877056812000722 (Accessed 6 October 2020).

[59] B.O.-C. in Podiatric Medicine and Surgery, undefined 2012, Total Ankle Replacement: A Historical Perspective, Podiatric.Theclinics.Com., n.d. https://www.podiatric.theclinics.com/article/S0891-8422(12)00096-1/abstract (Accessed 29 September 2020).

[60] C. Rushing, R. Law, C.H.-T.J. of F. and A. Surgery, undefined 2020, Early Experience with the CADENCE Total Ankle Prosthesis, Elsevier, n.d. https://www.sciencedirect.com/science/article/pii/S1067251620303252 (Accessed 7 October 2020).

[61] C. Gross, A. Palanca, J.D.-J.-J. of the, undefined 2018, Design Rationale for Total Ankle Arthroplasty Systems: An Update, Journals.Lww.Com., n.d. https://journals.lww.com/jaaos/FullText/2018/05150/Design_Rationale_for_Total_Ankle_Arthroplasty.3.aspx (Accessed 29 September 2020).

[62] A. King, N. Bali, A. Kassam, A. Hughes, … N.T.-F. and A., undefined 2019, Early Outcomes and Radiographic Alignment of the Infinity Total Ankle Replacement with a Minimum of Two Year Follow-Up Data, Elsevier, n.d. https://www.sciencedirect.com/science/article/pii/S1268773118300821 (Accessed 7 October 2020).

CHAPTER THREE

Interactions of bone with shoulder joint prostheses and possible failures

3.1 Introduction

Total shoulder arthroplasty (TSA) is increasingly employed to treat osteoarthritis, soft tissue severe deficiency at the shoulder joint, severe multi-fragmentary proximal humeral fracture, and any other musculoskeletal injury or disorder that incurs pain and limited functionally at the shoulder joint. TSA requires the highest joint arthroplasty after knee and hip arthroplasties. To enhance the clinical outcome, various shoulder replacement systems have been developed with different design concepts, functions, and performance. To treat osteoarthritis at the shoulder joint, total shoulder replacement (TSR) is be designed to be anatomically positioned at the glenoid and humeral head aspects of the scapula and humerus, respectively. In cases in which the rotator cuff muscle-tendons may not have adequate contraction functionality or may even be torn, anatomic total shoulder replacement (aTSR) is not a suitable prosthesis system for treating arthritis at the shoulder joint. In such conditions, reverse total shoulder replacement (rTSR) was developed to increase the stability of glenoid and humeral prostheses and reduce dependency on rotator cuff muscle-tendons. The shoulder replacement systems have design-based limitations that may cause loosening, migration, dislocation, fracture, and eventually failure of the system. The design and development of TSR systems have been improving based on observed complications and the growth of engineering technologies used to create design concepts and techniques. Prosthesis failures related to the design are not usually discussed or evaluated in clinical investigations that report follow-up clinical outcomes. This chapter provides a detailed discussion on the effectiveness of TSR systems for treating arthritis of the shoulder joint in view of bone—prosthesis interactions with a consideration of the influence of soft tissues. In the first section, the anatomy and physiologic condition of the musculoskeletal system at the shoulder joint is reviewed and effective technical parameters on interactions of the bone and TSR

Interactions of Bone with Orthopedic Implants and Possible Failures
ISBN: 978-0-323-95410-5
https://doi.org/10.1016/B978-0-323-95410-5.00005-0

prostheses are explored. Then, possible influences on the performance of TSR prostheses are discussed. In the next part, shoulder replacement systems are elaborated upon in depth regarding their design parameters and their interactions with bone tissue at the implanted site. In the next sections, reported failures and deficiencies of TSR prostheses reported in clinical studies are reviewed. The chapter attempts to provide a rigorous technical discussion to explain how such failures might occur and the effects of design improvements on reducing the failure rate. To explore these concepts, figures were selected or designed. The last section presents an overall discussion and summary of the reviewed and discussed concepts, as well as a future perspective and possible design concepts to enhance interactions of bone with TSR prostheses.

3.2 Shoulder joint

The shoulder or glenohumeral joint is a musculoskeletal joint that might be injured in a fall, a sports accident, or a vehicle accident. This joint is not considered to be a load-bearing joint as much as knee, hip, and ankle joints, because of a lack of body weight loading conditions. At the glenohumeral joint, the humerus at its head portion and scapula at its glenoid aspect are in direct contact or are articulated. The humeral head is shaped close to spherical and the glenoid is shaped with a concave surface, from which the shoulder joint is formed as a ball-and-socket joint at the upper limbs or extremities. The stability of the joint is established entirely by the ligaments and muscle-tendons. Rotator cuff muscles originate from the scapula and attach to the humeral head. The rotator cuff muscles are supraspinatus, infraspinatus, teres minor, and subscapularis and are connected from the superior aspect of greater tuberosity, the middle aspect of greater tuberosity, and the inferior aspects of greater tuberosity and lesser tuberosity, respectively (Fig. 3.1) [1]. The rotator cuff muscles connect to the back facet of the humeral head. This provides two functional advantages. In addition to facilitating upper limb rotation at the shoulder joint, the rotator cuff forces the humeral head articulate with the glenoid facet of the scapula. Such musculoskeletal loading conditions keep the humeral head in contact with the glenoid during movement of the humerus in various axes of rotations at this joint. The glenoid anatomy is structured as a shallow cavity featuring fibrocartilaginous labrum tissue at the entire edge. The labrum is a fibrocartilaginous tissue that acts as a flexible constraint to enhance articulation of the humeral head with the cartilage layer of the glenoid cavity. The labrum

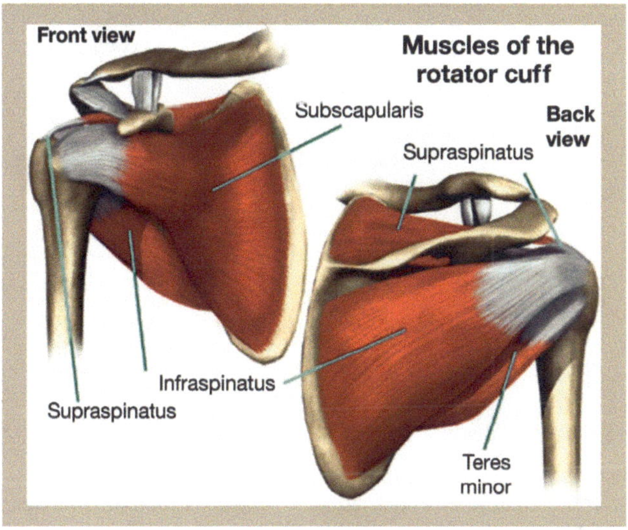

Figure 3.1 Illustration of rotator cuff muscle-tendons at the glenohumeral joint. The rotator cuff has a key role in the motion of the upper limbs at the shoulder joint. It consists of four muscle-tendons, as shown in this figure; they all originate from the scapula and attach at the back section of the humeral head at the greater and lesser tuberosities. All motions of abduction-adduction, flexion-extension, and internal-external rotation are provided through contraction of these muscles. Supraspinatus would be an abductor muscle-tendon, whereas the subscapularis and infraspinatus muscle-tendons operate flexion and extension motions at glenohumeral joint, respectively. The internal posterior is facilitated by the teres minor muscle-tendon and involves external motion. The successful rate of the anatomic TSA (aTSA) is strongly influenced by the rotator cuff muscle-tendons. Failure of these rotators to function causes various types of complications, and in severe conditions it necessitates revision surgery. Rotator cuff muscles might be overstretched or even torn during heavy physical activities. *Through the courtesy of S. Vollans, A.A.-S. (Oxford), undefined 2016, Rotator Cuff Tears, Elsevier, n.d. https://www.sciencedirect.com/science/article/pii/S0263931916000065. (Accessed 18 October 2020) and get permission from Elsevier.*

might be torn and necessitate replacement of the glenohumeral joint with total shoulder replacement (TSR).

In addition to labrum tissue, glenohumeral ligaments (Fig. 3.2) give strong stability to the humeral head for articulation with the glenoid in the anterior axis while the humerus is rotated anteriorly in flexion [2]. The coracohumeral ligament is another ligament superiorly supports the glenohumeral joint. The humerus can rotate in various axes of flexion-extension, abduction-adduction, and internal-external facilitated by the rotator cuff muscles. Rotation of the humerus at the shoulder joint combines

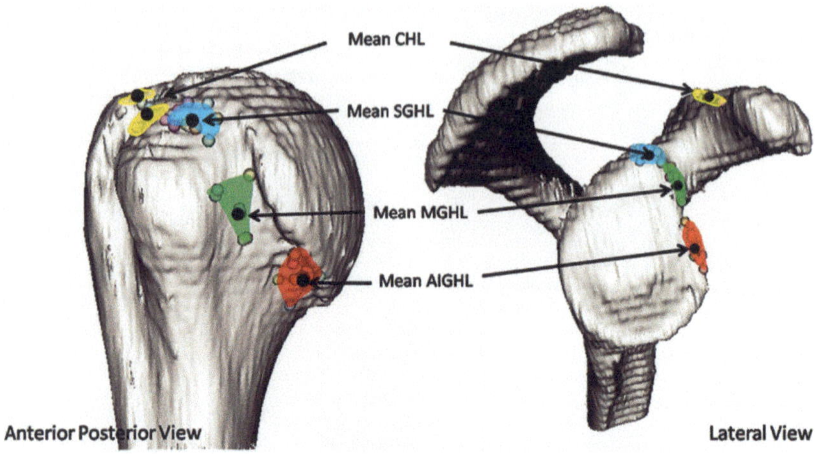

Figure 3.2 Illustration of the glenohumeral and coracohumeral ligaments at the glenohumeral joint. The glenohumeral ligaments are known as superior, middle and inferior glenohumeral ligaments. These ligaments provide relative stability to the humeral head and scapula at the shoulder joint. *AIGHL*, anterior band of the inferior glenohumeral ligament; *CHL*, coracohumeral ligament; *MGHL*, middle glenohumeral ligament; *SGHL*, superior glenohumeral ligament. *Through the courtesy of C. Yang, A. Goto, W. Sahara, … H.Y.-C., undefined 2010, In Vivo Three-Dimensional Evaluation of the Functional Length of Glenohumeral Ligaments, Elsevier, n.d. https://www.sciencedirect.com/science/article/pii/S0268003309002484. (Accessed 24 October 2020) and get permission from Elsevier.*

these three axes of rotation in daily activities or sports. From a mechanical point of view, the shoulder joint is a joint of soft tissues: labrum, muscles, tendons, and ligaments to hold the humerus in articulation with the glenoid facet of the scapula. A large amount of force is expectable in rotator cuff muscle-tendons during movement of the upper limbs (humerus, distal, ulna, and hand bones).

In addition to the weight, the generation of centrifugal force during motion of the upper limbs increases stretching and contraction forces of the rotator cuff muscles. In daily activities, centrifugal force might not be significantly introduced to the rotator cuff muscle-tendons; however, in specific activities such as those which athletes perform in hand sports (e.g., tennis, ball throwing, volleyball, swimming), the level of this type of force increases sharply owing to the high-speed motion of the upper limb ($F_{centrifugal} = \frac{mV^2}{r}$, where m, V, and r are the mass, velocity, and radius, respectively). Another factor that increases the effective force at the shoulder joint is when the upper limbs are abducted or flexed. Such a motion of the upper limbs generates a bending moment that increases the status at which the upper limbs are positioned, up to entirely straight at 90° abduction or flexion angle. An increase in the force as the result of upper limb weight, high-speed motion, and bending moment raises the transferred force to the muscle-tendons and ligaments at the glenohumeral joint. This increased force or loading condition cause soft tissue injuries and deficiency in their function, because of which the loading transferring pattern and mechanism at the shoulder joint between the humeral head and glenoid cavity is affected. This causes cartilage tissue biological activity to deteriorate and osteoarthritis to occur. The main effective force at the contact of the humeral head and glenoid is shear force owing to superior-inferior laxity of the humeral head in various axes. Applied transverse force in the transverse plane from the rotator cuff to enforce the humeral head through the central portion of the glenoid increases the level of shear force at bone-to-bone articulation. The inferior edge of the glenoid cavity bears a higher level of shear stress because of the generation of a greater amount of friction force at this zone at various angular positions of the humeral head. This shear stress affects cartilage or labrum tissue at the inferior area of the glenoid cavity. The performance of the supraspinatus muscle-tendon in holding the upper limbs from the superior facet and bearing gravity force has a significant influence on reducing the amount of shear stress at the inferior area of the glenoid cavity. Therefore, this muscle-tendon is at a higher risk for failure, tearing, or malfunction, in which localized stress at the inferior zone of the glenoid cavity results and cartilage and labrum tissues are affected. Deltoid muscle-tendon is another soft tissue that supports the movement of upper limbs at the glenohumeral joint. This muscle is large and is composed of three main muscles known as the posterior, lateral, and anterior heads, which act as the main abductor of the upper limbs for an abduction angle over 15°. The deltoid muscles support flexion and extension rotation of the upper limbs as well. Through an understanding of glenohumeral boundary and loading conditions, design concept of TSR can be carried out with enhanced effectiveness in long-term implantation, as be discussed in the next section. Another crucial and effective parameter in the design of the TSR is the anatomy of the scapula ending with the glenoid cavity. As illustrated in Fig. 3.3, bone stock in this zone is not large. The mineral density of trabecular bone at this zone is not as high that in metaphyseal portions of the distal femur or proximal tibia at the knee joint. Based on the anatomy of the glenoid section, the width of the inferior zone is higher than that of the superior zone [3]. Likewise, the inferior aspect of the glenoid cavity projects more laterally compared with the posterior aspect, from which a kind of

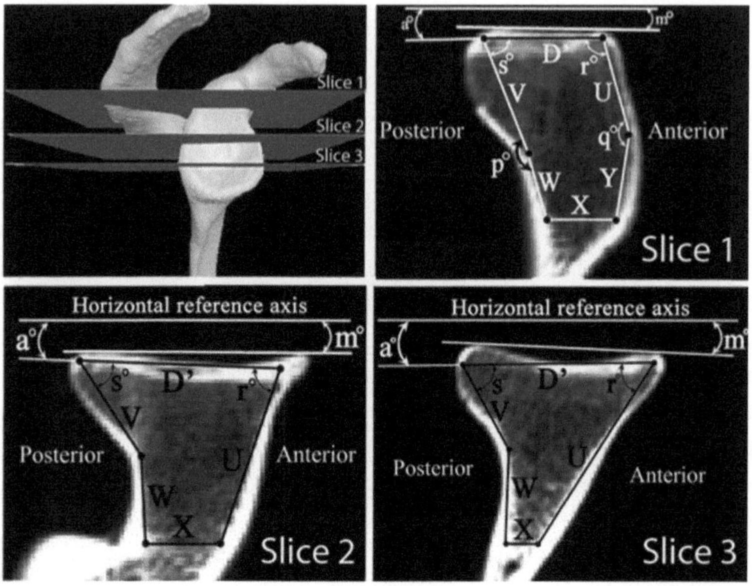

Figure 3.3 Radiographic illustration of bone cross-section and volume of bone stock in the transverse plane at the glenoid zone. As illustrated, the bone stock volume is limited for insertion of the glenoid prosthesis through the glenoid aspect of the scapula. The basic features of the glenoid prosthesis (e.g., keel, pegs, screws) need to be sized adequately to be optimal at this moderately narrow space compared with the metaphyseal section of long bones such as the femur or tibia. Such base features are best arranged in the axial axis in the frontal plane owing to the longer length of the glenoid aspect in the axial axis compared with the transverse axis. At the middle of the glenoid aspect of the scapula, the transverse length of bone stock is greater than that of the superior and inferior sections, which needs to be considered regarding a central peg or screw-peg in designing the base portion of the glenoid prosthesis. *Through the courtesy of G.B. Sharma, P.J. McMahon, D.D. Robertson, Structure modeling of the glenoid: Relevance to shoulder arthroplasty. Journal of Orthopaedic Research 32 (2014) 1471–1478. https://doi.org/10.1002/jor.22696 and get permission from Wiley.*

constraint is created to prevent increased laxity of the humeral head in the axial axis (or caudal-cranial axis) (Fig. 3.4).

3.3 Shoulder joint prostheses

When the cartilage or labrum tissues are significantly injured, torn, or degenerated, replacement of the shoulder prostheses is needed to relieve severe pain and inflammation at the injured zone and enhance functionality and range of motion at the glenohumeral joint. Shoulder replacement

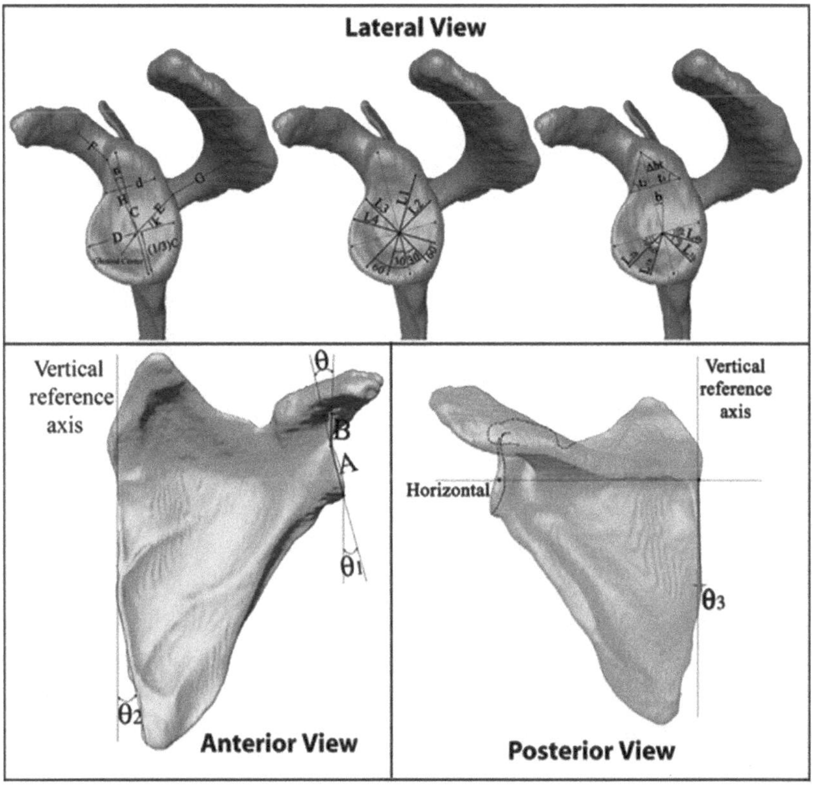

Figure 3.4 Illustration of scapula in various views. In the lateral view, the specific geometry of the glenoid aspect is seen. Its curved profile at the sidelong areas causes centric articulation of the humeral spherical head at the central area. The length of the inferior zone in transverse plane is greater to allow a higher contact surface of the humeral head at this zone, geometrically constraining humeral head motion in the axial axis toward the inferior direction. Constraint of the inferior zone can be observed in the anterior view, where is a relative angular axis from the inferior zone toward the superior zone (*AB line*, θ and θ_1 angles). *Through the courtesy of G.B. Sharma, P.J. McMahon, D.D. Robertson, Structure modeling of the glenoid: Relevance to shoulder arthroplasty. Journal of Orthopaedic Research 32 (2014) 1471–1478. https://doi.org/10.1002/jor.22696 and get permission from Wiley.*

involves a prosthesis assembly at the humeral side and a prosthesis assembly at the glenoid side. When reaming the cartilage tissue at the ends of the humeral head and glenoid cavity, it is necessary to replace the anatomically designed prostheses to dedicate the shape of the humeral spherical and glenoid cavity surfaces. The crucial clinical parameter contemplated for replacing the humeral and glenoid prostheses is preservation of all muscle-tendons and ligaments to maintain shoulder joint motion physiologically. As highlighted in Section 3.2, rotator cuff muscle-tendons provide musculoskeletal motion at the glenohumeral joint and force the humeral head to be positioned in the cavity of the glenoid, whereas the ligaments establish joint relative stability in various motions. Thus, the constraints created by these soft tissues may not be effectively provided by the specific design of the articulation prostheses and must be preserved in the surgical implantation procedure. Through this preservation, bone resection needs to be carefully managed so as not to intervene with areas at which soft tissues originated or are attached. This would create limited resection of the bone at the humeral head and glenoid cavity, and thus, the design of the prostheses would be associated with some limitations.

3.3.1 Design concept of anatomic shoulder replacement prostheses

Shoulder replacement prostheses have various designs [4]. In general, there are two categories of design concept: anatomic and reverse. Each category is designed with a long or short stem. A conventional design is a long-stem anatomic design concept that is in accordance with the anatomic shape of the humeral head and glenoid cavity. In this design, at the humeral assembly, the prosthesis needs to have a spherical surface at the proximal portion, which is connected to a median component to be stabilized on a long stem (Fig. 3.5A, E). Replacement of the humeral assembly through the proximal portion of the humerus begins by resection of the humeral head at the cartilage surface while preserving the attachment areas of the tendons and ligaments (e.g., greater and lesser tuberosity areas). The hemispherical component of the humeral assembly is positioned on the resected surface, whereas the stem is implanted through the prepared canal at the metaphysis and diaphysis sections of the proximal third portion of the humerus. On the other side, the glenoid assembly is a component manufactured from ultra-high molecular weight polyethylene (UMMWPE) polymer material. The glenoid prosthesis is thick (2–6 mm) (the values for the glenoid prosthesis are not accurate for all developed products; the highlighted values only address the scale of the normal glenoid polymeric prosthesis on the market). The glenoid prosthesis has a concave surface at the articulation surface with a spherical head of the humeral prosthesis and features a keel or pegs. The shape of the keel or pegs is designed to be well-inserted deep into the glenoid portion of the scapula for secure implantation of the glenoid prosthesis to bone. In this conventional shoulder replacement design, humeral and

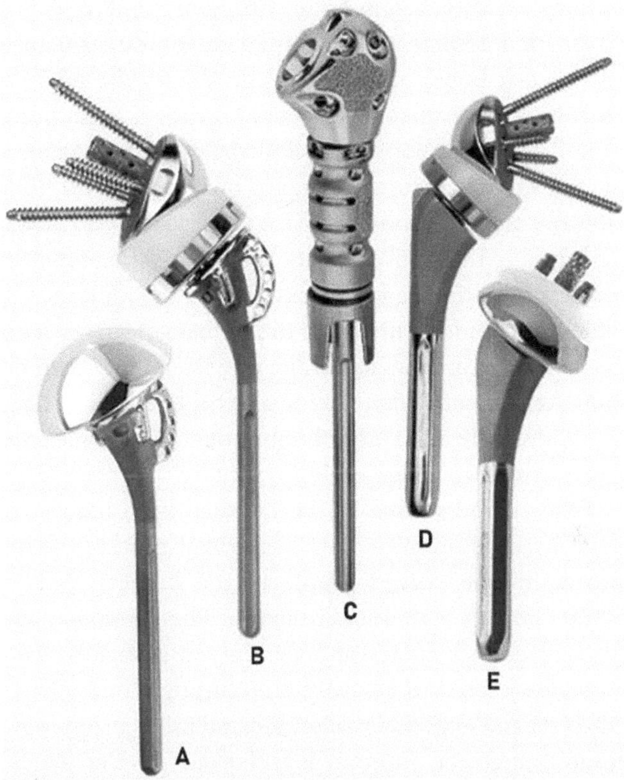

Figure 3.5 Illustration of various types of total shoulder replacements. (A, E) Anatomic total shoulder replacement (aTSR) used to treat shoulder arthritis without a deficiency or tear in the shoulder muscle-tendons. The aTSR is composed of a glenoid prosthesis that is made completely of ultra-high molecular weight polyethylene (UHMWPE) polymer or assembled from one UHMWPE insert and titanium/tantalum tray components. (B, D) Reverse total shoulder replacement (rTSR) used to treat shoulder arthritis with deficiency or malfunction of the rotator cuff muscle-tendons. Inverse articulation conditions are used to design an rTSR prosthesis, in which the glenoid prosthesis is metallic (CoCr and titanium alloys) with a convex profile and the humeral insert prosthesis is made of UHMWPE polymer with a concave profile. The overall stability of articulation conditions in rTSR is greater than that of aTSR to compensate for the deficient function of rotator cuff muscle-tendons. (C) Complete humeral prosthesis developed for replacement through the proximal portion of the humerus owing to severe bone loss at the humeral head. *Through the courtesy of A. Budeyri, R. Garofalo, S.K.-S. in Arthroplasty, undefined 2017, Convertible Implants: Simplifying the Revision from Anatomic Total Shoulder Arthroplasty to Reverse Total Shoulder Arthroplasty, Elsevier, n.d. https://www.sciencedirect.com/science/article/pii/S1045452717300895. (Accessed 24 October 2020) and get permission from Elsevier.*

glenoid prostheses are stabilized in bone with bone cement. As discussed in Chapter 2, cement interdigitating through the porous structuring bone, particularly trabecular bone, promotes integrity with the bone, whereas a rough or irregular surface feature (e.g., dent, groove, small protuberance) on the prosthesis enhances the mechanical shear strength at the cement—prosthesis interface. The important surgical point is that no bone stock is cut to position the glenoid prosthesis. This is because of the linkage of ligaments at the edges of the glenoid cavity (Fig. 3.2). Therefore, the glenoid prosthesis projects from the glenoid cavity, where the joint moves laterally. This necessitate nonspherical curving of the humeral head surface at contact with the articulation to compensate for the projected distance of the glenoid prosthesis from the glenoid cavity. (A worn or degenerated cartilage layer of the glenoid cavity is reamed and bone stock at the glenoid zone is completely maintained to enhance placement of the glenoid prosthesis.) The radius of the curved surface of the humeral head reduces the height of the humeral head component at the proximal portion of the humeral assembly—prosthesis (the curved profile of the humeral head might be designed in the form of a multiradii curve). The motion mechanism of the humeral head curved surface and the curved surface of the cavity of the glenoid prosthesis need to be analyzed using rigid body dynamics software or an analyzer and experimentally validated using an appropriate apparatus to ensure range of motion is achieved at the shoulder joint physiologically with no stiffness or generation of undesirable contact stress at the interface. Various types of TSR systems are shown in Fig. 3.5.

3.3.2 Design concept of reverse shoulder replacement system

Reverse TSA (rTSA) was introduced as a salvage surgical solution to treat severe shoulder joint injuries [5,6]. Such injuries occur after unsuccessful primary surgical treatment of the shoulder joint. Failure of rotator cuff muscle-tendons after primary treatment of shoulder joint arthritis is managed by rTSA. Likewise, the failure of proximal humerus fracture fixation is treated by rTSA. Even severe prosthesis loosening or dislocation after primary aTSA might be revised by rTSA. rTSA is also used as a primary surgical solution for shoulder joint arthritis along with significant rotator cuff muscle-tendon injuries. As discussed in the previous section, anatomic shoulder prostheses are designed in accordance with the anatomic shape of the humeral head and glenoid cavity. Such a shoulder prosthesis has adequate performance with an effective physiologic function of the muscle-tendons and ligaments at the glenohumeral joint. However, contraction of the muscle-tendon and stiffness of ligaments at the glenohumeral joint may not be adequate in elderly patients, which could reduce the success rate of the anatomic shoulder prosthesis or aTSA. This occurs as a result of the dependence of the humeral head on the rotator cuff muscle-tendons and glenohumeral ligaments to restrict the position of the humeral head component in close contact with the glenoid prosthesis and provide stability to the TSR prosthesis at the articulation interface in various motions of the upper limbs at the shoulder joint. During inappropriate contraction and function of soft tissues at the glenohumeral joint, soft tissues are irritated, creating pain to the patient. To treat osteoarthritis in the shoulder joint when there are soft tissue injuries or a malfunction, the shoulder replacement system needs to be designed to be lesser dependent for stability and functionality on the rotator cuff muscle-tendons and ligaments at the shoulder joint. In fact, stability of the humeral prosthesis at the articulation with glenoid prosthesis needs to be provided mechanically by greater contact constraints at the articulation interface compared with the conventional anatomic design of the shoulder prosthesis. Such a concept was introduced as the Grammont prosthesis in 2003 [7]. It was improved in a newer design concept (Fig. 3.6), which reverses concave–convex articulation as that of the aTSA. Thus, it is known as the rTSR, and the surgical procedure to implant this system is managed with rTSA. As illustrated in Fig. 3.5B, D, the glenoid prosthesis assembly is the convex component and the humeral prosthesis assembly is the concave component at the articulation interface. The convexity of the glenoid prosthesis at the articulation interface is spherical, in articulation with the concave spherical profile of the humeral prosthesis. The articulation component of the humeral assembly is positioned at the inferior aspect of the glenoid articulation component to stabilize contact between articulated components with minimal shear stress as a result of gravity force of the upper limbs. Because of this, less muscle force is needed to force the humeral prosthesis to the glenoid prosthesis. Muscle-tendon preservation of physiologic functionality is a significant challenge in rTSA surgery, particularly for the subscapularis and deltoid muscle-tendons [5–7]. By lateralizing the rTSR design, the warp of the deltoid muscle-tendon increases and contraction and stretching of the deltoid muscles significantly improve during a high range of motion at the shoulder joint (lateralization means the shifting of the articulation interface between glenoid and humeral articulated components toward the lateral aspect of the shoulder) (Fig. 3.6). Through the concept of lateralization, more bone stock is preserved at the proximal

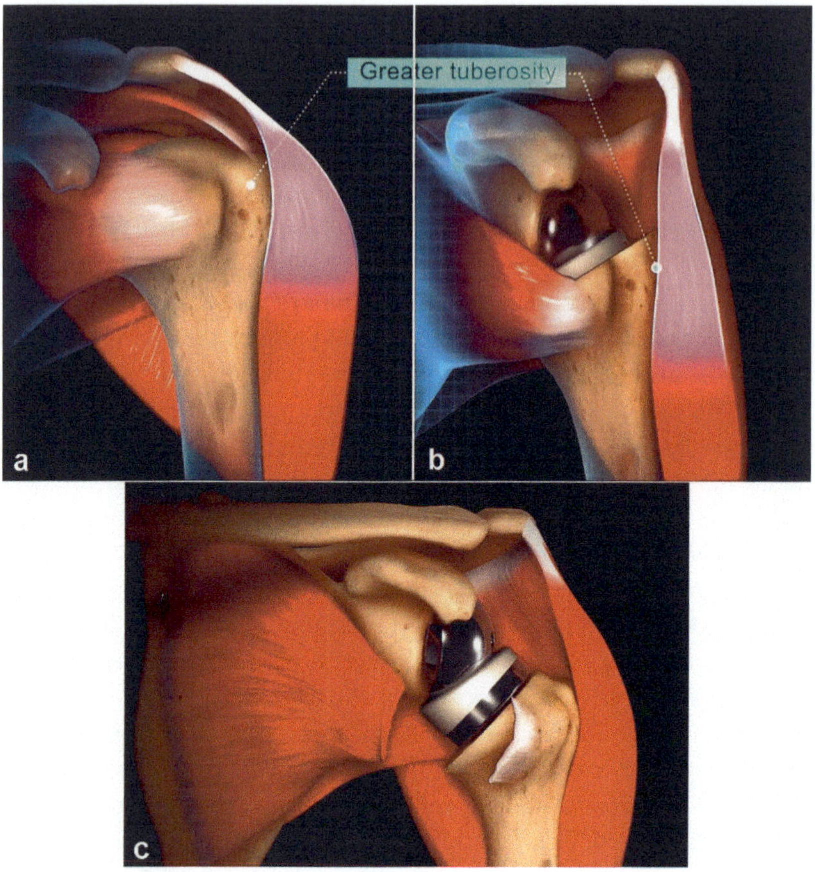

Figure 3.6 Illustration of deltoid warp advantage of lateralized reverse total shoulder replacement (rTSR). (A) Physiologic deltoid warp and attachment of subscapularis muscle-tendon through the tuberosity. (B) Low deltoid warp in old design concept of rTSR, which stabilizes the humeral prosthesis highly medial toward the glenoid. (C) Enhanced design concept of the rTSR by lateralizing the humeral prosthesis replacement, in which deltoid warp is significantly improved compared with the old design. The enhanced design of the rTSR is used to treat significant rotator cuff muscle-tendon deficiency, where the attachment zone of the subscapularis muscle-tendon at the tuberosity zone is better preserved. *Through the courtesy of J. Vourazeris, T. Wright, A. Struk, ... J.K.-J. of Shoulder and, undefined 2017, Primary Reverse Total Shoulder Arthroplasty Outcomes in Patients with Subscapularis Repair versus Tenotomy, Elsevier, n.d. https://www.sciencedirect.com/science/article/pii/S1058274616304384. (Accessed 25 October 2020) and get permission from Elsevier.*

portion of the humerus, particularly the area adjacent to the greater and lesser tuberosity. The glenoid prosthesis assembly is composed of multiple metallic components. The base component, which has a central peg, is replaced on the prepared glenoid by multiple screws. Then, the spherical component is inserted through the base component and is securely stabilized for effective articulation with the polymeric concave component of the humeral prosthesis assembly. The base component peg is tubular to enable insertion of the long peg of the spherical component through the base component. Loosening of the glenoid prosthesis is observed if stress shielding occurs at the interface of the screws and bone.

3.4 Interactions of bone with shoulder joint prostheses

Shoulder prostheses are not at high risk for loosening, migration, or dislocation compared with what occurs in knee, hip, and ankle prostheses. This is because there is not a large amount of body weight loading conditions to the prostheses or extensive rigor to the soft tissues at the glenohumeral joint. Likewise, subsidence of the prosthesis through bone tissue is not expected owing to the low level of compressive stress at the articulation surface, and it does not increase in obese or overweight patients compared with that which is observed in the body weight load-bearing musculoskeletal joints (e.g., knee, hip, ankle). However, other factors may influence the effective interaction of bone with prostheses. Owing to the important role of the soft tissues for the performance and function of the shoulder joint, a surgical plan for replacing TSR prostheses needs to be managed in accordance with the manufacturer's recommended procedure (the long-term experience of surgeons in TSA is beneficial for successfully managing the TSA). Appropriate and accurate guide and cutting instruments are needed for the precise preparation of prostheses replacement through bone tissue. The implanted prostheses have no displacement or dislocation after implantation and bone—prosthesis construct rigidity needs to be maintained in long-term implantation. Furthermore, prosthesis designs have to be compatible with physiologic loading conditions and the range of motion. Such agreement among prostheses is validated through various analytical, computational, and experimental biomechanical evaluation methods in which physiologic lading, motion, and boundary conditions of the shoulder joint are simulated in validated and calibrated computational and experimental setups.

3.4.1 Rocking horse effect

The rocking horse effect at the articulation of the humeral head and glenoid prostheses is the main reason for prosthesis loosening in shoulder arthroplasty that is evaluated biomechanically [8–12]. This effect occurs when the articulation contact becomes eccentric in the cavity of the glenoid prosthesis. As illustrated in Fig. 3.7, in such a condition, a bending moment is created with respect to the central axis of the glenoid prosthesis. This generated bending moment tends to deboned the glenoid prosthesis from its initial implanted position. With cyclic loading of this kind of undesirable mechanical phenomenon, the glenoid prosthesis may lose integrity with the bone.

3.4.2 Full conformity at articulation surface

Effective conformity of the humeral head and glenoid cavity is crucial in interacting between the shoulder prosthesis and bone tissue. Greater conformity enhances the contact surface at the articulation interface and reduces contact stress and ultimately the contact wear rate (in accordance with Archard's law, $\delta = k \cdot \sigma \cdot s$, in which the δ is the wear depth, k is the wear coefficient, σ is the contact stress, and s is the sliding distance). On the other hand, greater conformity reduces the laxity of humeral head motion in the cavity of the glenoid prosthesis, where physiologic motion at the shoulder joint becomes stiff or restricted. Such increased stiffness or restriction increases shear contact stress generated at the articulation interface in sidelong areas of the glenoid prosthesis. This may generates an eccentric bending moment and the rocking horse effect, which increases the risk for loosening at the interface of the bone and glenoid prosthesis. Therefore, it is essential to design the prosthesis at the articulation interface considering both wear and laxity.

3.4.3 Bone resorption at interface of glenoid prosthesis and bone

Because bone stock or volume at the glenoid section of the scapula (Fig. 3.3) is limited for replacing the glenoid prosthesis, bone resorption may occur at the bone–prosthesis interfaces. Glenoid replacement is managed by keel, peg, or screws known as glenoid base features. There are two types of loading conditions at the glenohumeral joint: compressive stress and axial stress. Axial stress at the articulation of the humeral head and glenoid prostheses transfers compressive stress at the interface of the glenoid base with bone. The loading axis and contact surface normally coincide under such conditions (Fig. 3.8). On the other hand, compressive force at the

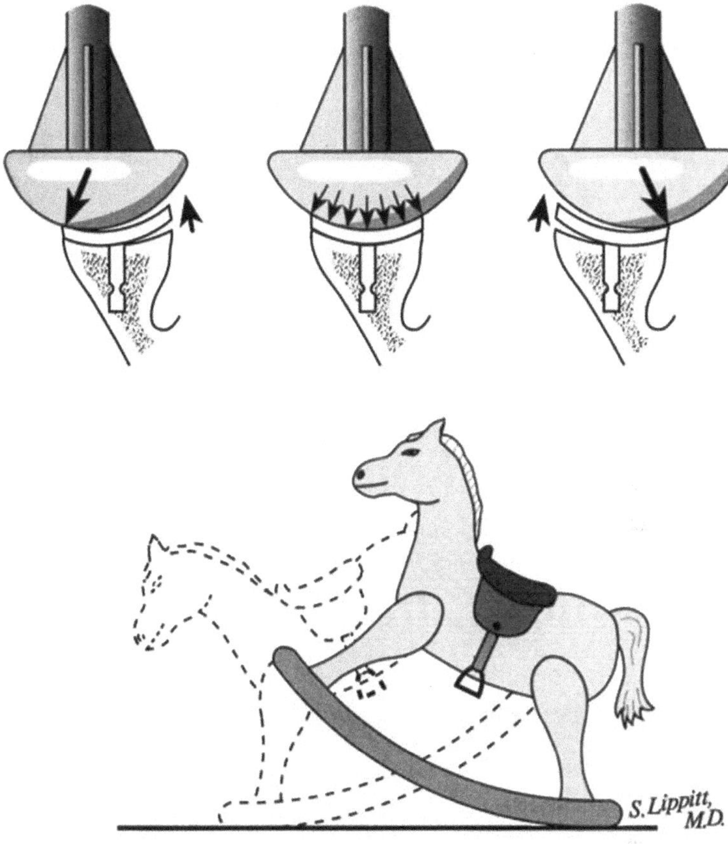

Figure 3.7 Illustration of the rocking horse concept. This mechanical concept is used to describe possible loosening of the glenoid prosthesis underneath the interface with the bone tissue. When loading concentrates at the inferior or posterior sidelong areas of the glenoid cavity of the prosthesis, a bending moment around the central axis of the glenoid cavity is generated because of which the integrity of the prosthesis–cement (in a cemented replacement) or bone–prosthesis (in a cementless replacement) deteriorates over time and loosens the prosthesis. Failure as a result of the rocking horse issue may be undetectable upon radiographic scanning owing to radiolucency of the UHMWPE polymer; however, it has been extensively evaluated biomechanically to enhance the design of the glenoid cavity profile to minimize occurrence of the rocking horse effect [12]. *Through the courtesy of L. Murphy, P. Prendergast, H.R.-J. of Shoulder and Elbow, undefined 2001, Structural Analysis of an Offset-Keel Design Glenoid Component Compared with a Center-Keel Design, Elsevier, n.d. https://www.sciencedirect.com/science/article/pii/S1058274601949766. (Accessed 25 October 2020) and get permission from Elsevier.*

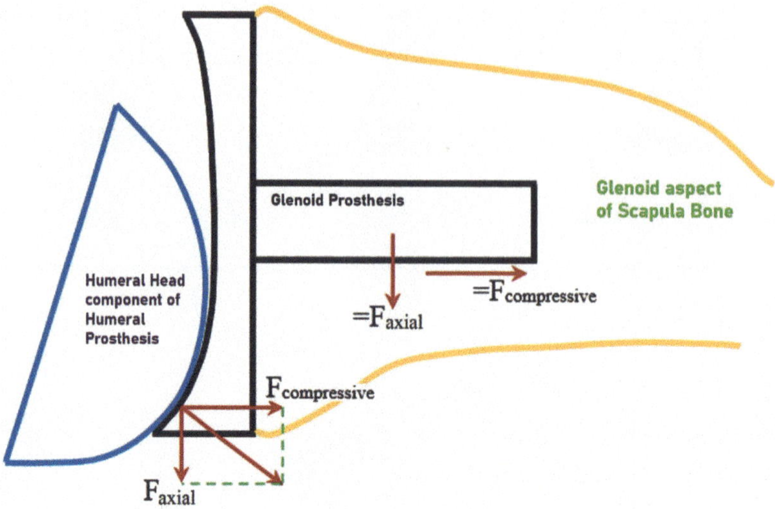

Figure 3.8 Illustration of articulation force and its transfer to bone and glenoid prosthesis interface at the base portion. Articulation force is generated at the contact surface of the humeral head component of the humeral prosthesis and cavity of the glenoid prosthesis. The generated force can be split into two perpendicular vectors: F_{Axial} and $F_{Compressive}$. F_{Axial} transfers at the interface of bone and the glenoid prosthesis at the base portion in the form of compressive force, whereas $F_{Compressive}$ transfers in the form of shear force at this interface. F_{Axial} transfers the loading condition through bone tissue to reduce the risk for bone tissue unloading and ultimately a reduction in bone resorption at this interface.

articulation interface is transmitted to shear stress at the interface of the glenoid base and bone. If the level of stress in a compressive loading condition is significant, mechanical stimulation for the bone apportion is created, and if it is in the optimal range for inducing osteogenic cells to form as mature bone cells, the long-term osseointegration of the glenoid prosthesis at a base feature occurs. If the base feature volume over available bone volume at the glenoid section is large, there is insufficient bone matrix to bear compressive loading conditions at the interface of the glenoid base feature and bone. The overall mechanical properties of bone volume between the base feature and cortical layer of bone are not high enough to act as a mechanical column to absorb the energy. This phenomenon cause the bone to overload and thus resorb. (Under such a condition, bone tissue attempts to increase its matrix compactness to counteract increased stress; however, over the time its biological activity is reduced and fibrous tissue might be created at the bone—prosthesis interface at the affected zone.)

3.4.4 Interactions of bone with humeral prosthesis

The humeral prosthesis is forced to be contact the glenoid cavity of the glenoid prosthesis. This is established through compressive force that rotator cuff muscles load to the humeral head prosthesis to stabilize it in articulation with the glenoid prosthesis, as physiologically observed in a healthy shoulder joint. Such compressive force in conjunction with ligaments constrained at the shoulder joint restrict stabilization of the humeral prosthesis at the glenoid cavity. On the other side or distal portion of the humeral prosthesis, the humeral stem is stabilized through the intramedullary canal of the proximal humerus. At a high range of motion from the physiologically natural position, the deltoid muscles are contracted to load a large amount of force for high angular abduction, flexion, or extension of the upper limbs. The attachment section of the deltoid tendons are at the proximal diaphysis section of the humerus, through which the deltoid muscle-tendon force tends to rotate the humerus and humeral prosthesis from its distal stem portion around the restricted constraint at the head portion. This generates a large amount of compressive stress at the interface of the stem portion of the humeral prosthesis and the cortical bone at the proximal third portion of the humerus. This compressive force tends to rotate the humeral prosthesis and creates a coupling bending moment between the most distal portion of the humeral stem (in contact with the cortical bone) and the most proximal portion of the humeral stem prosthesis that is in contact with the trabecular bone at the metaphysical section of the proximal humerus. With a consideration of constraint at the articulation of the humeral head with the glenoid cavity, this loading mechanism can be simulated as a cantilever beam. Fig. 3.9 shows how generated force at zone A could cause overloading through bone tissue, and in long-term implantation it may thicken the cortical layer or cause cortical hypertrophy. Increased engagement of the humeral prosthesis and cortical bone at the distal zone (zone A) reduces the lower engagement of the humeral prosthesis with the bone at the proximal zone with the trabecular bone and thus transfers the load to the proximal zone, causing stress shielding at the bone—prosthesis interfaces at this zone (zone B) [13]. As a result, bone resorption progresses in long-term implantation, as illustrated radiographically in Fig. 3.9. What is highlighted for the interaction of bone with a humeral prosthesis may be catastrophic in patients with low bone mineral density or for young patients with a high range of motion at the shoulder joint.

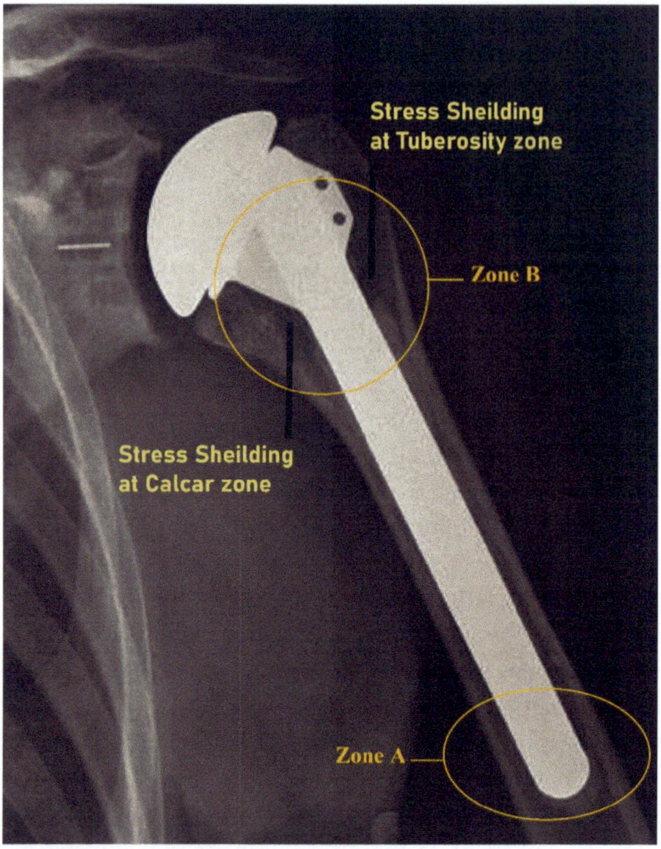

Figure 3.9 Radiographic illustration of anatomic total shoulder replacement. The UHMWPE polymer glenoid prosthesis is radiolucent; however, its base pegs inside the glenoid aspect of the scapula are less bright compared with bone tissue, and thus it is detectable in the radiographic image. Insignificant stress shielding can be seen at the tuberosity and calcar aspects of the proximal humerus after 4-years post-implantation. *Through the courtesy of B. Grogan, D. Song, C.J.-S. in Arthroplasty, undefined 2018, Cemented Humeral Shoulder Arthroplasty: Because it Works!, Elsevier, n.d. https://www.sciencedirect.com/science/article/pii/S1045452718300609. (Accessed 25 October 2020) and get permission from Elsevier.*

3.5 Possible failures of bone in interaction with shoulder prostheses

The loosening of shoulder prostheses has been reported in clinical studies [13–19]. Loosening of the glenoid prosthesis occurs for all types of complete polymeric, metal-backed, and completely metal glenoid prostheses. Failure of the glenoid is detectable on radiographic images as a radiolucent line or in areas of trabecular bone of the glenoid section of the scapula. The observation of radiolucent zones might not accompany symptoms at the initial stage, but when such zones grow, loosening, fracture, or dislocation of the glenoid prosthesis results. Failure of an implanted prosthesis through bone tissue occurs for various reasons such as infection, wear particles, improper surgical replacement, improper implant or prosthesis selection, improper preoperative study, and improper diagnosis of the indication. In addition to these reasons, failure may occur owing to design-related problems that create issues inherent in the interface of the bone and prosthesis. Determining such issues is a concept for enhancing prosthesis design and surface characteristics as the next generation of the prosthesis, suitable for elderly patients with a lower level of pain and for a higher level of functionality for young patients without loosening in long-term implantation. In this section, failure of the glenoid prosthesis that occurs for design-related reasons is reviewed with possible ineffective interactions with bone tissue. Normally, prosthesis failure is reported in clinical studies in which possible failures are not be discussed in depth. Thus, what is discussed in this section is in accordance with the fundamentals of technical concepts and prosthesis design parameters that were reviewed in Chapter 1 and in previous sections of this chapter, respectively. The conditions of patients reported in clinical investigations were also used to discuss the failures.

3.5.1 Loosening of the complete polymeric glenoid prosthesis

As highlighted in Section 3.3.1, the complete polymeric glenoid is used in the conventional design of the aTSR, which is implanted by the bone cement. Polymeric materials are highly bioinert and hydrophobic. Thus, the adherence and growth of osteogenic cells and other factors in bone formation on their surface is poor, and fibrous tissue is created at their interfaces with bone tissue. Because of the bone cement at the interface of the polymer implant and bone tissue, there are no direct contact surfaces with the bone tissue, and thus, fibrous tissue might not be created. However, the main matter is the adhesion strength between the bone cement and UHMWPE prosthesis. The only way to promote adhesion strength is to create irregular geometrical features on the surface of the polymer prosthesis (e.g., groove, hole, dent) (Fig. 3.10), where the shear strength at the cement–polymer prosthesis is increased and the risk for prosthesis dissociation is reduced [20]. This may not be a secure solution for long-term implantation owing

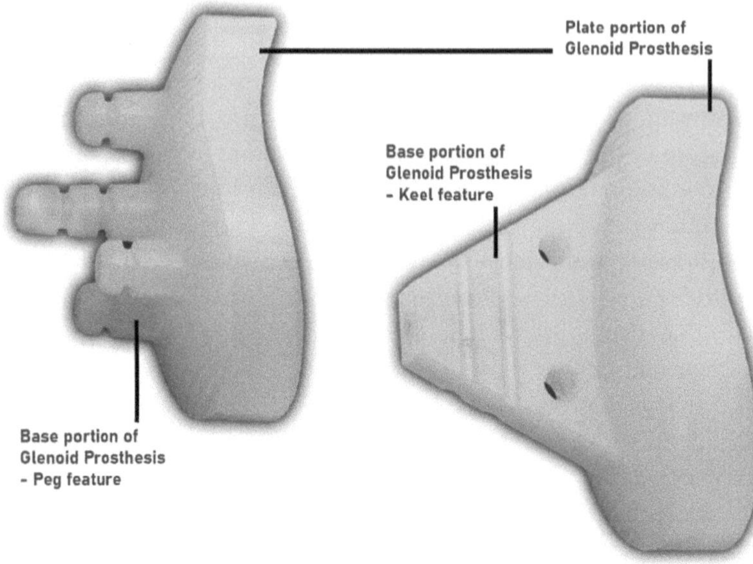

Figure 3.10 Illustration of complete polymeric glenoid prosthesis made of UHMWPE polymer. This type of glenoid prosthesis is used for an anatomic total shoulder replacement and is implanted with bone cement for increased stabilization in contact with bone tissue. The rocking horse effect occurs in loosening of the complete glenoid prosthesis at a higher risk level compared with a metal-backed glenoid prosthesis. The failure of integrity at the interface of bone–cement and cement–prosthesis at the base portion accelerates the rocking horse effect in loosening the complete polymeric glenoid prosthesis. *Through the courtesy of E.J. Strauss, C. Roche, P.-H. Flurin, T. Wright, J.D. Zuckerman, The glenoid in shoulder arthroplasty. Journal of Shoulder and Elbow Surgery 18 (2009) 819–833. https://doi.org/10.1016/j.jse.2009.05.008 and get permission from Elsevier.*

to high ductility and the brittleness of the UHMWPE polymer and cement, respectively. Unlike metallic prostheses, the UHMWPE polymer mechanical interaction with bone does not cause stress shielding at the periprosthetic bone. The use of multiple pegs or one keeled stem as the base portion of the glenoid prosthesis (Fig. 3.10) reduces the concentration of stress factor that is caused by a sharp reduction in the section area from the plate to base portion of the glenoid prosthesis.

A radiolucent line is observed beneath the glenoid plate portion, as illustrated in Fig. 3.11 [21]. Shear stress at the sidelong areas of the glenoid cavity increase, which reduce the conformity of the curved profile in the glenoid cavity, owing to significant mechanical mismatch between the UHMWPE polymer (for the glenoid prosthesis) and cobalt-chromium alloy (for the

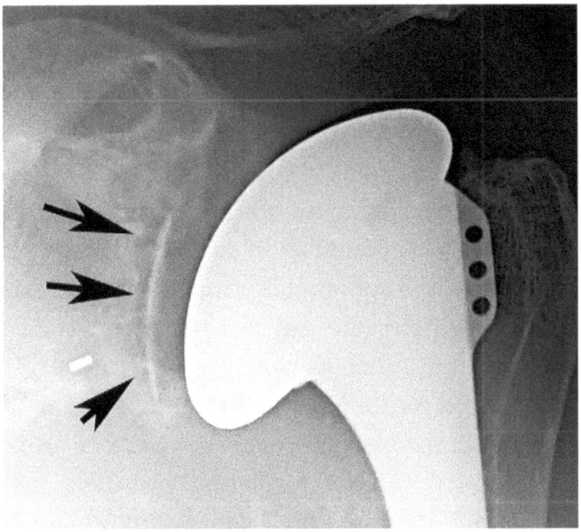

Figure 3.11 Radiographic illustration of radiolucent zone (shown by arrows) at the interface of bone and completely UHMWPE glenoid prosthesis on the underside surface. The radiolucency is related to the sidelong areas of the underside surface. At these areas, no peg, keel, or other feature applies effective loading conditions to bone tissue to induce bone formation. The low surface energy characteristics of the UHMWPE polymer prevent its effective and long-term integrity with the bone cement. These two factors may have caused the radiolucent zones in this radiographic scan, which shows areas in which bone mineral density was reduced and bone resorption occurred. *Through the courtesy of F. Matsen III, J. Clinton, J. Lynch, A.B.- JBJS, undefined 2008, Glenoid Component Failure in Total Shoulder Arthroplasty, Journals.Lww.Com, n.d. https://journals.lww.com/jbjsjournal/FullText/2008/04000/Glenoid_Component_Failure_in_Total_Shoulder.27.aspx. (Accessed 26 October 2020) and get permission from American Orthopedic Association.*

head component of the humeral prosthesis). This phenomenon occurs through wear or compressive plastic deformation of the polymer at the sidelong zones of the glenoid prosthesis (as the soft material) at the articulation surface with the CoCr humeral head (as a superhard material). This mechanical phenomenon concentrates the contact force at the articulation interface at the center of the glenoid prosthesis, where the base portion of the glenoid prosthesis inserts through the bone. Under such conditions, compressive force generated at the articulated surfaces is not effectively transferred underneath the bone tissue of the polymeric glenoid at the sidelong areas. This condition generates stress shielding owing to design deficiency and material failure at the interface of the humeral head and glenoid cavity. Overall, the complete polymeric glenoid component has inherent technical issues in the material and design aspects, with a potential effect on the progression of stress shielding areas beneath the surface of the glenoid prosthesis at the plate portion. Such a condition may become severe if a deficiency in the function of soft tissues at the shoulder joint progresses. The solution to reducing this potential design and material malfunction is to consider the peg at the sidelong areas. Although use of the pegs at the sidelong areas enhances the integrity of the glenoid prosthesis at these areas, it may not enhance the load transferring mechanism at the articulation surface. To overcome failure issues that are inherent and potentially inevitable for the completely UHMWPE glenoid prosthesis, the use of a metal tray such as that which is conceptually observed in a knee tibial tray is contemplated for the development of a metal-backed glenoid prosthesis to enhance the effectiveness of load transference beneath the glenoid prosthesis in direct contact with bone. The challenges of using a metal-backed glenoid prosthesis are reviewed and discussed in Section 3.5.2. Because of inherent failure issues observed for the completely polymeric glenoid prosthesis, a medium to high rate of radiolucency and loosening results at the interface of this glenoid prosthesis and bone. The rate of radiolucent lines at the interface of the completely polymeric prosthesis and bone is considerable (greater than 40%); however, many such cases do not progress to significant glenoid loosening that necessitates revision surgery.

3.5.2 Interaction of bone with metal-backed glenoid prosthesis

To improve the effectiveness of a completely polymeric glenoid prosthesis, the metal-backed glenoid was developed [12,22–24]. The use of metal-backed glenoids is similar to that for knee tibial, ankle tibial, or hip acetabular prostheses in which two components are used. One is the insert-bearing component that is polymeric and articulates with the other metallic component, and the other component is a tray component that is metallic and replaces bone as the base component to fix the polymeric insert. In the design of the metal-backed glenoid prosthesis, some advantages provided for improvement compared with the completely polymer glenoid prosthesis. First, biological interaction of metallic materials such as titanium (Ti) and tantalum (Ta) is much higher than that of polymeric materials, and such metallic materials are biocompatible. When the glenoid tray is fabricated from Ti or Ta materials, porous structuring of stem or peg features of the tray can use current manufacturing processes. Porous structuring of the base proton of the knee tibial tray, ankle talar, ankle tibial tray, knee femoral knee, and so forth and its design and process were used to develop the metal-backed glenoid prosthesis to enhance biological integration of the glenoid prosthesis to facilitate bone ingrowth through the prosthesis surface. A second advantage is easier machinability and processibility of metallic materials compared with polymeric materials. Through this advantage, the creation of various features on the surface and deep layers of metallic components can be achieved without significant residual stress or the deterioration of mechanical properties after processing. (Metallic materials are better relieved of stress in case of residual stress from a hard manufacturing process compared with polymers.) A third advantage is possible replacement of the metal prosthesis through bone without using bone cement. This facilitates direct contact between the bone and prosthesis to conduct bone matrix growth and eliminate integration failure at the cement–bone and implant–cement interfaces. A fourth advantage is the enhancement of load distribution beneath the surface of the polymeric insert of the glenoid assembly. As highlighted in Section 3.5.1, the load is concentrated at the central zone of the polymeric glenoid cavity, where sufficient loading conditions are not transferred beneath bone tissue at the sidelong areas, and where the risk for bone resorption and the observation of a radiolucent line and zone increases. With the use of a metal-backed glenoid, for a load concentrating at the central zone of the polymeric cavity, the load is distributed through larger areas of the metal tray beneath the component of the polymeric insert. (As a new idea, the design of the glenoid metallic tray could be enhanced by creating a gap at the central zone with direct contact at the sidelong areas for complete load transference from the sidelong areas to bone tissue underneath) (Fig. 3.12). This allows enhanced load transfer beneath bone tissue at the metallic glenoid tray and eliminates the risk for nonloading of the bone matrix at the sidelong areas of the glenoid cavity of the scapula. By enhancing the load transferring mechanism at the interface of the polymeric glenoid insert and metallic glenoid tray, loading conditions to bone tissue at the glenoid section of the scapula can be distributed by simulating physiologic loading conditions at the shoulder joint considering overall compressive and shear stresses and boundary conditions as well as soft tissue effects at the glenohumeral joint. Even the rocking horse effect might be prevented by an appropriate design of the interface profile between the glenoid insert and tray components. This issue could be a design concept strategy for a glenoid prosthesis with reduced rocking horse events.

Compared with a completely polymer glenoid prosthesis, metal-backed prostheses have a lower rate of loosening issues. However, the rate of revision surgery is greater in metal-backed glenoid prostheses [25]. Some metal-backed glenoid trays are fixed with screws through the glenoid aspect of the scapula. Screws at proximal and distal aspects of the glenoid are long

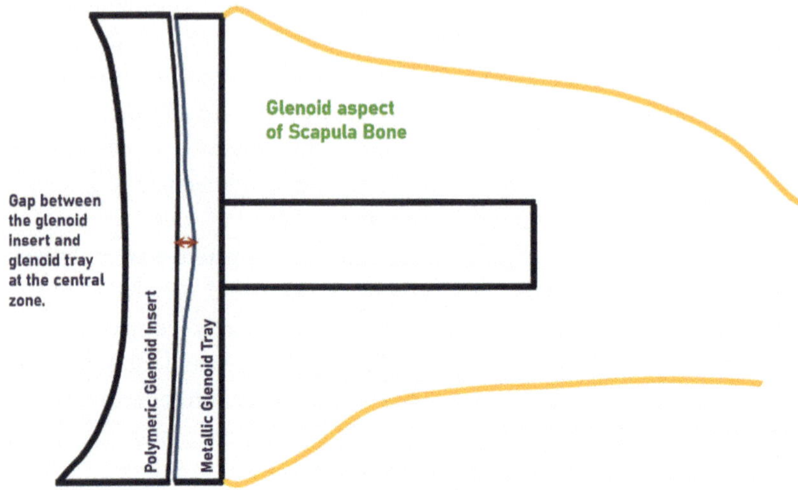

Figure 3.12 Illustration of a new design of metal-backed glenoid prosthesis, at which the design of the glenoid metallic tray could be enhanced by creating a gap at the central zone with direct contact at the sidelong areas for complete load transference from the sidelong areas to bone tissue underneath.

for more purchase through the bone and increased stability of the glenoid prosthesis. However, breakage of these screws was detected in clinical follow-up studies, in which the rate of hardware fracture in such metal-backed prostheses increased [26]. Screw fracture in the glenoid section of the scapula bone is not expected because of the lack of significant compressive stress from the superior aspect of the scapula at the implanted site of the screw. There may be an effect of loading of a bending moment to the long screw, which may result from axial stress generated at the articulation interface of the humeral head and glenoid prosthesis. Because of the long length of the screw with a low diameter in the threaded shaft, fatigue from bending stress may cause the screw to fracture in long-term implantation. Screw fracture occurs in patients with excellent bone mineral density, in whom the high rigidity of the bone—screw construct enhances load-bearing strength so that applied stress on the screw increases more than fatigue strength. On the other hand, in bone with poor mineral density, bone—screw integrity is affected by transferring of bending stress generated from axial stress (Fig. 3.8) at the articulation interface, where the screw suppresses the bone underneath and ultimately loosens. Fracture or loosening is reported at a significantly lower rate for a newer metal-backed glenoid tray at which screws are not used for fixation. In such glenoid titanium or tantalum trays, a central peg is tubular and three pegs in sidelong areas are structured in a porous form (Fig. 3.13). The midterm clinical outcome shows satisfactory

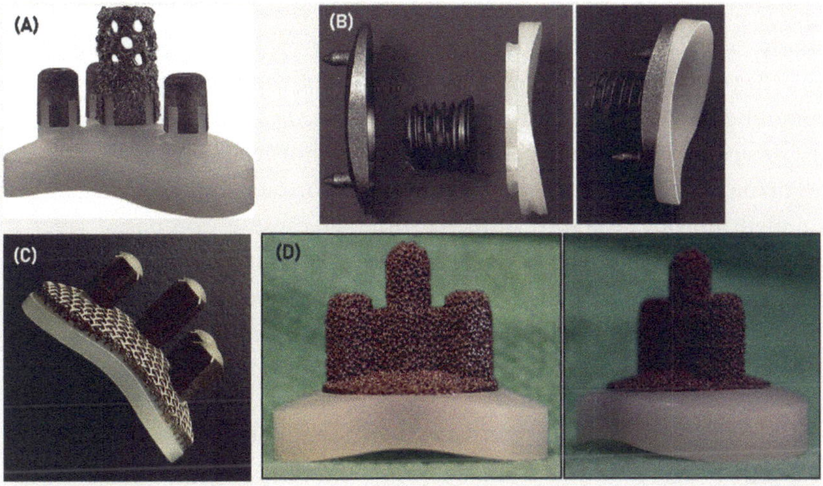

Figure 3.13 Illustration of designs of metal-backed glenoid prosthesis composed of two components of polymeric glenoid insert and metallic glenoid tray. (A) The polymer is injection molded through a central peg and three sidelong pegs made of titanium alloy. The central peg is bigger, with an internal hollow and a porous structured surface to conduct bone ingrowth. Through this, bone matrix is structured inside the central peg, and therefore the integrity of bone and the glenoid tray is well-achieved. Other sidelong pegs improve load transferring distribution from the articulation interface of the humeral and glenoid prostheses while reducing concentrated stress between the plate portion of the metallic tray and the central peg. (B) The polymer insert is interlocked to the metallic tray and holds the screw peg. The screw peg and metallic tray are fabricated from titanium alloy. (C) The central peg and three sidelong pegs are seen in this metal-backed glenoid prosthesis as well. The contribution of this prosthesis is injection molding of the polymer through a three-dimensional (3D) porous structured tray, which is osteoconductive for bone ingrowth beneath the surface of the plate portion of the metallic tray and reduces the risk for loosening in this area. (D) In this type of glenoid prosthesis, the polymer is injection molded to the completely 3D porous structured tantalum pegs. The plate portion is small and does not cover all of underside surface of the polymeric insert. The completely 3D porous structured peg is beneficial for capturing its whole volume with bone matrix when the pores are 200–500 μm. *(A) Through the courtesy of S. Kurtz, U.B. Handbook, undefined 2009, Ultra High Molecular Weight Polyethylene in Total Joint Replacement and Medical Devices, second ed., Academic, n.d. (B) Through the courtesy of F. Montoya, P. Magosch, . B.S.-J. of Shoulder and, undefined 2013, Midterm Results of a Total Shoulder Prosthesis Fixed with a Cementless Glenoid Component, Elsevier, n.d. https://www.sciencedirect.com/science/article/pii/S1058274612002947. (Accessed 28 October 2020). (C) Through the courtesy of S. Fucentese, J. Costouros, S.K.-J. of Shoulder and, undefined 2010, Total Shoulder Arthroplasty with an Uncemented Soft-Metal-Backed Glenoid Component, Elsevier, n.d. https://www.sciencedirect.com/science/article/pii/S1058274610000054. (Accessed 28 October 2020). (D) Through the courtesy of M. Budge, M. Kurdziel, ... K.B.-J. of Shoulder and, undefined 2013, A Biomechanical Analysis of Initial Fixation Options for Porous-Tantalum-Backed Glenoid Components, Elsevier, n.d. https://www.sciencedirect.com/science/article/pii/S105827461200290X. (Accessed 28 October 2020) get permission from Elsevier for all figures.*

results with the use of such a glenoid tray prosthesis [27]. As highlighted previously, a challenge with metallic implants or prostheses interacting with bone tissue is stress shielding at the interface of the implant and bone. As comprehensively discussed in Chapter 6 of *Osseoconductive Surface Engineering for Orthopedic Implants* (written by the lead author), the porous structure of the titanium implant at the surface and the fabrication of the implant in a complete porous structure from Ta material cause a composite layer to form, composed of bone matrix, and the implant material, acts as a mechanical transition layer between the bone and the bulk portion of the prosthesis. This was considered for the development of glenoid trays to reduce stress shielding and increase biological integration with bone tissue.

3.5.3 Failure of bone in interaction with metal-backed glenoid prosthesis

As highlighted in the previous section, metal-backed glenoid prostheses were developed to enhance load transference and biological integration of the glenoid tray component of the glenoid prosthesis through bone tissue at the glenoid aspect of the scapula. With such advantages, metal-backed glenoid prostheses have inherent or potential failures that progress or occur after implantation. A potential failure is the high rate of UHMWPE polymer wear at the articulation surface with the humeral head component [28]. This failure is significant for metal-backed glenoid prostheses owing to the low thickness of the polymer insert compared with a completely polymeric glenoid prosthesis. A reduction in thickness is inevitable because the polymeric insert component is supposed to interlock in the metallic tray component and available physiologic space is limited at the articulation of the overall glenoid prosthesis with the humeral head. Therefore, the overall mechanical strength of the polymeric insert in a metal-backed glenoid prosthesis is reduced and the level of density of energy absorbed in the overall volume of the insert is significantly increased. The increased level of absorbent energy accelerates the increase in contact stress at the articulation surface of the glenoid insert and humeral head where the level of wear rate increases. When this progresses, failure is catastrophic in long-term implantation, as observed for a retrieved metal-backed glenoid prosthesis [28]. Another failure occurs owing to increased stress concentration at the metallic glenoid tray between the plate and central peg portions. Such an issue is risky where the base portion of the glenoid tray features only a central peg (Fig. 3.13B). The interlocking mechanism of the polymer insert component to the metal tray component is the third mechanical factor that has an influence on the effectiveness of metal-backed glenoid prostheses in long-term implantation. Because of the increased level of axial stress at the articulation of the humeral head and glenoid prostheses (Fig. 3.8), the risk for polymer insert failure at the interlocking interfaces or zones with the metallic tray are significant, and under long-term cyclic physiologic loading conditions, polymer as the softer material is affected. The level of plastic strain at these interlocking interfaces or zones is increased and ultimately, debonding or dissociation of the glenoid polymer insert from glenoid tray components occurs (the mechanical properties of UHMWPE polymer are significantly lower than those of titanium alloy; for instance, the Young's modulus is about 1 GPa versus 110 GPa). These three highlighted design and material disadvantages of metal-backed glenoid prostheses are associated with the initial generation of glenoid prostheses and cause loosening, migration, or dissociation of the glenoid prosthesis. Under severe conditions, they increase the risk for humeral head superior dislocation owing to the ineffective constraint of the glenoid prosthesis at the articulation surface (Fig. 3.14). The design and construction of the glenoid prosthesis have improved with similar materials (UHMWPE and titanium alloy).

The use of a screw peg with an internal hole is effective in reducing the complication rate of old-generation metal-backed glenoid prostheses. This peg features a base portion of the glenoid tray that enhances the integrity of the bone tissue and pegs by grafting the internal hole of the peg to conduct internal growth of the bone matrix and biological integrity of the peg and bone tissue. However, conditions such as eccentric articulation of the humeral head and glenoid insert or tearing of the rotator cuff muscles introduce a large amount of shear stress at the cross-section of the screw peg and plate portion of the glenoid tray [23]. Such a severe failure of the metal-backed glenoid prosthesis with a hollow screw peg is displayed in Fig. 3.14.

3.5.4 Advantages of new metal-backed glenoid prosthesis in interaction with bone

In newer-generation metal-backed glenoid prostheses, pegs are designed as one central peg that is be longer and three sidelong pegs (one is in the superior section and two are in the inferior section). This enhances the load transferring mechanism between the glenoid tray and bone tissue. The generated axial and compressive loading conditions at the articulation of the humeral head and glenoid cavity of the glenoid insert (Fig. 3.8) are more distributed at a larger volume of bone tissue at the glenoid aspect of the scapula. This

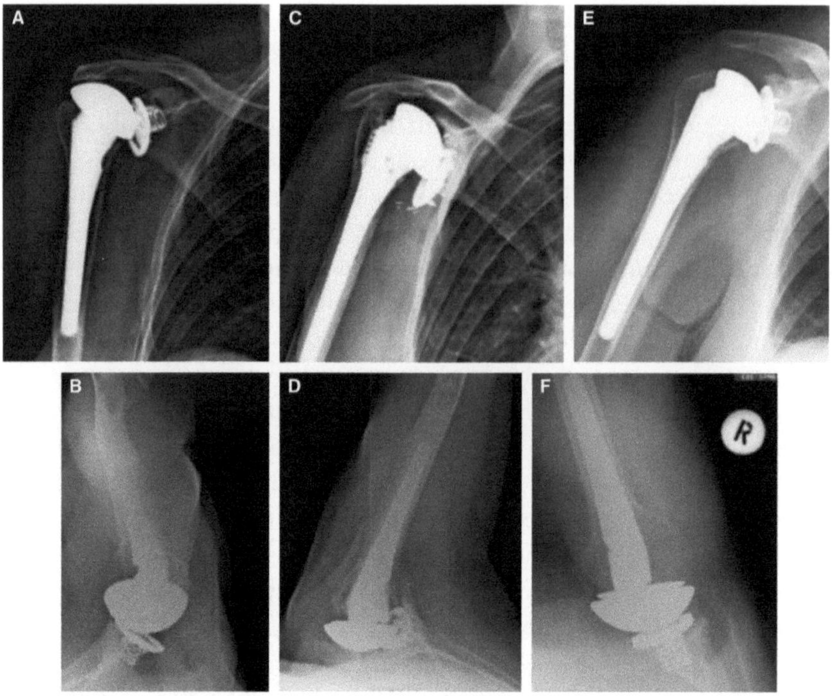

Figure 3.14 Radiographic illustration of metal-backed glenoid failure with a central peg screw (Fig. 3.13) in three patients. The progression of failure is seen as a fracture of the screw peg (A, B). In the other patient, fracture of the peg screw progressed to a more catastrophic situation as the extensive dislocation of the glenoid prosthesis and metal wear particles (C, D). UHMWPE wear can be observed as rigorous wear at 4 years after implantation (E, F) in the other patient, which is detectable as direct contact of the metal humeral head on the metal glenoid tray. *Through the courtesy of F. Montoya, P. Magosch, … B.S.-J. of Shoulder and, undefined 2013, Midterm Results of a Total Shoulder Prosthesis Fixed with a Cementless Glenoid Component, Elsevier, n.d. https://www.sciencedirect.com/science/article/pii/S1058274612002947. (Accessed 28 October 2020) and get permission from Elsevier.*

reduces the risk for concentrated stress at each peg compared to the status in the glenoid tray, which has only one central peg. Thus, loading conditions at each peg are controlled to prevent the overloading of bone tissue at the interface of the peg and bone and from bone resorption and loosening of the glenoid prosthesis. The three sidelong pegs prevent fracture of the glenoid tray from the connective zone between the plate and base portions. The other improvement in the metal-backed glenoid prosthesis is the porous structure of the pegs in their surface and bulk. Surface porous structuring is a general development strategy to enhance the osteoconductivity of orthopedic implants on their surfaces. It is carried out by techniques and processes such as three-dimensional (3D) printing, vapor deposition, and bead sintering. 3D manufacturing of the porous structure is another advanced technique in which the whole or partial volume of the implant is fabricated in a porous form with a pore size of 200–500 μm. Fabrication of the base portion of the glenoid tray from osteoconductive metals (e.g., titanium or tantalum) conducts bone formation through the pores of the porous structure and promotes glenoid tray integrity with bone tissue. A third advantage of new metal-backed glenoid prosthesis is the interlocking technique of the polymer insert to the metallic tray. Injection overmolding is used to fabricate a glenoid prosthesis composed of a UHMWPE glenoid insert and metallic glenoid tray. In this polymer engineering process, first the harder component with a significantly higher melting temperature is fabricated. Then, it is inserted into the injection mold. In the final step, polymer melt is injected into the harder component to solidify. With the internal geometrical features of the harder component (e.g., groove, dent, hole), the melt flows through deep layers, where the enhanced mechanical integrity of the two components is achieved compared with when mechanical integrity is managed by interlocking geometrical features on the surface of polymer and metallic components the old generation of metal-backed glenoid prostheses. Enhanced interlocking of the glenoid insert and tray prevents dissociation of the polymeric insert from the metallic tray owing to generated cyclic loading conditions in the axial axis (Fig. 3.8). It has other clinical advantages, such as easier replacement and a reduction in surgical time, which are out of the scope of this book. A main advantage of such enhanced interlocking is the significant reducing in wear rate at the articulation surface of the humeral head and polymeric insert. Injection overmolding fabricates a glenoid insert that is thicker, and the UHMWPE insert wear rate is considerably decreased, as discussed earlier. Increasing the insert thickness necessitates lowering the glenoid tray thickness in the plate portion. The fourth advantage of the new metal-backed glenoid prosthesis is the anatomic designing of the glenoid insert at the underside surface, which is designed with a curved profile similar to that of the cavity profile at the articulation surface (Fig. 3.13). This design consideration enables greater bone preservation at the glenoid aspect of the scapula and better physiologic load transference at the interface of bone and the underside surface.

3.5.5 Failure of bone in interaction with new metal-backed glenoid prosthesis

Although there are multiple potential advantages for metal-backed glenoid prosthesis, there are also potential failures for their long-term implantation in the treatment of shoulder joint arthritis. As illustrated in Fig. 3.13, metal-backed glenoid prostheses are developed in various designs and structures. A potential failure is debonding of the polymeric insert component from the metallic tray at the interlocked zone. As mentioned, bonding of two glenoid components is processed through injection overmolding. In this process, the polymer melt needs to penetrate the internal volume of the metallic component for high integrity. However, ingrowth melt flow may not be achieved because of the low thickness of the glenoid tray at the plate portion, in which the integrity of the glenoid components may be affected under cyclic physiologic axial stress in midterm or long-term implantation. Likewise, if the internal pathway of the melt inside the glenoid tray is not adequately designed, stress concentration and the risk for fracture at the initial section of the pathway increases and debonding of the insert and tray components of the glenoid prosthesis results. Debonding failure was reported for some cases, as demonstrated in Fig. 3.15 [29,30]. Another challenge for the new metal-backed glenoid prostheses is the glenoid tray with a 3D porous structured peg at the base portion. The growth of bone matrix is a superior advantageous for enhanced integrity of the glenoid prosthesis and bone tissue, but it could increase of concentrated stress at the connection or cross-section of the 3D porous peg and plate portion of the glenoid tray (Fig. 3.16). The generated strain of axial force at the articulation surface (Fig. 3.8) is more concentrated at the cross-section than distributes as a whole volume in the glenoid tray. The porous pegs and bone act as one integrated volume with no stress shielding or overloading at the interface of the pegs and bone tissue, such as that which occurs at the interface of solid pegs and bone tissue (in a glenoid tray with solid pegs, loading conditions are transferred in the whole volume of the metallic glenoid tray owing to the high level of molecular integrity between the plate and pegs of the tray as one integrated solid compared with when porous pegs lose integrity with the plate section as a result of the solidification of the bone matrix through the porous pegs).

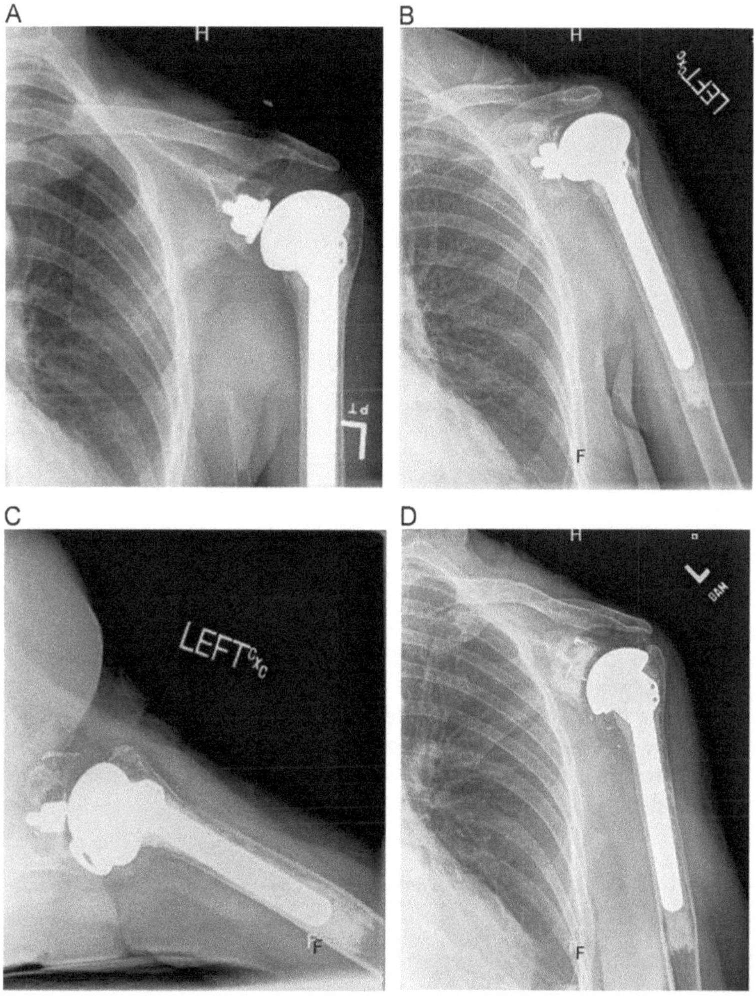

Figure 3.15 Radiographic illustration of metal-backed glenoid failure with three-dimensional porous structured peg 6 years after implantation (A–C). The case was reoperated by replacement of the completely UHMWPE glenoid prosthesis (D). *Through the courtesy of D. Gurin, W.S.J.-S. in Arthroplasty, undefined 2017, The Emerging Role of the Non-cemented Glenoid in Total Shoulder Arthroplasty, Elsevier, n.d. https://www.sciencedirect.com/science/article/pii/S1045452717300962. (Accessed 28 October 2020) and get permission from Elsevier.*

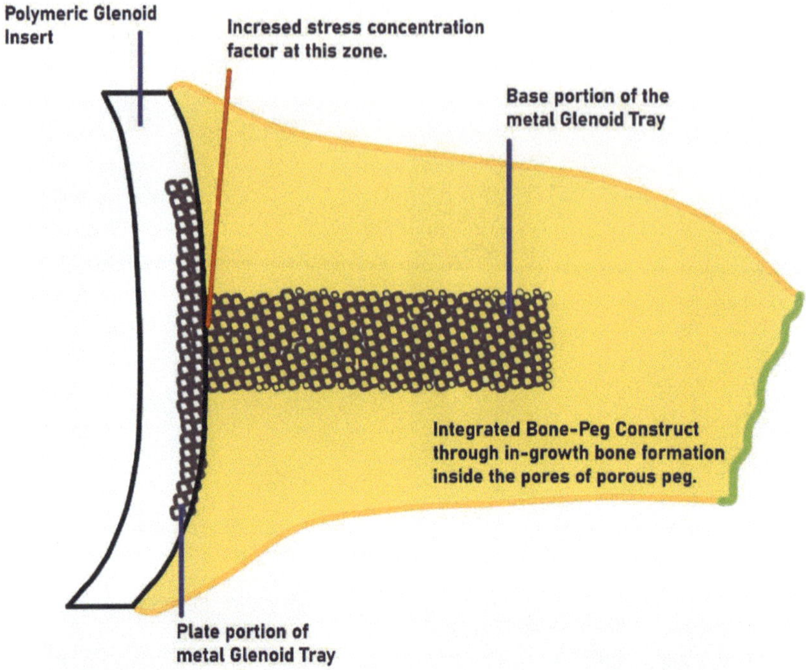

Figure 3.16 .Illustration of bone–prosthesis integrity in metal-backed glenoid with three-dimensional porous structured peg. The base portion of the metallic glenoid tray is biologically integrated with bone tissue through the growth of bone matrix through pores of the porous structured peg. This reduces the mechanical integrity of the glenoid tray between the base and plate portions, in which concentrated stress at the connective cross-section of these two portions increases. Thus, the risk for fracture at this section becomes higher.

3.5.6 Failure of bone with reverse total shoulder replacement

Failure of the rTSR has been reported as bone resorption at a greater tuberosity, calcar region, and left diaphysis of the proximal third portion of the humerus [31–35] and scapular neck at the inferior aspect of the glenoid [36–38]. As discussed in Section 3.4.4, transferred loading conditions through the tuberosity and calcar zones are not significant because of the design profile of the humeral prosthesis. In fact, the humeral stem is stabilized in the diaphysis portion of the humerus proximal third portion through the cortical bone, and the proximal portion of the humeral prosthesis is positioned at the humeral head zone through the trabecular bone. The proximal portion of the humeral prosthesis does not have a significant role in stabilization; thus, no loading conditions are transferred to bone tissue at this zone. The low level of loading conditions results in bone resorption, as illustrated in Fig. 3.17. Likewise, there are no loading conditions in the lateral aspect of the humeral stem in the diaphysis portion of the humerus proximal third zone. This is because of the transfer of loading conditions to the medial aspect of the humeral stem in abduction as a result of deltoid muscle-tendon force to the diaphysis proximal third. When there is angular rotation in the adduction axis from neutral position at the shoulder joint, the lateral aspect of the humeral stem is also stressed.

Another failure issue, mostly observed in the first generation of the rTSR, is severe bone resorption at the neck of the scapula and the inferior aspect of the glenoid. This phenomenon is known as scapular notch in the

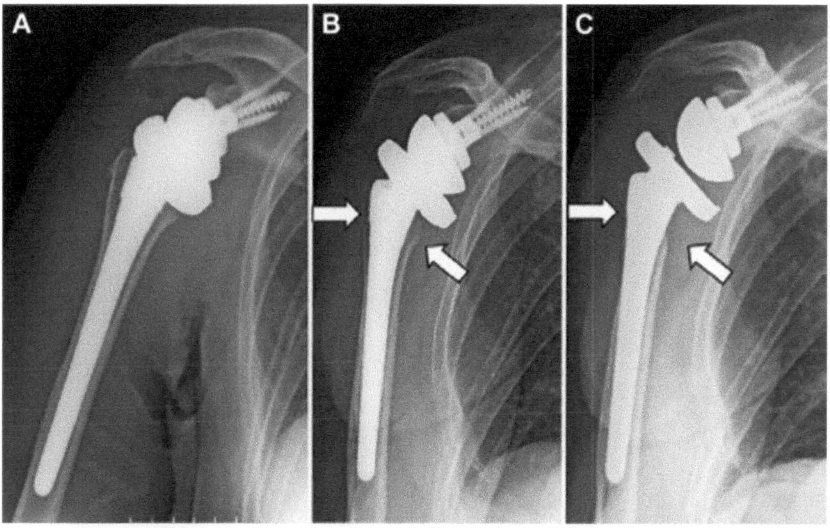

Figure 3.17 Radiographic illustration of bone resorption in greater tuberosity (left arrows), calcar zone (right arrows), and the lateral aspect of the humeral stem. (A) Immediately after implantation; (B) at 6 months; (C) at 1 year after implantation. Reverse total shoulder replacement (rTSR) for this patient was the onlay type, in which the neck of the humeral prosthesis at the proximal portion had a large difference with the tray component of the polymeric insert. This difference was lower in inlay-type rTSR because of the use of a conical tray component for the polymeric insert. The rate of bone resorption is lower in the inlay type compared with the onlay type [31,35]. *Through the courtesy of K. Inoue, N. Suenaga, N. Oizumi, … H.Y.-J., undefined 2020, Humeral Bone Resorption After Reverse Shoulder Arthroplasty Using Uncemented Stem, Elsevier, n.d. https://www.sciencedirect.com/science/article/pii/S246860261930172X. (Accessed 2 November 2020) and get permission from Elsevier.*

literature. As illustrated in Fig. 3.18, by impinging on the polymeric insert to the inferior scapula neck, the bone faces overloading conditions that cause gradual resorption of the bone at this zone. The initial design of the rTSR was managed medial to the glenohumeral joint and the polymeric insert at a low angular position in the abduction-adduction axis interfered with the scapular neck. In the newer lateralized design of the rTSR, the position of the glenohumeral joint is more lateral from the glenoid and thus eliminates prosthesis impingement with the scapular neck. This may be combined with internal angular bone sectioning to remove more bone stock at the glenoid section for implantation of the lateralized rTSR design. The scapular notch may loosen the glenoid tray under severe conditions, when rTSR revision would be required. Fig. 3.19 shows failure that occurred in the interaction of the glenoid prosthesis and bone in rTSA owing to the scapular notch.

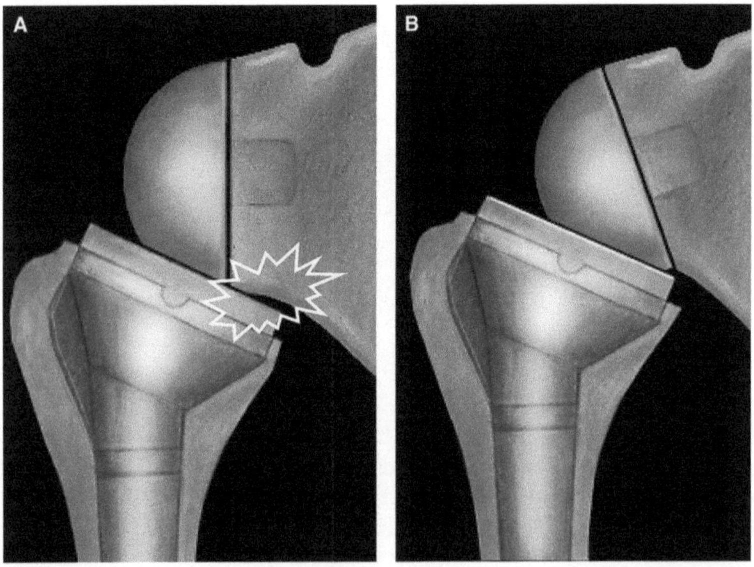

Figure 3.18 Illustration of the occurrence of the scapular notch. (A) The old medial design of reverse total shoulder replacement (rTSR) causes impingement of the polymeric insert with the scapula neck at the inferior aspect of the glenoid. (B) The newer lateral design of rTSR eliminates the risk for impingement and scapular notch. *Through the courtesy of T. Edwards, G. Trappey, C. Riley, … D.O.-J. of Shoulder and, undefined 2012, Inferior Tilt of the Glenoid Component Does Not Decrease Scapular Notching in Reverse Shoulder Arthroplasty: Results of a Prospective Randomized Study, Elsevier, n.d. https://www.sciencedirect.com/science/article/pii/S1058274611004204. (Accessed 2 November 2020) and get permission from Elsevier.*

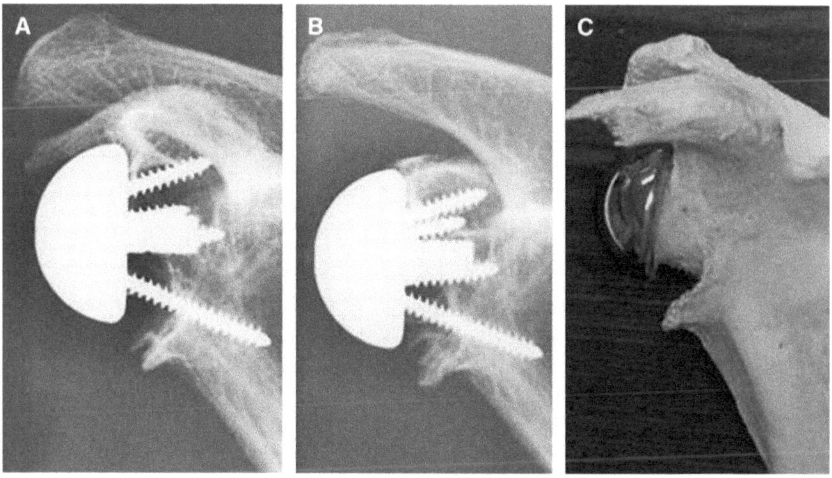

Figure 3.19 Illustration of scapular notch occurring at the interface of the glenoid tray and bone (A—C). As the result of this incident, bone resorption occurred at the inferior aspect of the glenoid and progressed proximally, in which the integrity of the bone and glenoid tray screws and peg deteriorated significantly and ultimately loosened the glenoid prosthesis. *Through the courtesy of C. Lévigne, P. Boileau, L. Favard, ... P.G.-J. of Shoulder and, undefined 2008, Scapular Notching in Reverse Shoulder Arthroplasty, Elsevier, n.d. https://www.sciencedirect.com/science/article/pii/S1058274608003467. (Accessed 2 November 2020) and get permission from Elsevier.*

3.6 Summary and more discussion

The failure of shoulder prostheses was reported in clinical investigations as a serious issue for patients with soft tissue deficiency at the shoulder joint. The shoulder joint highly depends on effective functionality of the muscle-tendons and ligaments at the glenohumeral joint, particularly the rotator cuff muscle-tendons. As highlighted in Section 3.2, the rotator cuff is composed of four muscles that facilitate motion at the shoulder joint and deltoid muscle to increase range of motion, particularly in abduction rotation. TSR prostheses (glenoid and humeral) were first designed in accordance with the anatomic shape of the glenoid cavity and humeral spherical geometry to establish the highest level of physiologic functionality of the glenohumeral joint with that of healthy joints. However, the placement prion of the glenoid prosthesis (base portion) had issues with long-term integrity with bone tissue. Osseointegration of the complete polymeric glenoid prosthesis was observed with the progression of radiolucent zones at the interface of the bone and prosthesis and an increase in the risk for loosening of the glenoid prosthesis. This issue was improved with the development of metal-backed glenoid prostheses to enhance the mechanical and biological integrity of the bone tissue and prosthesis while enhancing load transference from the articulation surface of the humeral head and glenoid insert cavity to the glenoid aspect of the scapula. Metal-backed glenoid prostheses have some technical issues, necessitating design and structural improvements. Screw breakage, stress shielding at the bone—prosthesis interface, increased stress concentration in the singular peg base portion, the accelerated wear rate of the polymeric insert, and so forth were discussed as design and structural issues that potentially and inherently existed in the initial generation of the metal-backed design. In the next generation of this type of glenoid prosthesis, design improvements enhanced its performance in long-term implantation. The design had a base portion with four pegs, in which the central peg is manufactured with an internal hole and a porous structured surface to enhance the biological integrity of the glenoid tray component of the metal-backed glenoid prosthesis with bone tissue. The thickness of the plate portion of the glenoid tray was reduced to increase the polymeric glenoid insert and significantly reduce the wear rate at the articulation interface of the TSR. Furthermore, the design or structurally enhanced metal glenoid tray was developed as a cementless prosthesis to be replaced without using bone cement; thus, the risk for dissociation at the interfaces of the cement—bone and cement—prosthesis are eliminated. Other issues have been reported for the newer generation of metal-backed glenoid prosthesis, such as dissociation of the polymeric insert from the metal tray and fracture of the metal glenoid tray at the cross-section between the base and plate portions (Fig. 3.15). With regard to failure of the humeral prosthesis of the aTSR, there are not significant reported data in clinical studies. The main issue associated with the humeral prosthesis is the progression of bone resorption at the greater tuberosity and calcar zone, which are not critical issues in many cases and do not cause loosening of the humeral prosthesis even in long-term implantation. In severely affected patients with osteoporosis or very low bone mineral density, grafting of the humeral stem at the humeral head portion or the use of stemless aTSR might be a solution. The design and structural improvements as well as material enhancement are undergoing design and development processing of the glenoid prostheses in aTSR systems.

Another type of TSR is known as rTSR, in which articulation conditions are reversely designed. The glenoid prosthesis is developed with a convex profile, whereas the humeral insert component is designed with a concave profile. Because of the lower dependency of the rTSR on soft tissue functionality to stabilize the humeral and glenoid prostheses, the complication rate of prosthesis loosening and dislocation is much lower than for aTSR. Articulation conditions between the humeral and glenoid prostheses depends on the stability of the glenoid tray fixation to the glenoid and humeral stem replacement through the humeral intramedullary canal. In revision cases in which aTSR prostheses fail, considerable bone resorption might occur, which increases failure of the rTSR as a revision solution for treating of such primary failed cases. Thus, the selection of the rTSR system in accordance with each condition is highly beneficial for the long-term success rate of the rTSR for revision surgery. Reported failures for the rTSR prostheses are bone resorption at the greater tuberosity, calcar, and scapula neck zones. Of these failures, bone resorption at the neck of the scapula at the inferior aspect of the glenoid may progress greatly and loosen the screws and peg of the glenoid tray component. Such a failure occurs in medialized rTSR in which the articulation interface is toward the glenoid zone and the humeral polymeric insert component direct interferes with the scapular neck. This issue is eliminated by lateralizing the articulation interface in the development of the newest rTSR. The newer design for the rTSR has other advantages, such as increased deltoid warping and enhanced physiologic motion at the shoulder joint, particularly high angular abduction motion.

This chapter reviewed interactions and failures of shoulder and glenohumeral prostheses with bone tissue and their relationships to the design parameters, features, and concepts of prostheses. However, there may be missed technical and related data on these topics, which will be updated in the next editions of this book. Furthermore, I have provided much discussion-based content based on my long-term scientific and industrial experience and studies, and hope they will be helpful and constructive for those who are interested in an in-depth and conceptual understanding of the contents in a more practical manner.

3.7 Remind and learn

In this section, some questions are provided as effective reminders of the contents reviewed and discussed in this chapter. Answering these questions enhances learning about the failures of the bone in interaction with TSR prostheses.

Q 3.1: Discuss the overall anatomy of the shoulder joint. Describe the rotator cuff muscles and the attached zone at the humerus. What are the function of the rotator cuff and labrum tissue in physiologic motion of the upper limbs at the shoulder or glenohumeral joint?
Q 3.2: Discuss the crucial role of the rotator cuff muscles in the stability and functionality of the upper limbs at the shoulder joint. Discuss the role of the glenohumeral and coracohumeral ligaments in enhancing physiologic motion and relative stability of the humeral head and scapula at the shoulder joint.
Q 3.3: Discuss loading conditions that might be introduced or transferred to the rotator cuff muscles. What is the effect of increased force to these muscles? What function of the supraspinatus muscle-tendon is crucial for effective loading transference between the humeral head and glenoid cavity in the glenohumeral joint?
Q 3.4: Describe the function of the deltoid muscles. What is bone stock of the scapula ends with the glenoid cavity? Discuss the anatomy of the glenoid cavity at the superior and inferior sections and what advantage is achieved in the inferior zones of the glenoid cavity. Considering Fig. 3.4, discuss the effective design of base features of the glenoid prosthesis (e.g., keel, pegs, screws).
Q 3.5: Discuss the TSA. What main parameter is contemplated for replacement of the humeral and glenoid prostheses, and why? Discuss bone preservation concerns in TSA.
Q 3.6: Discuss replacement of the humeral and glenoid prostheses in conventional TSR. What are the main concerns for replacing the glenoid prostheses? Discuss your answer considering the bone stock at the glenoid zone of the scapula and shifting of the articulation surface laterally. What design concept for the humeral prosthesis is required to compensate for joint shifting laterally? Discuss how the articulation surface of the humeral head component with the glenoid prosthesis may be analyzed and why it might be necessary.
Q 3.7: Referring to Fig. 3.6, discuss the various types of the TSR system. Compare the components of each type.
Q 3.8: What indications need to be treated by rTSA? Discuss why the aTSA might not be usable for these indications. Discuss why rTSA is known as reverse methods compared with aTSA.
Q 3.9: Discuss articulation conditions in rTSR and its benefits for enhancing the stability of prostheses and minimizing dependency on the function of muscles at the shoulder joint.

Q 3.10: What is lateralization in designing rTSR prostheses? Discuss the benefits of lateralization of the articulation interface in the function of deltoid muscles and preserving bone at the humeral head. Discuss the components of the glenoid prosthesis in rTSR.

Q 3.11: Discuss why the occurrence of failures such as migration, dislocation, subsidence, and even loosening of the TSR prosthesis are much less than what is observed in other joints such as the knee, hip, and ankle.

Q 3.12: Discuss the crucial role of the surgical procedure for effectively replacing TSR prostheses. What parameters or factors need to be carefully contemplated to manage TSA?

Q 3.13: What is the rocking horse phenomenon at the interface of the glenoid prosthesis and beneath bone tissue at the glenoid cavity? How does this phenomenon occur? Discuss what failures occur when this phenomenon progresses through cyclic physiologic loading at this interface. How can this effect be recognized on radiographic images? What is the current solution for reducing the risk for the rocking horse effect?

Q 3.14: What are the advantages and disadvantages of full conformity at the articulation interface between the humeral head and glenoid prostheses? Consider the effects of full conformity on wear and laxity at the articulation interface and the rocking horse effect at the interface of the glenoid prosthesis and beneath bone tissue at the glenoid cavity.

Q 3.15: Discuss the glenoid base feature and discuss how shear and compressive stresses at the interface of the glenoid base feature with bone are generated. What are the effects of these stresses to the bone–prosthesis interface at the glenoid zone? How does bone resorption progress when the glenoid base feature is not an optimal size with respect to the bone volume at the glenoid zone?

Q 3.16: What is the loading mechanism at the proximal third humerus where the humeral prosthesis is implanted? How do cortical hypertrophy and stress shielding occur at the distal and proximal zones of the humeral prosthesis and bone interface, respectively? Discuss what failures progressed owing to these inefficacies in interaction between the humerus and humeral prosthesis.

Q 3.17: What are the main reasons for failure in interactions of bone with TSR prostheses?

Q 3.18: What are the main challenges of a completely polymeric glenoid prosthesis in interfaces with bone tissue at the glenoid zone? Discuss why the bone–prosthesis interface might not interact effectively at the interface of the polymeric glenoid and bone. What is the overall benefit of multiple pegs or a one-keeled stem design of the glenoid base feature in developing a polymeric glenoid? How may the rocking horse effect progress at the polymeric glenoid prosthesis and glenoid cavity?

Q 3.19: What is the main reason for bone resorption underneath the interface of the polymeric glenoid prosthesis with the glenoid cavity? Why does plastic deformation or wear occur at the sidelong areas of the glenoid prosthesis at the articulation surface with the humeral head? Discuss the condition in which compressive contact force at the articulation surface of the humeral head and glenoid prostheses is concentered at the central zone and failure occurs. What is a possible design solution to reduce this failure in designing the polymeric glenoid prosthesis?

Q 3.20: What is the background development of the metal-backed glenoid prosthesis? What are the main advantages of the metal-backed glenoid prosthesis compared with the completely polymeric glenoid prosthesis? Discuss why the biocompatibility and biological integration of Ti and Ta metal-backed glenoid prostheses are greater than those of a polymeric prosthesis.

Q 3.21: What are the benefits of the metal-backed glenoid prosthesis compared with those of a polymeric prosthesis in view of manufacturing and the use of bone cement for replacement?

Q 3.22: Discuss how the contact force generated at the articulation surface of the humeral head and polymeric glenoid component in a metal-backed glenoid prosthesis would be better transferred to sidelong areas of the glenoid cavity of the scapula. What new design concept might be hypothesized to enhance this mechanical load transfer? (Refer to Fig. 3.13.)

Q 3.23: What types of metal-backed glenoid prostheses might be developed for the market? Discuss the manufacturing process of each type, with its design characteristics and benefits.

Q 3.24: What are the main failures of a metal-backed glenoid prosthesis that is fixed by a long screw through the glenoid zone of the scapula? Discuss your answer for both good and poor mineral density bone at the glenoid zone.

Q 3.25: Discuss the benefits of a metal-backed glenoid prosthesis with a central peg and three pegs in sidelong areas compared with a prosthesis fixed with screws to the glenoid aspect of the scapula.

Q 3.26: Why might the wear rate might be significantly greater in a metal-backed glenoid prosthesis compared with that in a completely polymeric glenoid prosthesis in articulation with the humeral head prosthesis? Discuss your answer in view of increased contact stress at the articulated surface in a metal-backed prosthesis. What failure might occur as the result of the increased wear rate at the articulated surface?

Q 3.27: What is a possible failure for a metal-backed glenoid tray with one central peg as the base feature? Why might debonding or dissociation of the glenoid polymer insert from the glenoid tray components occur?

Q 3.28: What are the main advantages and disadvantages of the metal-backed glenoid prosthesis with a central screw peg as the base feature? Discuss failure that might occur for this type of glenoid tray component and possible reasons (use Fig. 3.15 to discuss your answer).

Q 3.29: Discuss a metal-backed glenoid prosthesis with central and sidelong pegs as the base feature. What are the advantages of this newer design of the glenoid tray component?

Q 3.30: What are the benefits of the porous structure of the metal-backed glenoid prosthesis at the base feature of the glenoid tray component in interacting with bone? Which kinds of materials would be advantageous for fabricating a porous structured glenoid tray?

Q 3.31: Discuss injection overmolding in fabricating the newer metal-backed glenoid prosthesis with greater thickness of the insert polymeric component. What are the benefits of using this polymer engineering process to enhance the glenoid insert and tray integrity while reducing the wear rate at the articulation surface with the humeral head prosthesis?

Q 3.32: What is the advantage of the anatomic design of the glenoid insert at the underside surface in the interface with the glenoid cavity of the scapula?

Q 3.33: What is the challenge of effective interlocking between the polymeric glenoid insert and the metallic glenoid tray using injection overmolding? What failures might occur in midterm or long-term implantation of a metal-backed glenoid prosthesis manufactured with this process? (Consider Fig. 3.16 in discussing the failures.)

Q 3.34: Discuss why stress concentration might increase at the connection or cross-section of the 3D porous peg and plate portion of the glenoid tray in a completely porous structured glenoid tray. Why might mechanical integrity between the base and plate portion of such a glenoid tray component be reduced?

Q 3.35: What bone areas in contact with the humeral prosthesis in rTSR might be affected? What failures might occur in the affected zone? (Consider Fig. 3.18 in discussing the failures.)

Q 3.36: Describe the scapular notch phenomenon and under what conditions it might occur after rTSA. What failure might occur in interactions of the glenoid prosthesis and bone in rTSA as a result of this phenomenon? What design improvements have been implemented to eliminate this issue?

3.8 Think and challenge

In this section, some questions are provided for further thought about the reviewed and discussed contents of this chapter. The aim is to challenge the current facets of interactions and failures of the bone in interaction with TSR prostheses by answering these questions, which may enhance new design concepts for the future development of prostheses.

Q 3.37: As highlighted in this chapter, the interaction of the humeral stem with the humerus at the proximal portion may result in stress shielding at the proximal zone of the humeral prosthesis in interfaces with either trabecular or cortical bone tissue. What designing concept might be used to reduce or eliminate such failure significantly?

Q 3.38: As highlighted in this chapter, the concept for designing the glenoid prostheses is challenging owing to the small volume of scapula at the glenoid zone for effectively replacing the glenoid prosthesis. As reviewed, the risk for failure exists for all types of developed glenoid prostheses. Discuss whether other design concepts could be hypothesized to develop new glenoid prosthesis with a reduced risk for reviewed failure.

References

[1] S. Vollans, A.A.-S. (Oxford), undefined 2016, Rotator Cuff Tears, Elsevier, n.d. https://www.sciencedirect.com/science/article/pii/S0263931916000065. (Accessed 18 October 2020).

[2] C. Yang, A. Goto, W. Sahara, … H.Y.-C., undefined 2010, In Vivo Three-Dimensional Evaluation of the Functional Length of Glenohumeral Ligaments, Elsevier, n.d. https://www.sciencedirect.com/science/article/pii/S0268003309002484. (Accessed 24 October 2020).

[3] G.B. Sharma, P.J. McMahon, D.D. Robertson, Structure modeling of the glenoid: relevance to shoulder arthroplasty, J. Orthop. Res. 32 (2014) 1471–1478, https://doi.org/10.1002/jor.22696.

[4] A. Budeyri, R. Garofalo, S.K.-S. in Arthroplasty, undefined 2017, Convertible Implants: Simplifying the Revision from Anatomic Total Shoulder Arthroplasty to Reverse Total Shoulder Arthroplasty, Elsevier, n.d. https://www.sciencedirect.com/science/article/pii/S1045452717300895. (Accessed 24 October 2020).

[5] J. Vourazeris, T. Wright, A. Struk, … J.K.-J. of Shoulder and, undefined 2017, Primary Reverse Total Shoulder Arthroplasty Outcomes in Patients with Subscapularis Repair Versus Tenotomy, Elsevier, n.d. https://www.sciencedirect.com/science/article/pii/S1058274616304384. (Accessed 25 October 2020).

[6] R. Friedman, P. Flurin, … T.W.-J. of S. and, undefined 2017, Comparison of Reverse Total Shoulder Arthroplasty Outcomes with and Without Subscapularis Repair,

Elsevier, n.d. https://www.sciencedirect.com/science/article/pii/S1058274616304487. (Accessed 25 October 2020).
[7] S. Parada, P. Flurin, T. Wright, ... J.Z.-J. of S. and, undefined 2020, Comparison of Complication Types and Rates Associated with Anatomic and Reverse Total Shoulder Arthroplasty, Elsevier, n.d. https://www.sciencedirect.com/science/article/pii/S1058274620306224. (Accessed 25 October 2020).
[8] V. Sabesan, D. Lima, J. Whaley, ... V.P.-J. of Shoulder and, undefined 2019, The Effect of Glenohumeral Radial Mismatch on Different Augmented Total Shoulder Arthroplasty Glenoid Designs: A Finite Element Analysis, Elsevier, n.d. https://www.sciencedirect.com/science/article/pii/S1058274618308899. (Accessed 25 October 2020).
[9] L. Murphy, P. Prendergast, H.R.-J. of Shoulder and Elbow, undefined 2001, Structural Analysis of an Offset-Keel Design Glenoid Component Compared with a Center-Keel Design, Elsevier, n.d. https://www.sciencedirect.com/science/article/pii/S1058274601949766. (Accessed 25 October 2020).
[10] A. Karelse, A. Van Tongel, ... T.V.-J. of S. and, undefined 2015, Rocking-horse Phenomenon of the Glenoid Component: The Importance of Inclination, Elsevier, n.d. https://www.sciencedirect.com/science/article/pii/S1058274614006843. (Accessed 25 October 2020).
[11] C. Anglin, U. Wyss, D.P.-J. of Shoulder and Elbow Surgery, undefined 2000, Mechanical Testing of Shoulder Prostheses and Recommendations for Glenoid Design, Elsevier, n.d. https://www.sciencedirect.com/science/article/pii/S1058274600156733. (Accessed 25 October 2020).
[12] M. Budge, M. Kurdziel, ... K.B.-J. of Shoulder and, undefined 2013, A Biomechanical Analysis of Initial Fixation Options for Porous-Tantalum-Backed Glenoid Components, Elsevier, n.d. https://www.sciencedirect.com/science/article/pii/S105827461200290X. (Accessed 28 October 2020).
[13] B. Grogan, D. Song, C.J.-S. in Arthroplasty, undefined 2018, Cemented Humeral Shoulder Arthroplasty: Because it Works!, Elsevier, n.d. https://www.sciencedirect.com/science/article/pii/S1045452718300609. (Accessed 25 October 2020).
[14] D. Kovacevic, J. Hodgins, L.B.-S. in Arthroplasty, undefined 2014, Lucent Lines and Glenoid Components: What Do They Portend?, Elsevier, n.d. https://www.sciencedirect.com/science/article/pii/S1045452715000097. (Accessed 25 October 2020).
[15] S. Moulton, E. Cole, R. Gobezie, A.R.-J.O. Access, undefined 2019, Minimum 5-year Outcomes of Pegged Versus Keeled All-Polyethylene Glenoids, Elsevier, n.d. https://www.sciencedirect.com/science/article/pii/S2468602619300464. (Accessed 25 October 2020).
[16] B. Schoch, T. Wright, J. Zuckerman, P.F.-J.O. Access, undefined 2019, The Effect of Radial Mismatch on Radiographic Glenoid Loosening, Elsevier, n.d. https://www.sciencedirect.com/science/article/pii/S2468602619300476. (Accessed 25 October 2020).
[17] D.M. Kim, M. Aldeghaither, F. Alabdullatif, M.J. Shin, E. Kholinne, H. Kim, I.H. Jeon, K.H. Koh, Loosening and revision rates after total shoulder arthroplasty: a systematic review of cemented all-polyethylene glenoid and three modern designs of metal-backed glenoid, BMC Muscoskel. Disord. 21 (2020), https://doi.org/10.1186/s12891-020-3135-6.
[18] W. Aibinder, B. Schoch, C. Schleck, ... J.S.-J. of Shoulder and, undefined 2017, Revisions for Aseptic Glenoid Component Loosening After Anatomic Shoulder Arthroplasty, Elsevier, n.d. https://www.sciencedirect.com/science/article/pii/S1058274616303469. (Accessed 3 November 2020).

[19] J. Kany, J. Jose, D. Katz, J. Werthel, … P.S.-J. of S. and, undefined 2017, The Main Cause of Instability After Unconstrained Shoulder Prosthesis Is Soft Tissue Deficiency, Elsevier, n.d. https://www.sciencedirect.com/science/article/pii/S1058274617300605. (Accessed 3 November 2020).
[20] E.J. Strauss, C. Roche, P.-H. Flurin, T. Wright, J.D. Zuckerman, The glenoid in shoulder arthroplasty, J. Shoulder Elbow Surg. Board Trust 18 (5) (2009) 819–833, https://doi.org/10.1016/j.jse.2009.05.008.
[21] F. Matsen III, J. Clinton, J. Lynch, A.B.- JBJS, undefined 2008, Glenoid Component Failure in Total Shoulder Arthroplasty, Journals.Lww.Com, n.d. https://journals.lww.com/jbjsjournal/FullText/2008/04000/Glenoid_Component_Failure_in_Total_Shoulder.27.aspx. (Accessed 26 October 2020).
[22] S. Fucentese, J. Costouros, … S.K.-J. of Shoulder and, undefined 2010, Total Shoulder Arthroplasty with an Uncemented Soft-Metal-Backed Glenoid Component, Elsevier, n.d. https://www.sciencedirect.com/science/article/pii/S1058274610000054. (Accessed 28 October 2020).
[23] F. Montoya, P. Magosch, … B.S.-J. of Shoulder and, undefined 2013, Midterm Results of a Total Shoulder Prosthesis Fixed with a Cementless Glenoid Component, Elsevier, n.d. https://www.sciencedirect.com/science/article/pii/S1058274612002947. (Accessed 28 October 2020).
[24] S. Kurtz, U.B. Handbook, undefined 2009, Ultra High Molecular Weight Polyethylene in Total Joint Replacement and Medical Devices, second ed., Academic, n.d.
[25] A. Papadonikolakis, F.M.I.- JBJS, undefined 2014, Metal-Backed Glenoid Components Have a Higher Rate of Failure and Fail by Different Modes in Comparison with All-Polyethylene Components: A Systematic Review, Journals.Lww.Com, n.d. https://journals.lww.com/jbjsjournal/Fulltext/2014/06180/Metal_Backed_Glenoid_Components_Have_a_Higher_Rate.12.aspx?__hstc=215929672.82af9c9a98fa600b1bb630f9cde2cb5f.1528588800314.1528588800315.1528588800316.1&__hssc=215929672.1.1528588800317&__hsfp=1773666937. (Accessed 28 October 2020).
[26] S. Martin, D. Zurakowski, T.T.- JBJS, undefined 2005, Uncemented Glenoid Component in Total Shoulder Arthroplasty: Survivorship and Outcomes, Journals.Lww.Com, n.d. https://journals.lww.com/jbjsjournal/fulltext/2005/06000/uncemented_glenoid_component_in_total_shoulder.14.aspx. (Accessed 26 October 2020).
[27] D.M. Kim, F. Alabdullatif, M. Aldeghaither, M.J. Shin, H. Kim, D. Park, E. Kholinne, I.-H. Jeon, K.H. Koh, Do modern designs of metal-backed glenoid components show improved clinical results in total shoulder arthroplasty? A systematic review of the literature, J. Sagepub. Com. 8 (2020), https://doi.org/10.1177/2325967120950307.
[28] R.J.S.C.C.G.W P. Boileau, Arthroplasty of the shoulder, J. Bone Joint Surg. Br. 88 (2006) 562–575, https://doi.org/10.1302/0301-620X.88B5.16466.
[29] D. Gurin, W.S.J.-S. in Arthroplasty, undefined 2017, The Emerging Role of the Non-cemented Glenoid in Total Shoulder Arthroplasty, Elsevier, n.d. https://www.sciencedirect.com/science/article/pii/S1045452717300962. (Accessed 28 October 2020).
[30] S.T. Watson, G.K. Gudger, C.D. Long, … J.T.-J. of Shoulder and, undefined 2018, Outcomes of Trabecular Metal–Backed Glenoid Components in Anatomic Total Shoulder Arthroplasty, Elsevier, n.d. https://www.sciencedirect.com/science/article/pii/S1058274617306328. (Accessed 28 October 2020).
[31] K. Inoue, N. Suenaga, N. Oizumi, … H.Y.-J., undefined 2020, Humeral Bone Resorption After Reverse Shoulder Arthroplasty Using Uncemented Stem, Elsevier, n.d. https://www.sciencedirect.com/science/article/pii/S246860261930172X. (Accessed 2 November 2020).

[32] S. Shah, A. Roche, S. Sullivan, B. Gaal, ... S.D.-J., undefined 2020, The Modern Reverse Shoulder Arthroplasty & an Updated Systematic Review for Each Complication: Part II, Elsevier, n.d. https://www.sciencedirect.com/science/article/pii/S2666638320301262. (Accessed 3 November 2020).

[33] P. Boileau, J. Raynier, M. Chelli, ... J.G.-J. of S. and, undefined 2020, Reverse Shoulder—Allograft Prosthesis Composite, with or Without Tendon Transfer, for the Treatment of Severe Proximal Humeral Bone Loss, Elsevier, n.d. https://www.sciencedirect.com/science/article/pii/S1058274620302767. (Accessed 3 November 2020).

[34] F. Lam, D. Bhatia, S. Mostofi, ... K.V.R.-C., undefined 2007, Biomechanical Considerations of the Normal and Rotator Cuff Deficient Shoulders and the Reverse Shoulder Prosthesis, Elsevier, n.d. https://www.sciencedirect.com/science/article/pii/S0268089006001617. (Accessed 3 November 2020).

[35] J. Gagliano, S. Helms, ... G.C.-J. of S. and, undefined 2017, A Comparison of Onlay Versus Inlay Glenoid Component Loosening in Total Shoulder Arthroplasty, Elsevier, n.d. https://www.sciencedirect.com/science/article/pii/S1058274617300599. (Accessed 3 November 2020).

[36] M. Kowalsky, L. Galatz, ... D.S.-J. of Shoulder and, undefined 2012, The Relationship Between Scapular Notching and Reverse Shoulder Arthroplasty Prosthesis Design, Elsevier, n.d. https://www.sciencedirect.com/science/article/pii/S1058274611003983. (Accessed 3 November 2020).

[37] T. Edwards, G. Trappey, C. Riley, ... D.O.-J. of Shoulder and, undefined 2012, Inferior Tilt of the Glenoid Component Does Not Decrease Scapular Notching in Reverse Shoulder Arthroplasty: Results of a Prospective Randomized Study, Elsevier, n.d. https://www.sciencedirect.com/science/article/pii/S1058274611004204. (Accessed 2 November 2020).

[38] C. Lévigne, P. Boileau, L. Favard, ... P.G.-J. of Shoulder and, undefined 2008, Scapular Notching in Reverse Shoulder Arthroplasty, Elsevier, n.d. https://www.sciencedirect.com/science/article/pii/S1058274608003467. (Accessed 2 November 2020).

CHAPTER FOUR

Interactions of bone with hip joint prostheses and possible failures

4.1 Introduction

The hip joint is one of the most challenging part of the musculoskeletal system for tretament if injured. The anatomy and connective soft tissues at this joint are highly specific for the transference of body weight loading from the upper part of the body through the lower extremities while providing rotational motion in various axes. Surgical treatment of the hip joint is required for injuries such as osteoarthritis, rheumatoid arthritis, multifragmentary femoral head fracture, and acetabular wall fracture. Management of implant replacement at the hip joint is difficult because of multiple main muscle-tendons that originate from or are attached to this joint. Total hip arthroplasty (THA) is managed in most severe cartilage or bone injuries at the hip joint, in which two group of prostheses (femoral and acetabular) are replaced through the proximal femur and acetabulum, respectively. In accordance with the severity of the injury, muscle-tendon functionality, bone stock conditions, and level of patient-demanded activity, various prostheses might be used. The stability of total hip replacement (THR) prostheses has failed shortly, in the midterm, and long after implantation owing to inadequate interaction between the prosthesis and bone as well as other clinical issues such as infection, inappropriate surgical technique, and an ineffective or improper postoperative treatment plan. Instability of the acetabular prosthesis was reported to occur at a higher rate compared with the femoral prosthesis because of the difficulty of stabilizing or fixing the acetabular shell component inside the acetabulum cavity. Increased contact force at articulation of the femoral head and acetabular liner has a significant effect on loosening and dislocation of the acetabular prosthesis even when it is fixed to the pelvis using multiple screws. On the other hand, femoral stem stabilization or fixation through the diaphyseal section of the proximal third femur provides mechanical stability. However, in the midterm or long term, various zones of the bone at the implanted site

Interactions of Bone with Orthopedic Implants and Possible Failures
ISBN: 978-0-323-95410-5
https://doi.org/10.1016/B978-0-323-95410-5.00004-9

are affected by overloading or stress shielding, during which cortical hypertrophy or bone resorption occurs, respectively.

This chapter reviews and examines these challenges and concepts. Because of the direct relation of prosthesis design parameters with the interaction of bone tissue with THR prostheses, the main types of acetabular and femoral prostheses are discussed according to their design features and benefits, as well as their effectiveness in treating hip injuries. For each reviewed type of prosthesis, the extent of failure of bone tissue interacting with the THR prostheses at the implanted site is discussed in depth. Furthermore, the effects of surface parameters of the hip prostheses are highlighted as a crucial factor for the enhanced integrity of bone and prostheses at their interfaces.

4.2 Hip joint

The hip joint links the upper section of the body to the lower limbs. Because of the pelvis at this joint, hip injuries are critical and incur serious musculoskeletal dysfunctionality and complications for patients. The pelvis articulates with the femur through the acetabulum cavity. The proximal section of the femur is shaped with spherical geometry (known as the femoral head) positioned at the spherical cavity of the acetabulum. The femoral head extends through a neck to the main body of the femur. The femoral neck is a risky zone to be fractured under severe loading or in osteoporotic patients with low mineral density bones. The geometry of the femoral head and neck are specific with a high loading conditions at the hip joint. A narrower neck compared with the femoral head allows a higher range of motions in various axes. General conditions of the hip joint are similar to those in the shoulder and knee joints. The structure of articulation at the hip joint is not similar to the shoulder joint because of highly compressive force at the hip joint compared with a low amount in the shoulder joint. This is also true for the knee joint, owing to a higher degree of motion in the abduction-adduction axis. Therefore, the hip joint is a median joint between the shoulder and knee joints in view of the range of motion. The loading conditions at the hip joint generate compressive and bending stresses to the femoral neck necessitating strong muscle-tendons and ligaments to reduce the severe catastrophic effect of such concentrated loading conditions at the femoral neck zone. Muscles such as piriformis, obturator externus and internus, pectineus, adductors (brevis, longus, and magnus), iliopsoas, iliacus, and gluteus (medius and minimus) have opposite force or moment relative to the body weight force and moment at the femoral neck (it act similar to an engineered river bridge stabilized by upper cables in a long zone. By adding vehicles passing from the bridge, the opposite force generated at the cables increases to counteract the weight load of the vehicles). Thus, the effectiveness of muscle function at the hip joint has a considerable effect on biomechanical strength of the femoral neck while positioning the femoral head through the acetabulum cavity.

Attachment of the muscles to the femur mostly originate from the pelvis. They apply compressive force to various portions of the femur to provide stability to the femur in various angular motions of the femur while preventing concentrated stress and bone fracture as a result of increased compressive and bending stress at the hip joint. Range of motion at the hip joint is categorized as flexion-extension, abduction-adduction, and internal-external rotation. Hip joint arthritis occurs through the wear of cartilage on the surface of the femoral head or acetabulum cavity (the cartilage articular surface is shaped like a horseshoe at the acetabulum cavity). Hip osteoarthritis necessitates hip arthroplasty, in which the cartilage surface of the acetabulum is reamed, whereas the femoral head and neck portions of the proximal femur are cut, in which some ligaments between the pelvis and femur are cut and completely removed (Fig. 4.1) [2]. In most cases, both portions of the cartilage articular surfaces at the femoral head and acetabulum are affected. Thus, it is necessary to replace prostheses at both sides of the hip joint. Prosthesis replacement at the hip joint is known as THA and the replaced prostheses is called THR. In the next section, prostheses of the THR are discussed according to their design characteristics and features.

4.3 Interactions of bone with hip joint prostheses

Hip joint prostheses are femoral and acetabular. Femoral prostheses are composed of components including a femoral stem and ball, whereas acetabular prostheses consist of an acetabular shell and liner components.

4.3.1 Interactions of femur bone with femoral prosthesis

The femoral stem is designed anatomically to dedicate the geometry of the femur from the trochanter zone through the diaphysis portion of the femur at the proximal third; it is shaped like a neck in the most proximal portion [3]. The design of the femoral stem in the frontal plane is managed with an extended width at the proximal portion and gradually narrows to the distal portion for replacement through the intramedullary canal of the diaphysis.

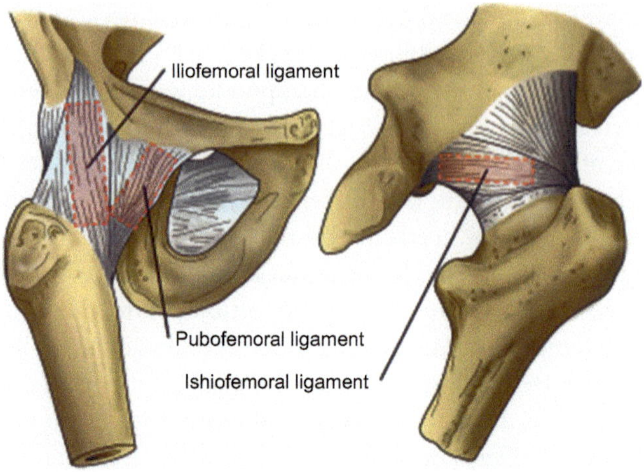

Figure 4.1 Illustration of ligaments at the hip joint. Some interconnective ligaments between the pelvis and femur such as the iliofemoral and pubofemoral ligaments might be cut in total hip arthroplasty. This creates specific conditions for the long-term stability of total hip replacement prostheses. Such a condition is not observed in other joint arthroplasties such as knee, shoulder, and ankle arthroplasties. Thus, the design of femoral and acetabular prostheses and their condition at the articulation interface are challenging, and the rate of failures, such as dislocation, loosening, fracture, and subsidence, are greater compared with other joint arthroplasties. *Through the courtesy of S. Schleifenbaum, T. Prietzel, C. Hädrich, R. Möbius, F. Sichting, N. Hammer, Tensile properties of the hip joint ligaments are largely variable and age-dependent — an in-vitro analysis in an age range of 14–93 years (2016). https://doi.org/10.1016/j.jbiomech.2016.09.001 and permission from Elsevier.*

The extended width portion is at the trochanter zone (Fig. 4.2). This design concept of the femoral stem at the trochanter and diaphysis zones facilitates load transfer from the articulation interface through the femur. Such a design concept of the femoral stem component is not completely physiologic. However, because the current concept of the femoral stem inserts this prosthesis through the femur bone, such a design might be the best solution to model the femoral stem prosthesis. In a healthy condition, transferred compressive force from the articulation interface at the hip joint is borne by trabecular bone at the head, neck, and trochanter zones, whereas the transferred bending moment is borne by cortical bone at the diaphysis zone of the femur. The extended width of the femoral stem at the trochanter zone is completely solid and made of a metallic material; it is different in view of mechanical properties and structure from that of bone tissue. The narrow width portion of the femoral stem at the diaphysis zone has similar

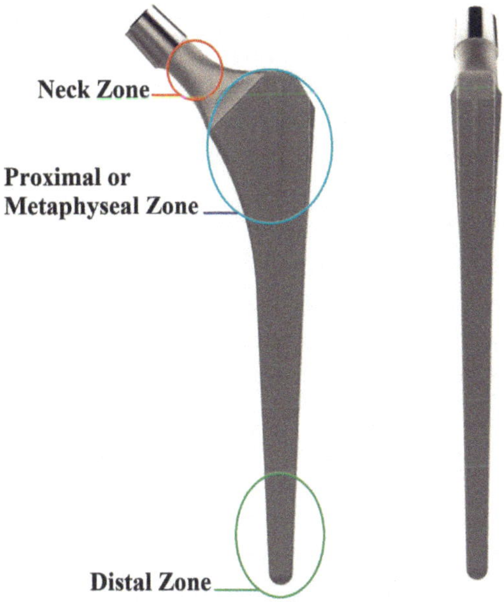

Figure 4.2 Illustration of femoral stem prosthesis. This prosthesis is inserted through the proximal third femur for articulation with the acetabular prosthesis. The femoral stem comprised of various parts as neck, metaphyseal, and diaphyseal. The neck portion is dedicated to the femoral neck of the femur whereas the metaphyseal and diaphyseal portions are designed to be anatomically matched to the anatomy of the proximal third bone at the trochanter and diaphyseal sections. The femoral stem is normally fabricated from titanium alloy using hot forging and features a porous osteoconductive layer at the circumference of the metaphyseal portion. The basic principles of the femoral stem design are similar among all orthopedic implant manufacturers; however, specific features and design concepts are implemented to enhance its effectiveness in long-term implantation, as discussed in this chapter. *Through the courtesy of C. Park, H. Eun, S. Oh, H. Kim, … S.L.-T.J. of, undefined 2019, Femoral stem survivorship in Dorr type A femurs after total hip arthroplasty using a cementless tapered wedge stem: a matched comparative study with type B femurs, Elsevier. (n.d.). https://www.sciencedirect.com/science/article/pii/S0883540318311173 (accessed November 17, 2020) and permission from Elsevier. The names of femoral stem portions have been added to the original figure.*

conditions. From a mechanical view, the strength of the trochanter and diaphysis zones against the transferred bending moment is significantly increased by replacing the femoral stem through these zones of the proximal third of the femur, which causes stress shielding at these zones. This increases stiffness at the hip joint during various activities such as running or walking, in which flexibility of the femur under physiologic loading is eliminated.

The design concept of a metallic femoral stem through the middle zone of the femur was established many years ago. The aim of this prosthesis development was to introduce a salvage treatment to reduce pain and provide a medium level of satisfaction of functionality and movability at the hip joint. The introduction of a bending moment in healthy conditions to the femur creates a reaction force by contracting the muscles (as highlighted in Section 4.2) and maintains muscle strength in over the long term. Therefore, the effect of increased stiffness after THA at the hip joint does not involve muscle and bone tissue in load bearing of the bending moment. (The anatomy or geometry of the proximal femur was structured to generate a large amount of bending moment and involves the diaphysis zone and associated attached muscles along the length of the femur). Replacement of the metallic femoral stem at the femur proximal third intervene with the load distribution along the whole diaphysis portion of the femur from the proximal third to the distal third. Because of the high bending modulus of the metallic femoral stem, energy transferred from the bending moment is borne by the prosthesis, and thus the bone and attached muscles are in poor contraction. This also causes stress to concentrate at the diaphyseal section of the femur under the most distal section of femoral stem prosthesis, which might be a reason for the bone fracturing at this section in long-term post-implantation of the femoral stem. To discuss why this traditional design concept of the femoral stem is still used to develop the femoral stem prostheses, some general facts need to be provided that are associated with implant design. First, based on the developed technologies of materials engineering in the past, most components are fabricated in a solid form. Second, artificial prostheses for joint replacement applications need to be replaced in the middle canal of the cortical bone shell through a long-stem design feature or through trabecular bone by extended features such as pegs or screws. This necessitates mechanical stabilization of the prosthesis inside the bone for long-term functionality. The technique of prosthesis positioning in such conditions is inevitable among joint prosthesis developers. The ideal solution is the development of a femoral stem from a material with significantly lower mechanical properties with adequate fatigue or long-term performance under bending stress. For instance, 30% carbon fiber—reinforced polyetheretherketone is an advanced thermoplastic polymer composite with mechanical properties close to those of cortical bone; it is an alternative material fabricating the femoral stem (the long-term efficacy of this material has not been effectively validated for use in developing a femoral stem). Another solution is to structure the femoral stem in three-dimensional (3D) porous form to allow bone ingrowth through the interconnective pores and establish the material integrity of the prosthesis and cortical shell of the diaphysis. Such integrity would be biologically achieved and highly dependent on the biological reaction of the patient's body and strengthen newly formed bone matrix inside the porous structured femoral stem. Another issue with this concept is the concentration of stress at the section where the femoral stem prosthesis is not implanted inside bone. Such a potential cause of failure was observed for a metal-backed glenoid prosthesis with 3D porous structured pegs, as reviewed in Chapter 3 and illustrated in Fig. 3.16. In the next sections, design modifications of a femoral stem from traditional models to newer ones are reviewed and their effectiveness in enhancing interactions with the femur are discussed. The second component of the femoral prosthesis is shaped like a ball; it is a complete sphere with an internal hole to be placed through the neck section of the femoral stem component. The design point in this component is the significantly lower diameter of the ball compared with the femoral head. The ball is manufactured from a high mechanical strength material such as cobalt chromium (CoCr) alloy. The aim of the design is to lower the overall weight of the THR. Because of the large size of the femoral head, if the ball component is manufactured from a metallic material, its weight would be high and increase the stiffness of the femoral prosthesis. Likewise, a metallic ball with a similarly sized femoral head would necessitate the fabrication of an acetabular shell from a polymeric material as the articulated conjunct component (the conjunction of metal and polymer materials, particularly CoCr-ultrahigh molecular weight polyethylene [UHMWPE] has the lowest wear rate compared with metal-on-metal or polymer-on-polymer articulations).

4.3.2 Interaction of pelvis bone with acetabular cup prosthesis

The femoral ball is articulated with the acetabular liner, which is designed to be hemispherical with an internal hemisphere cavity for contact with the ball. The femoral head is modeled in the form of a hemisphere connected to the femoral stem neck through a ball. The hemisphere component is made of polymeric materials. Because the hemispheric femoral head is made from a polymeric material, the acetabular shell component acts as a tray component to link the liner component to the pelvis at the acetabulum zone. To stabilize the acetabular liner completely, it is entirely positioned through the acetabulum zone; thus, the metallic acetabular shell is thin and hemispherical. Replacing the acetabular shell inside the acetabulum is managed through tabs or is fixed by screws. Articulation of femoral and acetabular prostheses is achieved by rotating the ball component in the hemispheric cavity of the polymeric acetabular liner component. The integrity of the acetabular shell and femoral stem with the bone is crucial for the long-term functionality and performance of the THR. The acetabular shell needs to be biologically integrated with bone or integrated using bone cement. Traditionally, the use of bone cement was extensive for replacing the acetabular shell where cement penetrates bone tissue (especially the pores of trabecular bone) at the cement—bone interface and is interdigitated through irregularly geometric features (e.g., grooves, holes) of the acetabular shell in the prosthesis—cement interface. This type of acetabular shell is known as a cemented prosthesis. The acetabular shell is processed with an osteoconductive layer on its contact surfaces with the bone at the acetabulum zone. Various types of material engineering techniques such as bead sintering are used to enhance bone ingrowth on the surface of the acetabular shell to establish biological integration in conjunction with mechanical integration, which would are normally managed to replace the acetabular shell. Replacing the acetabular shell through biological integration with bone is known as cementless replacement. Cemented and cementless replacements of the femoral stem were also developed by implant developers. In next section, various techniques of biological integration of the acetabular shell and femoral stem are reviewed and discussed.

The surface of the enhanced acetabular cup is processed through plasma spraying of titanium as the first layer followed by plasma spraying of hydroxyapatite (HA) as the second and top layer. Plasma spraying of pure titanium in a vacuum environment under optimized processing conditions creates a macrostructured and mildly porous coating layer on the metallic (stainless-steel or titanium alloy) acetabular shell component. The titanium underlay layer acts as an osteoconductive host layer to enhance bone ingrowth through the surface of the shell component. Because of the poor mechanical strength of the plasma-sprayed HA coating layer, the underlay of titanium promotes the shear and compressive strengths of the HA-coated layer and thus the long-term effectiveness of the processed layer on the surface of the acetabular cup compared with a single HA layer. Regarding this topic, the lead author has provided a detailed discussion on the effectiveness of plasma spraying to enhance the osteoconductivity of orthopedic implants in Chapter 2 of *Osseoconductive Surface Engineering for Orthopedic Implants*. Another factor for enhancing clinical outcomes is the significant metallurgic modification of the UHMWPE. UHMWPE is manufactured by compression molding in a controlled environment to prevent the penetration of contaminants (even microsize particles) during solidification of the polymer resin in the mold. In addition, molded UHMWPE (in the shape of a sheet or rod) is cross-linked through gamma irradiation at multiple periods with a controlled gamma dose to eliminate free radicals at the surface or a bulk structure of the UHMWPE. Furthermore, the oxidative resistance of the UHMWPE sheet, rod, or final component is enhanced through its doping in the antioxidant agent (e.g., vitamin E). Another improvement factor, machining of UHMWPE to be shaped in the final design, was significantly improved by developments in milling machines, cutting tools, cooling techniques, and so forth (enhancements of UHMWPE have had a significant effect on reducing the wear rate on all articulation interfaces of joint prostheses). Overall, metallurgic (materials and surface engineering techniques) and design enhancements of the acetabular cup are effective in promoting clinical outcomes, as reported by Gaillard et al. [4].

4.4 Failures of pelvis bone in interaction with acetabular cup prosthesis

As highlighted in Section 4.3.2, the acetabular cup prosthesis is positioned in the acetabulum cavity of the pelvis. Its integration is provided mechanically and biologically. Because of the large amount of contact stress at the articulation interface of the femoral ball and acetabular liner prostheses, the interface of the acetabular shell and bone is affected, and loosening of the acetabular cup results [5]. Therefore, the acetabular cup has different designs and structural parameters or features [6]. The conventional acetabular cup is a cup-shaped shell and liner that is replaced using bone cement. To obtain more stability of the acetabular cup in the acetabulum, the shell is fixed with multiple screws, and then the liner is press-fit inside the shell cup. In some cases, the shell is manufactured as a cage (a cup with many peripheral holes on it entire surface). The cage is embedded with grafting materials to facilitate bone formation through the holes of the shell and thus enhanced biological integrity with bone. In another type of acetabular cup prosthesis, the articulation interface is developed at two interfaces known as a dual mobility (DM) acetabular cup. Shell contact surfaces with the acetabulum might have pores to establish an osteoconductive layer to provide biological integrity to

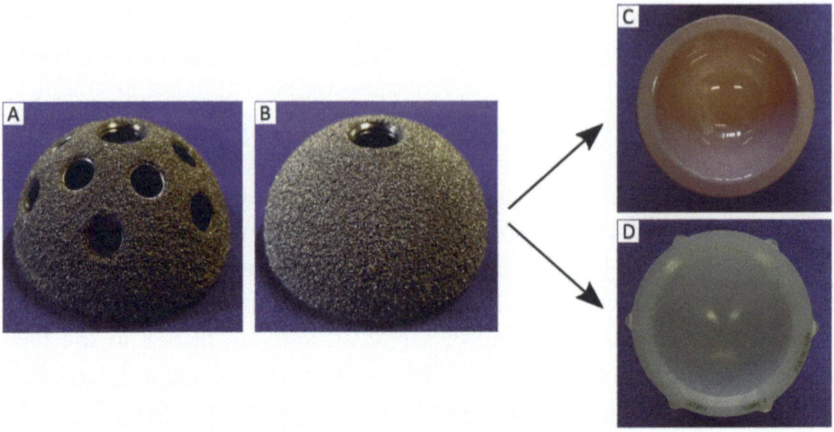

Figure 4.3 Illustration of acetabular cup prosthesis composed of two components: an acetabular shell (A, B) and liner (C, D). The acetabular shell is normally manufactured from stainless-steel or titanium alloy materials, and the acetabular liner is manufactured from ultrahigh molecular weight polyethylene polymer (D) or ceramic (e.g., alumina) (C). Both components of the acetabular prosthesis are hemispheric to be positioned securely at the cavity of the acetabulum. The acetabular shell has various design features and benefits, such as multiple holes (A), a single hole (B), and a porous structured surface (A, B). The acetabular liner is developed with press-fit features as peripheral projections on its outer surface (D). Other types of acetabular components are developed for specific injuries and patient conditions, some of which are reviewed in this chapter. *Through the courtesy of P. Messer-Hannemann, G.M. Campbell, M.M. Morlock, Deformation of acetabular press-fit cups: influence of design and surgical factors, Clin. Biomech. 69 (2019) 96–103. https://doi.org/10.1016/j.clinbiomech.2019.07.014 and permission from Elsevier.*

the acetabular shell with bone without using bone cement. Fig. 4.3 illustrates two types of acetabular shell and liner.

4.4.1 Interactions of pelvic bone with dual-mobility acetabular cup and possible failures

The acetabular cup is normally designed as a single-mobility interface in which the ball component rotates in the cavity of the acetabular liner component. In this condition, alignment of the ball and liner cavity is essentially centric. With any deficiency at the interface of the liner and ball, articulated components tend to be eccentric, in which contact stress at this articulation increases and contact wear accelerates. Although rotation or motion of the ball component is constrained by the hemispheric cavity of the acetabular liner, contact force between these components is best maintained in a specific range to prevent accelerated wear. Deficient muscle function after implantation or malpositioning of prostheses during surgical implantation generates increased contact stress at the articulation interface, suppressing the softer component or material (here, the softer component is the acetabular liner, which is made of UHMWPE polymer), where the level of contact stress and rigidity at the articulation interface are undesirably increased. Likewise, the patient's preoperative musculoskeletal condition might be incompatible with the selected THR system for effective alignment of the femoral and acetabular prostheses at the articulation interface with no or minimal abundant contact loading. Another point is the relative size of the articulation interface in the THR, may be much less than that of a healthy hip joint. Large loading conditions at the hip joint are transferred through a small contact interface between the femoral and acetabular components, which are not mechanically adequate in the design of the joint between two components and concentrate the load and energy at such a relatively small joint. Therefore, there is a need to improve loading transference and reduce joint stiffness at the interface. As highlighted in Section 4.2, the size of the articulation interface may not be effectively achieved by increasing the ball diameter when it is manufactured from a metallic material. One improvement idea developed by implant designers is dual mobility, in which two articulation interfaces are considered in the design of the THR to reduce increased contact stress and associated potential failures that might result [7—10]. In this developed concept, the first articulation interface is between the acetabular shell and liner components, whereas the second is between the liner and ball components (Fig. 4.4). The partial range of rotation of the femoral prosthesis around the acetabular prosthesis is transferred to the interface of the liner and shell interface, where the hip joint is thought to be more flexible and smooth in physiologic motion at this joint, with better durability of the polymeric liner component in long-term implantation.

The long-term effectiveness of DM acetabular cups was observed in a clinical study of 167 patients, all of whom had an enhanced DM cup prosthesis [4]. Clinical outcomes were reported for the greater than 10-year follow-up study, in which no loosening and dislocation was detected, and for which range of motion was appropriate (Fig. 4.5). The DM design concept is a better alternative for patients with a risk for instability in the case of conventional THR replacement [12]. By enhancing motion between the femoral and acetabular prostheses at the dual articulation interfaces, prosthetic dislocation of the femoral head from the acetabular liner is reduced. The articulation interface of the femoral head and acetabular liner was improved in the designs of DM cups by capturing more than

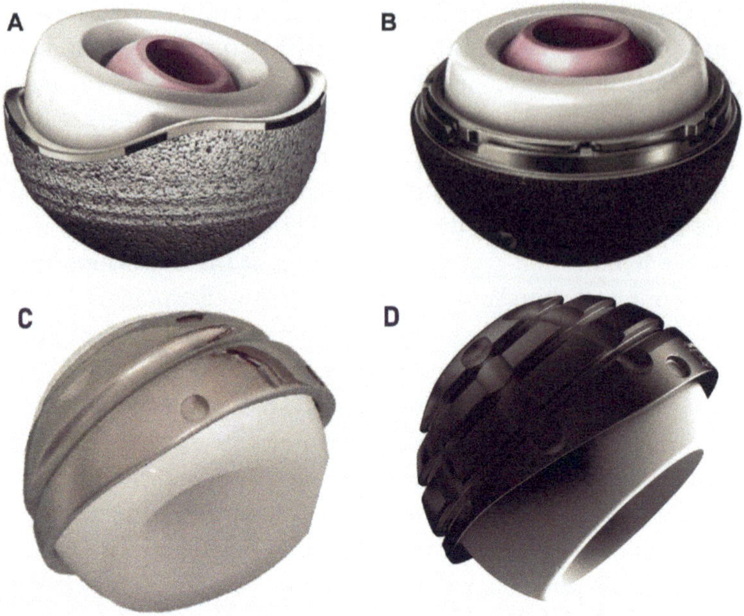

Figure 4.4 Illustration of dual-mobility acetabular cup prosthesis, in which the acetabular liner has rotational motion inside the acetabular shell (A–D). The femoral head is press-fit through the cavity of the liner so that the head is constrained at the cavity from dissociation, with rotational motion. Dual-mobility is good for patients with abductor muscle deficiency and for indications in which there is a high risk for acetabular shell loosening or a high wear rate at the femoral head-acetabular liner articulation. The anatomic design of the monolithic acetabular shell in the DM cup (A, D) enhances the stability of the liner inside the shell during physiologic motions. *Through the courtesy of (C, D) Rowan et al. R. Fe, S. Aj, L. Jk, W. Gh, Dual-mobility vs fixed-bearing total hip arthroplasty in patients under 55 years of age: a single-institution, matched-cohort analysis, n.d. https://www.sciencedirect.com/science/article/pii/S0883540317304059 (accessed February 15, 2021). N. Lebeau, M. Bayle, R. Belhaouane, M. Chelli, E. Havet, B. Brunschweiler, P. Mertl, Total hip arthroplasty revision by dual-mobility acetabular cup cemented in a metal reinforcement: a 62 case series at a minimum 5 years' follow-up, Orthop. Traumatol. Surg. Res. 103 (2017) 679–684. https://doi.org/10.1016/j.otsr.2017.04.009 permission from Elsevier for both images.*

50% of the femoral head with the cavity of the liner. This results in a lower range of motion at this articulation, but it is compensated for at the articulation interface of the acetabular liner and shell components. DM cups have better clinical outcomes for obese patients [13]. This might be because of the better stability of DM cups in various rotational motions physiologically generated at the hip joint. In obese patients, the impact of increased body weight loading has a significant influence on contact stress at the articulation

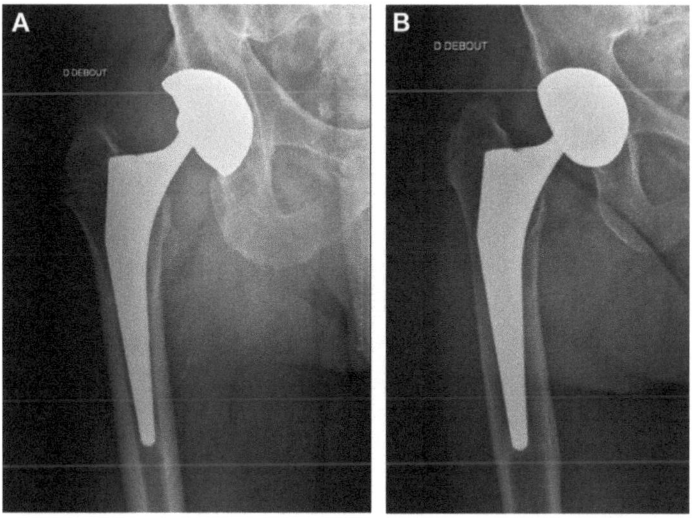

Figure 4.5 Radiographic illustration of dual-mobility acetabular cup prosthesis in total hip arthroplasty in a patient at (A) 11 years and (B) 15 years after implantation. No failure was reported for this patient. A higher range of mobility at the hip joint, a lower risk for acetabular dislocation, a lower potential risk for impingement of the femoral stem neck and acetabular prosthesis, and less jumping of the femoral head from the liner cavity were reported for the dual-mobility cup compared with the conventional cup [11]. *Through the courtesy of R. Gaillard, R. Kenney, J.L. Delalande, C. Batailler, S. Lustig, Ten- to 16-year results of a modern cementless dual-mobility acetabular implant in primary total hip arthroplasty, J. Arthroplasty 34 (2019) 2704–2710. https://doi.org/10.1016/j.arth.2019.06.051 and permission from Elsevier.*

surface, which increases the risk for wear at the articulation interface. Increased contact stress increases contact force at the interface of the acetabular cup prosthesis and acetabulum bone, where shear stress at this bone–prosthesis is significantly increased. This mechanical interaction between the bone and prosthesis causes an undesired effect through periprosthetic bone at the acetabulum zone as bone resorption. When this failure progresses, the acetabular cup gradually loosens or migrates [14]. The DM acetabular cup reduces increased contact loading at the interface of the acetabular prosthesis and acetabulum cavity and thus reduces the risk for acetabular failure in obese patients.

In DM cups, contact at the articulation surface is increased owing to the presence of two articulation surfaces. Thus, localized contact stress and contact force are generated at a lower range, in which shear stress at the interface of the acetabular shell and acetabulum is reduced and a lower rate of acetabular cup dislocation and loosening results [11]. Dissociation of the acetabular liner from the acetabular shell in the DM design is a risky problem. The acetabular shell is designed with a deeper spherical cavity compared with the conventional design of the acetabular shell. This allows greater constraint of the movable acetabular liner when it is positioned inside the acetabular shell. The design of the acetabular shell could be further enhanced by extending the medial side to feature an acetabular shell with higher constraint of the acetabular liner, preventing dislocation or dissociation of the liner from the shell component. Constraint of the femoral ball inside the cavity of the acetabular liner is another design improvement to enhance the function of the DM cup in long-term implantation. Through this, the interface of the ball and liner might increase to above 80% of the external surface of the ball (Fig. 4.4A and B). The femoral head is impacted through the cavity of the acetabular liner to be positioned entirely inside the acetabular liner. The femoral head can still rotate inside the liner cavity while it is constrained from separating from the liner. However, based on reported clinical outcomes [12], the femoral head might separate from the liner. This is known as intraprosthetic dislocation; it is a type of failure unique to DM cups.

The suitability of the DM cup was observed with promising results in a short-term study [12] for revision THA (rTHA) cases. In these cases, bone loss or muscle insufficiency might exist, increasing the risk for instability of the prostheses of revision THR (rTHR) after implantation. Adequate function of the abductor muscles is required to hold the femur and femoral prosthesis construct in an anatomic or physiologic position at the articulation joint between the acetabular liner and femoral head components of the THR. In conjunction with other muscles (as reviewed in Section 4.2), the abductor muscles at the hip joint establish an upward axial load to counteract the downward force of the body weight load and provide equilibrium at the hip joint. This force tends to hold the femur, in which the transfer of excessive downward force at the interface of the acetabular shell and acetabulum cavity is prevented. In patients with poorly functioning abductor muscles (known as muscle insufficiency or deficiency in the clinical literature), the stability of the femoral prosthesis at the articulation interface with the acetabular prosthesis may be ineffective. The DM cup with or without a cage-cup stabilization component is an alternative with a lower rate of acetabular dislocation or loosening in primary or revision surgeries [15]. Another scenario is bone loss in patients who have undergone rTHR. In unsuccessful primary THA with progressive bone loss at the interface of the acetabulum cavity and prosthesis, management of the new hip prostheses replacement is challenging in rTHA. Use of the modular DM cup to reduce increased contact force at the hip articulation joint, and the cup-cage (with a bone graft) could enhance interaction between the bone and prosthesis at the acetabulum cavity of the pelvis (further details of cup-cage component are presented in Section 4.4.4) [10,16]. Fig. 4.6 shows a failed primary THA using a DM cup revised with a cup-cage prosthesis.

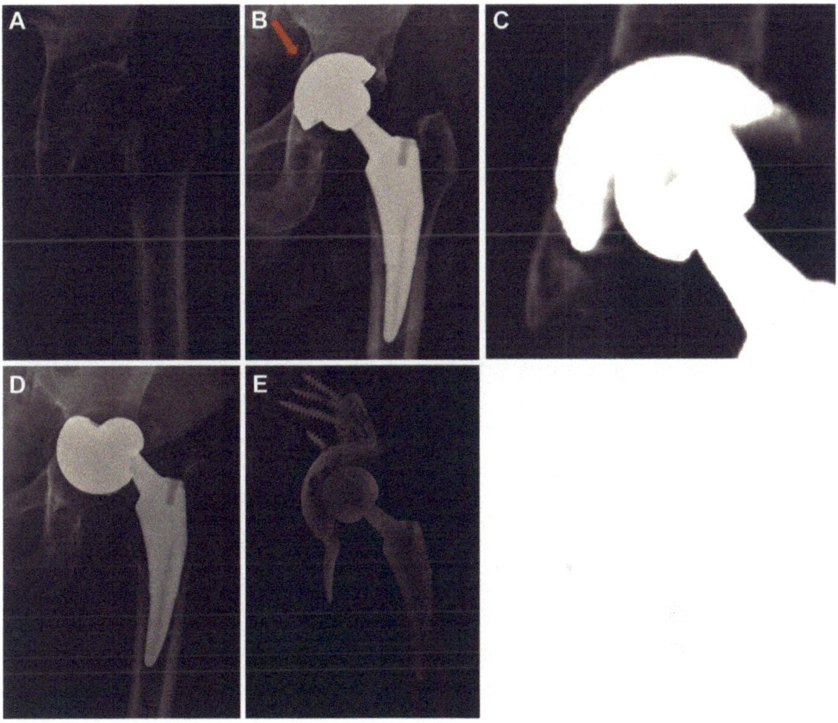

Figure 4.6 Radiographic illustration of acetabular wall fracture (B, C) in an elderly patient. A dual-mobility acetabular cup prosthesis was used in the primary total hip arthroplasty (THA). The patient was given the diagnosis of a dislocated fracture of the femoral neck (A), which was managed by total hip replacement with a dual-mobility acetabular cup and short-length femoral stem prosthesis. Loosening of the acetabular cup progressed until catastrophic dislocation of the cup occurred (D). The patient had low bone mineral density. Thus, dual-mobility replacement with no fixation screw or cup-cage is not a secure solution for such a patient. The monolithic nature of a conventional dual-mobility cup without a hole at the acetabular shell necessitates modular dual-mobility cups for these patients. The revised THA was managed with a cup-cage reconstruction prosthesis. *Through the courtesy of P.D. Sunilkumar, K.J. Oh, H.W. Cho, S.M. Kim, Monolithic dual mobility cup total hip arthroplasty has high complication rates with surgical fixation in elderly with femur neck fracture, J. Arthroplasty. (2020). https://doi.org/10.1016/j.arth.2020.06.091 and permission from Elsevier.*

4.4.2 Interactions of pelvic bone with multiple-hole porous structured acetabular cup and possible failures

Acetabular shells may have multiple holes with a porous structure surface at the convex surface [17–22]. Such acetabular shells are advantageous for the cementless replacement of an acetabular prosthesis through the acetabulum cavity. The holes facilitate the insertion of cancellous screws through bone for mechanical integration of the shell component with the pelvis, whereas the porous structured surface acts as an osteoconductive layer for bone ingrowth for biological integration. The multiple-hole porous structured acetabular shell is beneficial for patients with a high level of osteoporosis at the acetabulum cavity and muscle abductor deficiency at the hip joint. In such patients, adequate stability of the THR prostheses might not be effectively achieved in long-term implantation. The acetabular shell may lose integrity with the acetabulum cavity and ultimately dislocate. The porosity of the porous structured surface is significant; the level of porosity ranges from 60% to 80%, and the interconnective pore network, thickness, and roughness of the porous structure are beneficial on the surface of the acetabular shell. By fabricating the multiple-hole porous structured acetabular shell from titanium alloy, a porous titanium structure on its surface is possible with a high level of adhesion strength for the porous layer. Titanium alloy has superior osteoconductivity compared with stainless steel and cobalt chromium (CoCr) alloys. Stainless steel is normally used to fabricate acetabular shells because of the greater mechanical modulus compared with titanium alloy, in which a thinner acetabular shell can be developed. However, in the cementless replacement of an acetabular shell, the technical priority is osteoconductivity of the acetabular shell in contact with bone. Titanium acetabular shells are manufactured in two portions in the form of bulk and porous structures. The bulk portion is fabricated by forging for the highest mechanical properties, and the porous portion is made by sintering titanium beads on the surface of the forged bulk portion [17–19,21]. Alternatively, both bulk and porous portions can be processed through additive manufacturing methods such as selective laser melting (SLM) [20]. (Processing of the porous structured layer on the surface of the acetabular shell was completely reviewed in Chapter 6 of *Osseoconductive Surface Engineering for Orthopedic Implants*. Likewise, the effectiveness of the porous layer on the osteoconductivity of joint prostheses [e.g., acetabular shell] was discussed in depth in that chapter). Fig. 4.7 illustrates a failure in which a multiple-hole acetabular shell was used in primary THA.

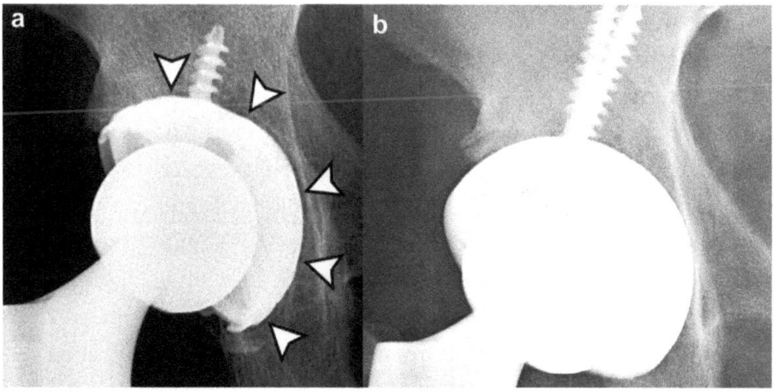

Figure 4.7 Radiographic illustration of cementless multiple-hole acetabular shell loosening with porous titanium structure at its outer surface, in contact with the acetabulum cavity after 6-month implantation (A). The patient felt hip pain as the result of loosening of the acetabular cup. As shown in the left radiograph image, a radiolucent line can be observed at the interface of the shell and acetabulum (B). Although the shell featured a porous engineered surface, bone ingrowth was not achieved in almost all interface zones. The body mass index (BMI) was 32.8 kg/m^2 for this 68-year-old woman. For the early mechanical fixation of the shell, one short screw was used. The high BMI and insufficient fixation stability of the shell did not facilitate early mechanical stability for the effective osseointegration or biological integration of the shell and acetabulum in this cementless replacement total hip arthroplasty. As highlighted by Long et al. [23], an ineffective porous structure of the engineered layer with low porosity might be another reason why bone matrix did not form through the shell surface. Revision surgery was managed using a tantalum highly porous acetabular shell and three screws for fixation. *Through the courtesy of W.J. Long, S. Nayyar, K.K. Chen, D. Novikov, R.I. Davidovitch, J.M. Vigdorchik, Early aseptic loosening of the Tritanium primary acetabular component with screw fixation, Arthroplast. Today. 4 (2018) 169–174. https://doi.org/10.1016/j.artd.2017.11.009 and permission from Elsevier.*

A multiple-hole porous structured acetabular shell is an alternative for managing an rTHA in which greater stability of the rTHR at the articulation joint is required [22]. For the revision management of the THA, various replacement systems were developed that are intended for different conditions. When sufficient bone stock at the acetabulum and sufficient hip abductor muscles efficiency exists, the use of a multiple-hole porous structured acetabular shell is advantageous to provide greater stability compared with a conventional cemented or cementless replacement for the acetabular shell. Failure of this type of acetabular shell might occur [23,25], as illustrated in Figs. 4.7 and 4.10.

4.4.3 Interactions of pelvis bone with press-fit acetabular cup and possible failures

Press-fit replacement of the acetabular cup for primary THA is an alternative technique for cemented or cementless types of acetabular cup replacement [26–29]. In this technique, the acetabular shell is fit through the acetabulum cavity, where the recoil elastic behavior of bone facilitates the primary stability of the acetabular cup prosthesis inside the acetabulum cavity (to press-fit the acetabular shell, the diameter of the reamed acetabulum cavity is slightly lower than that of shell spherical diameter). The press-fit replacement is good for a patient with adequate bone stock with good mineral density at the acetabular zone for the effective replacement of the acetabular shell with an enhanced level of primary stability and integrity in contact with bone. The developed acetabular cup for press-fit replacement is known as the press-fit acetabular cup. The press-fit acetabular cup is designed to be hemispherical with no extended fins or pegs to eliminate inadequate seating of the cup into the acetabulum cavity. If the notch created in the acetabulum cavity to position the fin or peg is slightly deviated, overall seating of the acetabular shell inside the acetabulum cavity may cause inadequate distribution of the press-fitting force on the whole surface of the acetabulum cavity and thus excessive press-fitting force to the bone and prosthesis loosening to occur in a short time. Likewise, this type of acetabular cup does not have holes peripherally for better press-fitting; thus, no fixation screw can be used to increase the stability of the press-fit acetabular cup to the pelvis. The cementless press-fit acetabular cup is an alternative prosthesis for young patients who need a high range of motion. The use of bone cement and fixation screws may cause acetabular loosening to progress in midterm implantation owing to excessive transferred loading in at a high range of motion. The primary stabilization of the press-fit cementless cup relies on effective and complete engagement of the acetabular shell contact surface with the acetabulum cavity. Therefore, there must not be gaps at the interface of the bone–prosthesis for the highest press-fit of the acetabular shell with bone at the acetabulum cavity (a gap might be observed at the polar zone of the cup after implantation of the acetabular shell [24]) (Fig. 4.8).

A press-fit acetabular cup is essential for mechanical integrity with pelvis bone at the whole surface. This is achieved by creating a porous structured surface at the contact surface of the acetabular shell with bone at the acetabulum cavity. Micromotions of various types of porous structured surfaces with different wall thicknesses of the acetabular shell were investigated

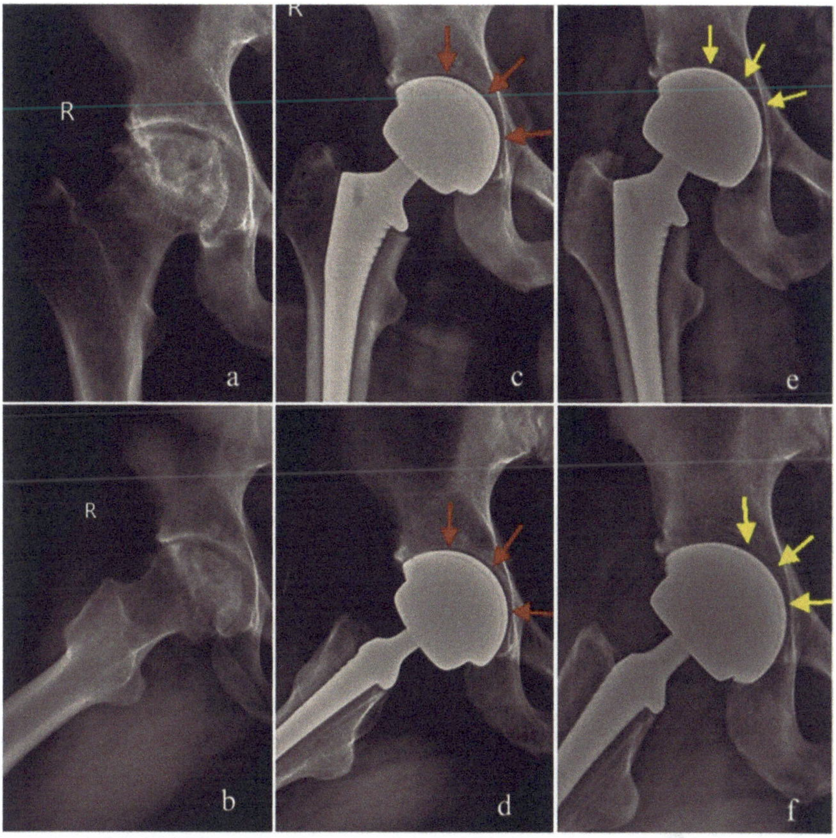

Figure 4.8 Radiographic illustration of press-fit acetabular cup replacement through the acetabulum cavity. Because of press-fitting of the acetabular cup, all contact surfaces of the acetabular shell component might not be fit at the acetabulum cavity without a gap. Normally, at the polar zone of the shell-acetabulum, an interface gap is observed in the press-fit replacement (C, C) (*red arrows*). As highlighted by Agarwala et al. [24], bone ingrowth is achieved after 5–14 months after surgery. In this case, bone ingrowth is seen in radiograph images taken after 10 months (*yellow arrows*). Bone ingrowth can be conducted by the porous structure of the acetabular shell surface at contact surfaces with the acetabulum cavity. The high adhesion strength of the porous layer is crucial to prevent the removal or release of particles from the coated or structured layer. *Through the courtesy of S. Agarwala, C. Jadia, M. Vijayvargiya, Incorporation of screwless press-fit acetabular cups and disappearance of polar gaps, J. Clin. Orthop. Trauma. 11 (2020) 85–90. https://doi.org/10.1016/j.jcot.2018.11.012 and permission from Elsevier.*

biomechanically [27]. Acetabular shell stability after press-fitting to bone was assessed through in vitro evaluation. The test was managed by press-fitting the shell through artificial or cadaver bone. The bone—prosthesis construct was then stabilized in a hip simulator apparatus in which physiologic loading conditions were simulated in accordance with ISO 14242 or an enhanced newer testing protocol. The optimum level of porosity resulted in lower elastic micromotion. A higher range of porosity enhanced engagement of periprosthetic bone tissue (especially trabecular bone), in which the flexibility of the porous structured layer and bone was greater than when the porosity level was at the optimum range, and the level of micromotion at the bone—prosthesis interface increased. Conversely, at a lower range of porosity, the engagement of the bone and porous structured surface was insufficient to endure applied loading conditions at the bone—prosthesis interface; hence, a greater level of micromotion was observed compared with the optimum range of porosity. Greater wall thickness of the acetabular shell had less elastic micromotion of the prosthesis at the interface with bone. The porous structured surface of the acetabular shell at the contact interface with bone might be implemented through plasma spraying, bead sintering, or SLM [24]. The titanium alloy might be selected as a superior material for fabricating the acetabular shell for better adhesion strength in the created porous structured layer from titanium alloy powder or beads. In addition, because of the need for a high level of wall thickness (4—6 mm) for the press-fit acetabular cup, the use of titanium alloy is advantageous to reduce the acetabular shell weight compared with stainless steel, which is traditionally used to fabricate this metallic prosthesis.

The press-fit acetabular cup has advantages and disadvantages, and it is suitable for young patients with adequate bone stock at the acetabulum zone and appropriate functioning of the hip muscles (especially the abductor muscles). Because no fixation features are used, such as fins, pegs, or screws, to replace the press-fit acetabular cup through the acetabular cavity (not even bone cement), the loosening rate might be significant and patient selection for such a prosthesis replacement is rigorously limited. Strategies such as a larger diameter of the femoral head compared with the conventional range (48 mm vs. the conventional size of 28 mm) or a thick porous structured layer (0.5—1.0 mm) may increase the effectiveness of this acetabular cup size for primary and secondary stability in contact with bone tissue at the acetabulum cavity of the pelvis.

4.4.4 Interactions of pelvis bone with cup-cage acetabular cup and possible failures

The cup-cage component is an alternative solution for treating severe or massive bone loss at the acetabulum involving the acetabular rim (Paprosky 3B bone defect at the acetabulum aspect of the pelvis) [25,30–32]. This component is used with an acetabular shell and liner components as extra components to establish the acetabular prosthesis complex with greater stability in rTHA [33]. Because of the bone loss created in failed primary THA with or without involvement of the acetabulum discontinuity, more space of the acetabulum might be reamed, and thus the constraint range of the acetabulum cavity is significantly reduced. Thus, it is necessary to fix the acetabular complex or assembly to the pelvis by multiple screws. Such fixation uses the cup-cage component with multiple extended flanges seated on the ilium and ischium aspects of the pelvis. The flanges have holes for purchase of the cancellous screw through the pelvis at the ilium and/or ischium aspects. The shell component is first placed at the extra reamed acetabulum with a greater hemispheric diameter compared with the conventional shell component used in primary THA, followed by placement of the cup-cage component over the shell component and its fixation by screws. The cup-cage has holes in the cup portion and extended flanges to facilitate placing the screws through the pelvis. The acetabular liner is stabilized using bone cement over the cup-cage. To reduce fixation stress at the interface of the cup-cage screws and bone, the DM cup-cage and liner are used to treat massive bone loss at the acetabulum [14], in which the cup-cage and shell components are one component, whereas the cup portion of the cup-cage component does not have holes (if the DM cup mechanism concept is desired, no hole and screws are used in the cup portion of the cup-cage owing to the articulation of the acetabular liner with the cup-cage. Otherwise, the wear rate of the polymeric acetabular liner might accelerate as a result of the impingement of the pole zone of the liner with the edges of holes or the screw head at the cup zone of the cup-cage). Using of the porous structured acetabular shell enhances the biological integration of the acetabular complex to bone, in which the stability of the acetabular complex at the acetabulum cavity is promoted. Fig. 4.9 demonstrates the possible failure of THA using the cup-cage component. Fig. 4.10 illustrates the use of the cup-cage component in an rTHA.

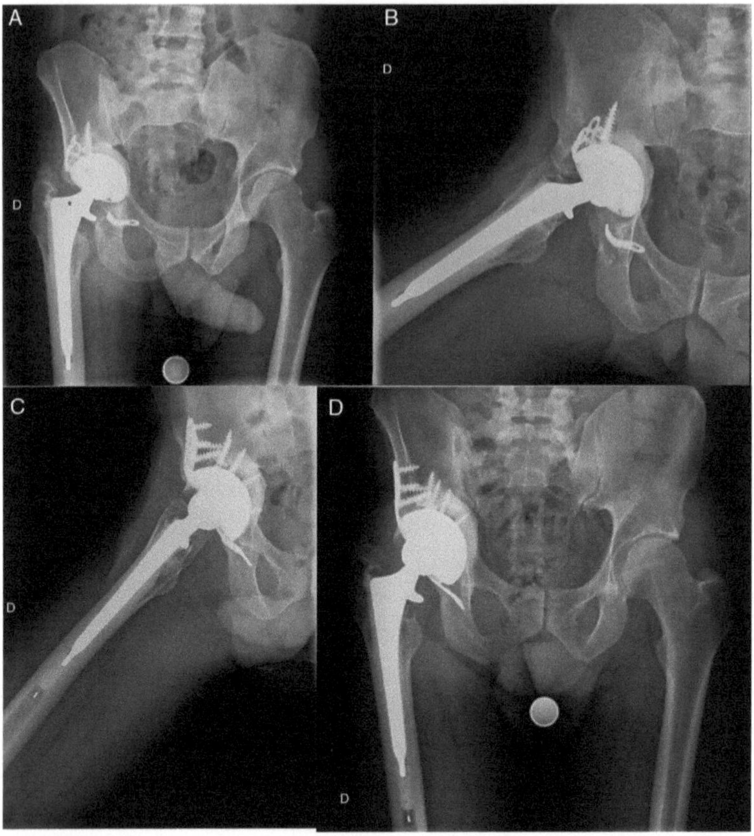

Figure 4.9 Radiographic illustration of total hip arthroplasty using cup-cage prosthesis. This type of acetabular cup is used in patients with a fracture at the acetabular wall or another pelvic discontinuity at this zone. It is also used for severe bone defects or loss at this zone (other types of acetabular cup might not be technically reliable for such severe conditions). The radiograph images are those of a 40-year-old patient at the time of first surgery in 2003. After several revision surgeries, a revision operation was managed using a cup-cage in 2012, which failed owing to fracture of the prosthesis flange at the ischial portion (A, B). The patient was then operated on again using a new cup-cage prosthesis with larger flanges at the iliac portion with more purchased screws and a multiple-hole acetabular shell to insert multiple screws through the acetabular wall and iliac zones (C, D). Reconstruction of pelvic fracture at the acetabular wall and severe bone loss at this zone are challenging and the scenario might be complicated, as seen in this patient. The combination of a cup-cage with flanges and a separate multiple-hole acetabular shell with a porous structured surface as well as the use of a bone graft at the implanted site are a secure surgical solution for such an indication. *Through the courtesy of A. Fraile Suari, F. Marqués López, M. Cuenca Llavall, M. Tey Pons, A. León García, Reconstruction for pelvic discontinuity and massive acetabular defects, Rev. Esp. Cir. Ortop. Traumatol. 64 (2020) 64–73. https://doi.org/10.1016/j.recot.2019.06.004 and permission from Elsevier.*

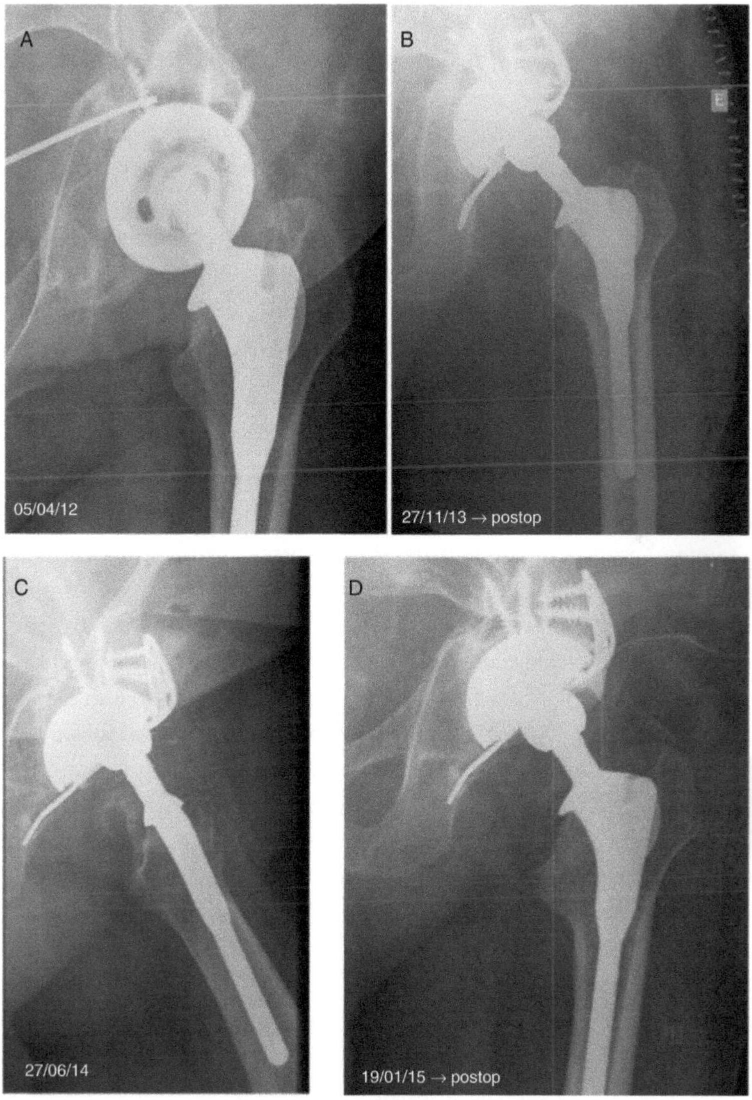

Figure 4.10 Radiographic illustration of acetabular wall fracture reconstruction using a cup-cage prosthesis. Fracture nonunion and acetabular collapse are seen. The fracture was managed using a multiple-hole acetabular shell for screw fixation of the fracture (A). The failed acetabular prosthesis was managed by a cup-cage reconstruction prosthesis (B). Dissociation of the polymeric acetabular liner was observed 6 months after implantation of cup-cage prosthesis (C). The liner was exchanged in revision surgery (D). *Through the courtesy of A. Fraile Suari, F. Marqués López, M. Cuenca Llavall, M. Tey Pons, A. León García, Reconstruction for pelvic discontinuity and massive acetabular defects, Rev. Esp. Cir. Ortop. Traumatol. 64 (2020) 64–73. https://doi.org/10.1016/j.recot.2019.06.004 and permission from Elsevier.*

4.4.5 Other types of acetabular cup

Acetabular cups also have other specific types for more complicated acetabulum conditions. The acetabular liner might be fixed to the acetabular shell by screws [34] to increase the stability of the liner inside the shell component. A stemmed acetabular cup and double cup are other types of acetabular cup that are beneficial for severe acetabular bone loss [35] and the management of advanced Paprosky 3A and 3B acetabular defects [36], respectively. However, clinical studies of these acetabular cups are scarce. In the next edition of this book, this section will be more complete regarding the types of acetabular cup prostheses with sufficiently reported clinical outcomes.

4.5 Failures of femur bone in interaction with femoral stem prosthesis

General aspects of the femoral stem prosthesis in interaction with femur bone are highlighted in Section 4.3. Initially, the attempt was to design the components of this prosthesis anatomically. However, based on reported complications in clinical investigations [37—43], there was a further focus on the physiologic design of the femoral stem. The target was to enhance interaction between the prosthesis and the proximal portion of the femur, where the femoral stem is implanted. The femoral stem is the key component of the femoral prosthesis, and improving or enhancing the design involves the entire component. Femoral stem interactions with the metaphyseal and diaphyseal zones of the proximal third femur are studied to enhance load transference and distribution from the femoral stem through bone tissue. In the following sections, various types of femoral stem designs with specific features and design improvements are reviewed and their effectiveness in enhancing THR functionality in long-term implantation is discussed. Figs. 4.11 and 4.12 show failures at periprosthetic bone after replacement of the femoral stem prosthesis in THA.

4.5.1 Interactions of femure bone with tapered femoral stem and possible failures

The femoral stem component is enhanced with a taper in its design in the mediolateral (frontal plane) or posteroanterior (sagittal plane) axis [3,44—47]. The taper is designed by creating an angular direction in the femoral stem, as illustrated in Fig. 4.13. In the sagittal plane (posteroanterior

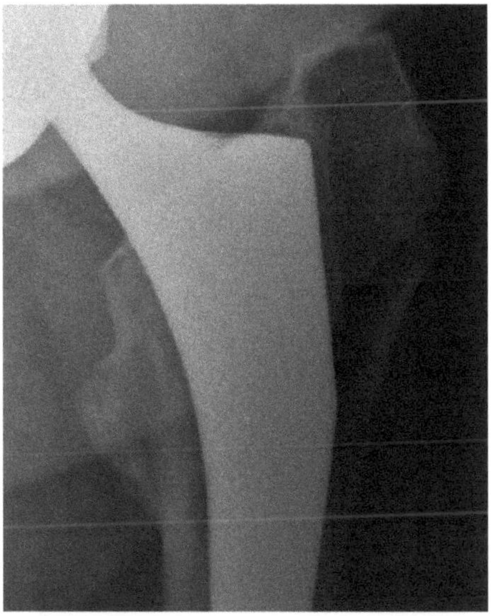

Figure 4.11 Radiographic illustration of resorption of periprosthetic bone at the contact surface of the femoral stem with metaphyseal proximal third femur bone in the trochanteric zone (blue arrow). In long femoral stem replacement, such bone resorption progresses laterally to the trochanteric zone over long-term implantation, whereas in ultrashort and short femoral stems, such failure is observed less often. *Through the courtesy of R. Gaillard, R. Kenney, J.L. Delalande, C. Batailler, S. Lustig, Ten- to 16-year results of a modern cementless dual-mobility acetabular implant in primary total hip arthroplasty, J. Arthroplasty 34 (2019) 2704–2710. https://doi.org/10.1016/j.arth.2019.06.051 and permission from Elsevier.*

[AP] axis), tapering might be implemented in either the metaphysis or diaphysis portion of the femoral stem. Tapering in metaphyseal portion enhancing the stability of the proximal section of the femoral stem. It is a mechanical factor to reduce the risk for femoral stem subsidence through the proximal femur, whereas it promotes rotational stability against a torsional or bending moment generated during physiologic motion at the hip joint (articulation of the femoral head and acetabular liner joint). Another taper in the AP axis is used at the diaphyseal portion of the femoral stem to reduce the risk for overloading to the cortical shell while providing an effective fit of the femoral stem through the canal of diaphyseal portion of the proximal third femur. The thickness of the cortical layer at the distal and proximal sections of the proximal third femur are slightly different (it increases from the proximal to distal portions), in which the risk for overfitting the distal portion of the femoral stem increases, ultimately causing cortical

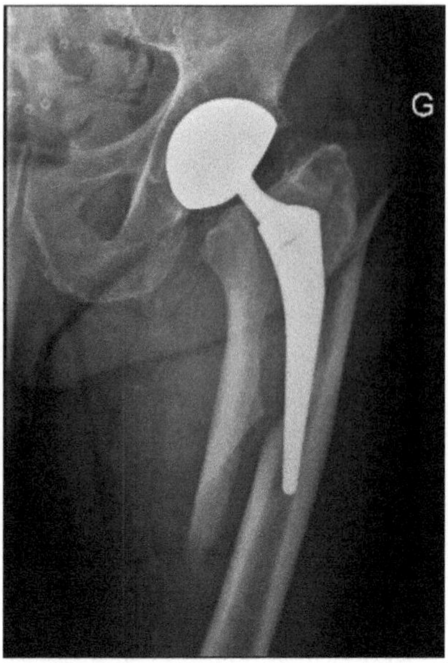

Figure 4.12 Radiographic illustration of femoral fracture at the diaphyseal section of the proximal third femur. Possible reasons were not discussed in Gaillard et al. [4]. Load transference and distribution using the short tapered femoral stem is better than a long and untapered femoral stem, in which the stress shielding upon weakening of the cortices at the diaphyseal zone is less. Such a situation occurs in elderly patient with a body mass index of >30 kg/m^2 and low bone mineral density. Although the combination of a dual-mobility acetabular cup and short femoral stem are beneficial for such a patient, the risk for femoral fracture or another type of failure remains high [41–43]. *Through the courtesy of R. Gaillard, R. Kenney, J.L. Delalande, C. Batailler, S. Lustig, Ten- to 16-year results of a modern cementless dual-mobility acetabular implant in primary total hip arthroplasty, J. Arthroplasty 34 (2019) 2704–2710. https://doi.org/10.1016/j.arth.2019.06.051 and permission from Elsevier.*

hypertrophy at this section (Fig. 4.14). Such a scenario is seen when the most distal portion of the femoral stem is replaced too tight, in which the proximal portion may not be entirely inserted through the femur. Another taper of the femoral stem is featured in the medial-lateral (ML) axis to enhance anatomical fitting at the metaphyseal and diaphyseal sections of the proximal third femur. The tapered design in this section reduces overloading at the diaphyseal section and enhances bone conservation at the metaphyseal section. In the ML axis or frontal plane, tapering at the lateral aspect is helpful for bone conservation at the trochanter zone, whereas there is a reduction in the progression of stress shielding effect at the trochanter zone owing to its morphometric design concept (Fig. 4.16) [48,50,51]. In the old design, the femoral stem has no taper in the ML and AP axes in the diaphyseal or distal portion. The current conventional femoral stem is tapered in the ML axis at metaphyseal section and may have two tapers (one is in the ML axis and the other in the AP axis, known as a double-tapered femoral stem). Furthermore, tapering of the femoral stem might have three tapers, one in the ML axis and the other two in the AP axes, as described earlier (known as a triple-tapered femoral stem) [52]. In some models, tapering might be added to the transverse plane to reduce subsidence or dislocation of the femoral stem in both the axial and ML axes. In general, tapering of the femoral stem increases its stability at the metaphyseal section and reduce overfitting as well as the progression of cortical hypertrophy at the most distal portion of the femoral stem. It also enhances effective load transference at the interface of the femoral stem and cortices at the diaphyseal portion of the proximal third femur, in which the enhanced interaction at this zone would facilitate a higher level of transferred load through bone. Fig. 4.13 displays a double-tapered femoral stem in various axes.

4.5.2 Interactions of femur bone with short-stem femoral stem and possible failures

Femoral stem components are traditionally developed with a long length, as illustrated in Fig. 4.2. As a result of the long-term use of long-length femoral stems, major overloading at the most distal section and stress shielding at the interfaces with the metaphyseal zone were observed (stress shielding effect also occurs at the proximal portion of the stem in the diaphyseal zone). As highlighted in the previous section, tapering of the stem is the alternative design concept extensively used to enhance the interaction of the proximal third femur with the prosthesis. However, complications such as the ineffective insertion of the femoral stem in femur Dorr type A morphology or geometry, the progression of stress shielding at the metaphyseal zone (particularly at the trochanter zone), and periprosthetic bone fracture are to be potential failures of the long femoral stem. Owing to such failures, particularly in younger patients, orthopedic implant developers introduced the short stem design concept to reduce these complications over short or long-term implantation [49,53–57]. The short femoral stem is inserted through the proximal femur metaphyseal zone with short contact interfaces with the diaphyseal zone (the length of the stem is different among the implant manufacturers; however, for all designed short stem components, the goal is to have optimal engagement with the diaphyseal bone). Engagement of the stem type prosthesis, such as the humeral stem, tibial tray, and

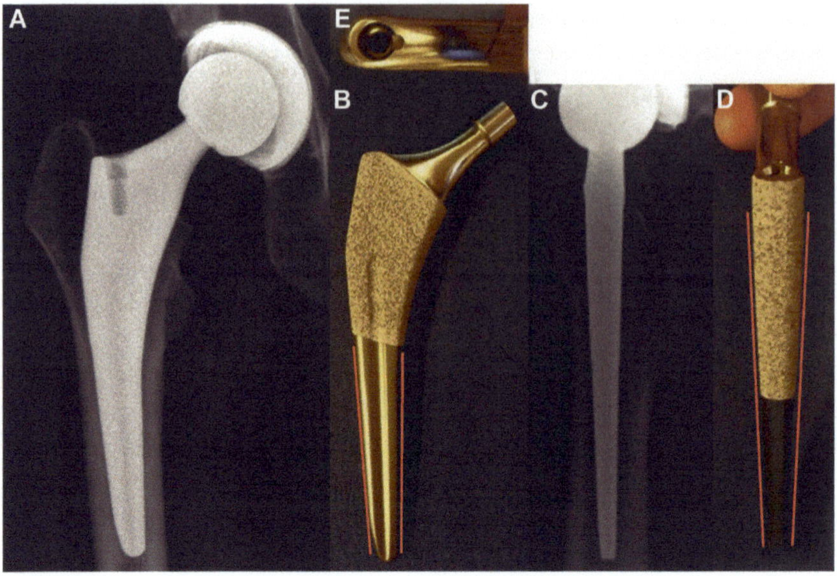

Figure 4.13 Illustration of short-length double-tapered femoral stem in anteroposterior (AP) (A, B), lateral (C, D), and top (E) views, with a rough surface at the metaphyseal zone and a coating of titanium-nitride to reduce ion release in contact with bone while enhancing wear resistance at the neck in contact with the femoral head. Tapering in the ML and AP axes (or in the frontal and sagittal planes) is highlighted with *red lines*. The longitudinal groove feature (observable in B) prevents overfitting of the femoral stem through the cortices, in which the risk for cortical hypertrophy is reduced, particularly in a Dorr type A morphology femur with highly thick bone cortices at the diaphyseal section. Fluting of the femoral stem tip at the most distal portion eliminates the risk for cortical hypertrophy caused by overloading at this zone. *Through the courtesy of L. Warth, T. Grant, N. Naveen, … E.D.-T.J. of, undefined 2020, Inadequate metadiaphyseal fill of a modern taper-wedge stem increases subsidence and risk of aseptic loosening: technique and distal canal fill matter!, Elsevier. (n.d.). https://www.sciencedirect.com/science/article/pii/S0883540320301674 (accessed November 17, 2020) and permission from Elsevier.*

femoral stem, in diaphyseal bone facilitates stability against physiologic loading in various axes of rotation. The press-fit or scratch-fit stem-type prosthesis with cortices at the diaphyseal zone of long bones in either cementless or cemented replacement (in cemented replacement, the fitting advantage is achieved using bone cement at the interface of the bone and prosthesis) is a secure and reliable strategy for implanting the stem-type prosthesis. However, because of the way in which the cortical bone is tapered and the diversity of cortical thicknesses at the proximal third femur, the risk for failure occurring is high. Thus, low engagement of the femoral stem with the diaphyseal section of the proximal third femur is beneficial to enhance load transference with the metaphyseal bone to eliminate the risk for cortical hypertrophy and fracture at the diaphyseal proximal third femur and significantly reduce the progression of stress shielding and bone loss at the trochanteric zone. Owing to these advantages, the short femoral stem is a superior solution for primary THA cases with a high risk for cortical hypertrophy, periprosthetic femur bone fracture, and bone loss at the trochanter zone, in addition to specific postoperative medical treatment planned and followed up by the associated surgeons and clinicians. The use of the short femoral stem has technical advantages for rTHA, particularly in younger patients. Normally, in revision surgery of joint replacements, there are bone loss zones caused by the progression of stress shielding at periprosthetic bone tissue. When removing the prosthesis, even more bone loss might occur, particularly in a primary cemented prosthesis, in which the integrated bone matrix at the cement is removed when the cement is removed. The scenario is challenging when the remaining bone cement is inside the canal of the diaphyseal section of the proximal third femur and its removal necessitates narrowing the cortical thickness at this zone, where the tight replacement of the new femoral stem may not be achieved effectively and the risk for femoral fracture at the narrowed zone increases. A key factor in rTHA is the conservation of bone stock as much as possible for rerevision surgery in case of failure of the revision prostheses. Thus, replacing the prostheses with minimal bone intervention is required for successful management of the rTHA. A short femoral stem is an important solution for conserving bone stock at the diaphyseal section of the proximal third femur.

A challenge in using a short femoral stem is its stability. The risk for instability and subsidence is greater in short compared with long femoral stems soon or in the midterm after implantation. Biomechanical factors such as abductor and adductor muscle deficiency and low mineral density of bone tissue at the metaphyseal zone of the proximal third femur accelerate the failure of the short femoral stem. Compared with the long femoral stem, the stability of the short stem is highly influenced by the effective integration of bone and the prosthesis at the metaphyseal section of the proximal third femur. In addition to the taper for its effective mechanical replacement through the metaphyseal zone, biological integration with bone tissue is crucial for long-term stability (the advantages of biological integration at the metaphyseal section of the proximal third femur and associated surface engineering methods to enhance the femoral stem at this zone are discussed

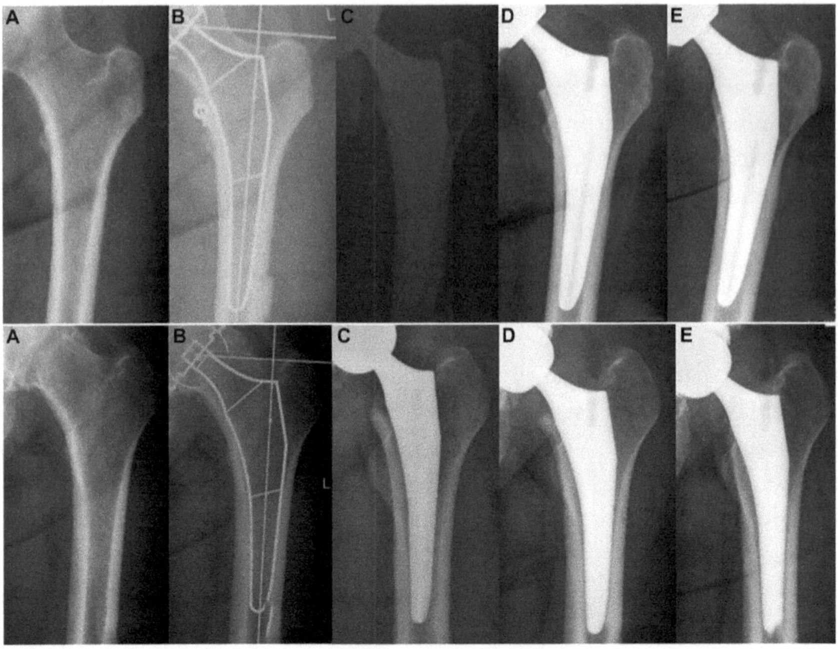

Figure 4.14 Illustration of short double-tapered femoral stem replacement through proximal third femur (A) in two patients. The top and bottom series of images are for a patient without and with prosthesis subsidence, respectively. In the bottom series, based on femur templating (B), the femoral stem size was determined. However, inadequate filling of the diaphyseal canal is observed in the immediate radiograph image (C). As a result, stem subsidence is observed at 1 month (D) and 1 year (E) after implantation. Adequate sizing of the short femoral stem is challenging for optimal fitting of the stem without overfitting and possible cortical hypertrophy in long-term implantation with no occurrence of subsidence caused by less fitting. *Through the courtesy of L. Warth, T. Grant, N. Naveen, ... E.D.-T.J. of, undefined 2020, Inadequate metadiaphyseal fill of a modern taper-wedge stem increases subsidence and risk of aseptic loosening: technique and distal canal fill matter!, Elsevier. (n.d.). https://www.sciencedirect.com/science/article/pii/S0883540320301674 (accessed November 17, 2020) and permission from Elsevier.*

in Section 4.5.3). The occurrence of subsidence as a result of ineffective sizing of the femoral stem is illustrated in Fig. 4.14.

In addition to the short stem as an advantage for young or active patients, the ultrashort femoral stem was developed as an alternative cementless replacement [49] (Figs. 4.15 and 4.16). Distal stabilization of the femoral stem in the diaphyseal section of the proximal third femur is important for long-term stability. A long-term clinical study using an ultrashort femoral stem reported good clinical outcomes without loosening, radiolucent lines,

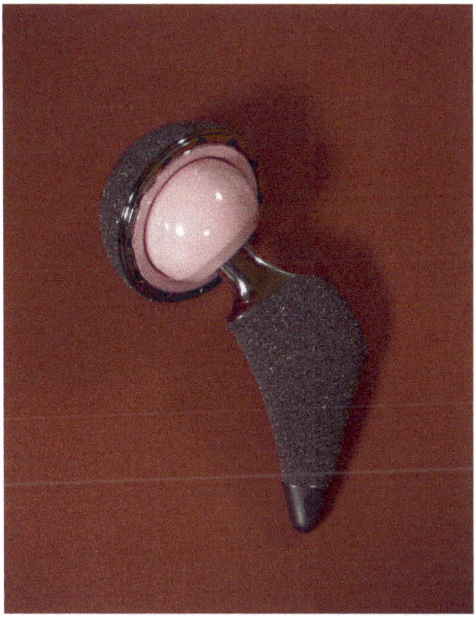

Figure 4.15 Illustration of ultrashort femoral stem. The design is highly anatomic and physiologic, dedicating the profile of the trochanteric profile at the lateral aspect. This section of the femoral stem is not completely anatomic in the long-term or short-term types because of the emphasis on engaging the stem with the diaphyseal bone to enhance stability. In the ultrashort stem, optimal contact surface of the prosthesis with the trabecular bone at the metaphyseal zone is required to increase the stability and integrity of the prosthesis and bone. *Through the courtesy of Y. Kim, J.P.-T.J. of arthroplasty, undefined 2019, Long-term outcomes of ultra-short metaphyseal-fitting anatomic cementless femoral stem in total hip arthroplasty with ceramic-on-ceramic articulation for, Elsevier. (n.d.). https://www.sciencedirect.com/science/article/pii/S0883540319304048 (accessed November 17, 2020) and permission from Elsevier.*

bone resorption, dislocation, subsidence, or significant stress shielding at the metaphyseal zone of the proximal third femur [49]. Although good outcomes were highlighted in that study for this specific femoral stem prosthesis, other factors such as a meticulous surgical procedure, the use of alumina-on-alumina (a ceramic-on-ceramic articulation at the joint of total joint replacements), the use of a larger femoral head, highly porous structured layering, surface flare structuring (a method for significantly enhancing surface roughness and scratch-fit of the joint prosthesis), and anatomic design were required to maintain the long-term efficacy of interactions of the ultrashort length femoral stem with bone tissue. The advantages of ultrashort

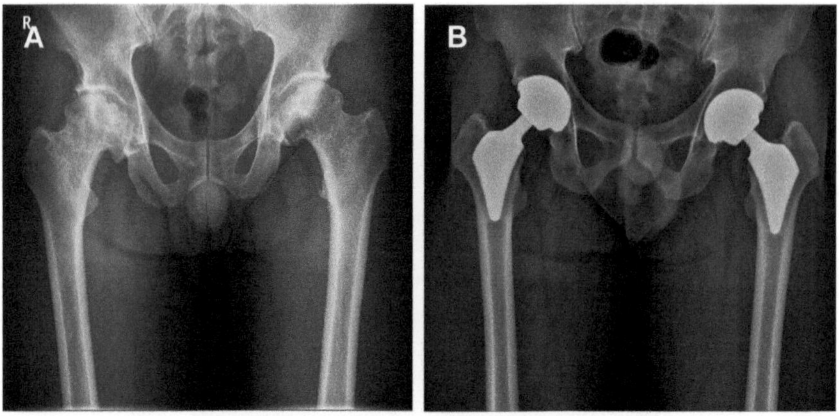

Figure 4.16 Radiographic illustration of ultrashort femoral stem replacement through the proximal third femur. This type of femoral stem is inserted through the metaphyseal section of the proximal third femur with minimal engagement with diaphyseal section. In this case, osteonecrosis is observed at the femoral head (A). The femur is categorized as Dorr type A. Good replacement conditions were reported at 16 years after surgery (B). The only issue is slight bone resorption at the calcar zone in both hips, which did not have a significant effect on clinical outcomes after 16 years. *Through the courtesy of Y. Kim, J.P.-T.J. of arthroplasty, undefined 2019, Long-term outcomes of ultra-short metaphyseal-fitting anatomic cementless femoral stem in total hip arthroplasty with ceramic-on-ceramic articulation for, Elsevier. (n.d.). https://www.sciencedirect.com/science/article/pii/S0883540319304048 (accessed November 17, 2020) and permission from Elsevier.*

or short femoral stems are achievable in long-term implantation for young patients with a high level of activity when other effective enhancement strategies are contemplated in the design and replacement of THR prostheses.

4.5.3 Interactions of femur bone with surface porous structured femoral stem and possible failures

Porous structuring of the femoral stem at the metaphyseal section is a well-established strategy to enhance biological integration with the trabecular bone at the metaphyseal zone of the proximal third femur [58–60]. Various surface engineering methods are used to create a porous structured layer on the surface of the femoral stem at this zone. The femoral stem is fabricated from titanium alloy; thus, the creation of a titanium porous structured layer is advantageous to ensure adequate adhesion strength for the formed layer to the femoral stem prosthesis. Ti porous layering occurs by vacuum plasma spraying and bead sintering as pioneering methods to create a thick layer (over 100 μm) in the orthopedic implant industry. The technical challenges with these processes were completely reviewed and discussed in Chapters 2 and 6 of *Osseoconductive Surface Engineering for Orthopedic Implants*, written by the lead author. Another alternative method for creating the porous structured surface is additive manufacturing methods known as 3D printing or rapid prototyping processes. As discussed in Chapter 6 of *Osseoconductive Surface Engineering for Orthopedic Implants*, SLM is used to fabricate bistructural prostheses composed of integrated solid and porous components. This enables the predesign of the porous structure in CAD modeling software to provide a regular pore design profile in the whole porous portion of the prosthesis, which is formed to be circumferential with the femoral stem at the metaphyseal portion in contact with the trabecular bone zone of the proximal third femur. Surface porous structuring features the femoral stem prosthesis with osteoconductive surface layering to promote bone ingrowth through the deep layer of the prosthesis surface. Ingrowth bone formation establishes the biological integration of the prosthesis with bone tissue, in which the stability of the prosthesis is enhanced. Biological integrity is highly beneficial; however, other factors such as the anatomic design, effective load transference at the bone—prosthesis interface, and soft tissue functions have a prominent effect on the long-term stability of the prosthesis at the implanted site.

4.6 Summary and discussion

Interaction of the bone with THR prostheses and possible failures were studied in this chapter based on the fundamentals and reported clinical outcomes in published clinical studies. In order to effective understanding of the interactions of the bone with THR prostheses, concepts such as stress shielding, bone resorption, overloading, bone hypertrophy, increased contact force, increased contact stress at the bone—implant interface, muscle-tendon deficiency, bone stock or mineral density, patient-demanded activities, body weight or BMI value, and adequate implant selection need to be considered. To some extent, these concepts are contemplated by the implant developers and orthopedic surgeons, better clinical outcomes would be resulted in long-term implantation. Normally, a high survival rate is reported in clinical studies; however, with an emphasis on a narrower selection of injured patients with specific conditions (e.g., patients with highly demanding activity, poor bone stock or bone mineral density, obesity, muscle-tendon deficiency, specific bone morphology), the rate of failure and complications significantly increases and the rate of revision surgery becomes considerable. Therefore, focusing study on possible the failures of THA is helpful for highlighting further areas for development or ideas for implant designers and developers, and it provides technical guidance for selecting adequate prostheses for treating hip injuries. An important factor in the developing new prostheses is their cementless replacement through the acetabulum cavity, femur proximal metaphyseal zone, and femur intramedullary canal. The traditional method of prosthesis fixation or stabilization through bone cement will soon be eliminated, particularly for those with severe bone loss (currently, in many countries or hospitals, cemented replacement techniques are the dominant injury management procedure for patients without complex or specific conditions). Cementless replacement techniques provide or facilitate treatment for some complicated cases that might not be effectively managed by cemented replacement. For instance, severe bone loss or poor bone stock at the acetabular wall would best be managed by a cup-cage prosthesis combined with the use of a bone graft at the affected zone. Various aspects of interactions of bone with a THR prosthesis and possible failures were reviewed and discussed in this chapter. Acetabular and femoral prostheses are designed with different features and benefits. Some traditional design concepts are still used to develop new prostheses, whereas enhanced features are contemplated to increase the effectiveness of the prosthesis after implantation. Acetabular cup prostheses are designed with a cup or hemispheric shape as press-fit, DM, cup-cage, multiple-hole, porous coated or structured, anatomic, and so forth. On the other hand, the femoral stem is designed as untapered, double-tapered, triple-tapered, short stem, ultrashort stem, anatomic, porous structured surface, and so on. In one developed acetabular or femoral prosthesis, one or multiple types of design features and benefits might be used to enhance clinical outcomes. In the acetabular prosthesis, the DM feature enhances the stability of the THR in the high range of motion at the hip and reduces the risk for contact stress and shear stress at the articulation and acetabular shell—acetabulum interfaces. The advanced anatomic design of the acetabular shell enhances the stability of the acetabular liner while moving or constraining inside the acetabular shell, particularly in the medial inferoposterior aspect. In addition, the multiple-hole acetabular shell provides insertion of cancellous screws through the iliac portion of the pelvis to enhance acetabular prosthesis stability in specific cases (e.g., in patients with fracture at the acetabular wall). Furthermore, the cup-cage component is best for buttressing the iliac, acetabulum, and ischial aspects of the pelvis in complicated cases with severe bone loss and discontinuity at the acetabulum. In the femoral prosthesis, cortical hypertrophy at the diaphyseal zone of the proximal third femur is significantly reduced using double-tapering of the femoral stem in both the sagittal and frontal planes. In addition, short and ultrashort femoral stems are beneficial to reduce bone resorption at the metaphyseal zone of the proximal third femur. In particular, the advanced anatomic and physiologic design of the femoral stem at the proximal lateral aspect enhances load transference at the trochanteric zone to reduce bone resorption progressing at this zone in long-term implantation. Overall, the effective design and manufacture of THR prostheses with advanced features and benefits combined with the adequate selection of the prosthesis based on the severity of the injury and the conditions of bone and soft tissue at the injured zone are and vitally important for successful THA. In this edition of the book, the main conventional and state-of-art aspects of the interactions of bone with THR prostheses were reviewed and discussed; further details regarding this topic will be added in the next edition of the book.

4.7 Remind and learn

In this section, some questions are provided for effective reminders of the reviewed and discussed contents in this chapter. Answering these questions enhances the learning about the interactions of bone with THR prostheses.

Q 4.1: Describe the overall physiologic conditions of the hip joint. Compare the physiologic motions at the hip joint with those at the knee and shoulder joints.

Q 4.2: Discuss the effective loading conditions at the neck zone of the proximal femur. Discuss the benefits of the geometry of the femoral head and neck. How would the equilibrium of loading conditions be achieved physiologically at the hip joint to counteract the increased effect of the body weight load on the proximal zone of the femur? Discuss the significant effects of the muscles at the hip joint on the function of the THA.

Q 4.3: Discuss the THA and which ligaments would be cut. What physiologic challenges might occur by cutting these ligaments in designing and developing THR prostheses?

Q 4.4: Discuss the prostheses included in THR. What are the design characteristics of the femoral stem compared with the anatomic shaping of the bone at the proximal femur from the trochanter to the proximal third diaphyseal zones? Discuss why replacing the metallic femoral stem might not be physiologically adequate for insertion through the bone. What failures might occur as a result?

Q 4.5: What is the effect of the current metallic femoral stem on loading between the muscles and femur? What a reason for preprosthetic bone fracture at the diaphyseal section of the femur under the most distal section of femoral stem prosthesis? Discuss why the femoral stem has been designed and fabricated in the current form for many years. What are possible solutions to reduce the ineffective effects of the femoral stem in the current design, structure, and material? Discuss the advantages and disadvantages of the solutions.

Q 4.6: Discuss the articulation conditions of the THR prostheses between the polymeric acetabular liner and femoral ball components. Provide a comprehensive comparison of articulation conditions between the femur and pelvis in a healthy hip joint with the articulation conditions in THR prostheses.

Q 4.7: Discuss the general characteristics of acetabular prosthesis components (acetabular liner and shell) in view of their design, structure and material. What types of replacement methods might be used to stabilize the acetabular prosthesis at the acetabulum cavity? Discuss various types of acetabular prostheses.

Q 4.8: What are the disadvantages of articulation conditions between the acetabular liner and femoral ball in conventional THR? What is the overall design concept of the DM acetabular cup to reduce the ineffective effect of the conventional acetabular cup at the articulation interface? Discuss the mechanism of articulation interfaces in DM design and the mechanical advantages that result in long-term implantation of the THR at the hip joint.

Q 4.9: Refer to Section 4.3.2: What coating processes are beneficial to be performed on the metallic acetabular shell to enhance its osteoconductivity at the acetabulum cavity? What metallurgical processes are used to enhance UHMWPE polymer for use as an acetabular liner? Discuss the benefits of these material engineering improvements in the performance of THR prostheses.

Q 4.10: Discuss how the DM acetabular cup prosthesis may reduce the risk for failure in obese patients or in those with a deficiency in the hip muscles.

Q 4.11: What is the main functional challenge of DM at the hip joint? What design improvements might be helpful to reduce the ineffective function of this acetabular cup in long-term implantation? (Discuss three design improvements in your answer).

Q 4.12: What issues might be observed in failed primary THA? Discuss the benefits of the DM cup for rTHA.

Q 4.13: Refer to Fig. 4.6: What is the main challenge of using a DM cup for a patient with low bone mineral density? What design solution is considered for the effective use of the DM cup when the risk for acetabular wall fracture is high?

Q 4.14: Discuss the overall characteristics of the multiple-hole acetabular shell. For which patients might this type of shell be beneficial? What are the benefits of porous structuring of the multiple-hole cup? How might the porous structured layer be processed on the surface of the multiple-hole cup? What material is good to use to manufacture the multiple-hole cup?

Q 4.15: Refer to Fig. 4.7, What are the main challenges of using multi-hole acetabular shell in THA? Discuss your answer in view of the number and length of utilized screws for fixation of the multi-hole cup to the acetabular wall and effective porous structured layer on the its contact surfaces with acetabulum cavity.

Q 4.16: Under what physiologic conditions is the multiple-hole acetabular shell usable for rTHA?

Q 4.17: Discuss the overall characteristics and implantation procedure of the press-fit acetabular cup. Discuss why no extended features such as fin, peg, or even screw holes are designed for this type of acetabular cup. Why might the cemented press-fit cup be inadequate for young patients? What is the reason for creating a polar gap in the replacement of cementless replacement of a press-fit cup? What is the challenge of creating such a gap at the interface of the prosthesis and bone? How might such an issue be solved in cementless replacement of the press-fit acetabular cup?

Q 4.18: Discuss why a high or low level of porosity of the porous structure on the surface of the press-fit acetabular shell may increase the extent of elastic micromotion. What issue might result when the porosity level is not optimized? What is the effect of press-fit acetabular shell thickness on the extent of micromotion at the bone—prosthesis interface?

Q 4.19: Discuss the overall characteristics of the cup-cage acetabular component and its surgical implantation and fixation to the acetabulum cavity and pelvis. What types of injuries at the acetabulum cavity are indicated for use of the cup-cage component?

Q 4.20: What are the technical considerations for using the cup-cage component with a DM acetabular cup in rTHA? What failures might occur if the DM cup is used with a conventional cup-cage component with holes at the cup portion?

Q 4.21: Refer to Fig. 4.9: Under what conditions might the THA or rTHA fail in the reconstruction of massive bone loss or fracture at the acetabular rim? What strategies enhance the replacement and fixation of the acetabular prosthesis to treat such severe injuries?

Q 4.22: What types of failure at the periprosthetic bone occur after replacing femoral stem prostheses in THA? Discuss the possible reasons for such failures.

Q 4.23: In what axes or planes of the femoral stem component might tapering be considered? What are the advantages of femoral stem tapering in the metaphyseal and diaphyseal portions of the proximal third femur?

Q 4.24: What is cortical hypertrophy at the most distal portion of the femoral stem? How might such an issue occur? What is a Dorr type A femur and why is the risk for overfitting of the femoral stem in such a type of bone morphology high? Discuss how tapering, longitudinal groove, and fluting of the distal tip of the femoral stem could reduce or eliminate this type of failure.

Q 4.25: Discuss the overall characteristics of the short femoral stem. What are the benefits of the short femoral stem compared with the conventional long stem? What are the benefits of the short femoral stem for primary THA in view of bone conservation in rTHA?

Q 4.26: What are the failures of the short femoral stem? Discuss why effective stability and bone–prosthesis integrity at the metaphyseal portion of the proximal third femur are vital to the success of the short femoral stem in long-term implantation. Refer to Fig. 4.14: Discuss the optimal conditions for replacing the short femoral stem and the effect of adequately sizing of this component in THA.

Q 4.27: Discuss the overall characteristics of the ultrashort femoral stem. What are the benefits of this type of femoral stem in THA? What development strategies might be necessary for the effective stability of the ultrashort femoral stem in the metaphyseal portion of the proximal third femur?

Q 4.28: Discuss the overall characteristics of the femoral stem with a porous-structured surface at its metaphyseal zone. What are the benefits of this type of processed femoral stem in THA? What surface engineering methods might be used to create a porous structured layer around the femoral stem at the metaphyseal portion?

4.8 Think and challenge

In this section, a question is provided for further thought about the reviewed and discussed contents of this chapter. The aim is to challenge current facets of interactions of bone with THR prostheses by answering this questions, which may enhance design concepts for the future development of prostheses.

Q 4.29: As highlighted in this chapter, the interaction of the humeral stem with the humerus at the proximal portion results in stress shielding at the proximal zone of the humeral prosthesis in interfaces with trabecular and cortical bone tissue. What design concept might be used to reduce or eliminate such a failure?

References

[1] Q. Liu, X. Cheng, D. Yan, Y. Zhou-J. of orthopaedic translation, undefined 2019, Plain radiography findings to predict dislocation after total hip arthroplasty, J. Orthop. Translat. Elsevier. (n.d.). https://www.sciencedirect.com/science/article/pii/S2214031X18301402 (accessed December 2, 2020).

[2] S. Schleifenbaum, T. Prietzel, C. Hädrich, R. Möbius, F. Sichting, N. Hammer, Tensile properties of the hip joint ligaments are largely variable and age-dependent – an in-vitro analysis in an age range of 14–93 years, J. Biomech. (2016), https://doi.org/10.1016/j.jbiomech.2016.09.001. Elsevier.

[3] C. Park, H. Eun, S. Oh, H. Kim, et al. S.L.-T.J. of, undefined 2019, Femoral stem survivorship in Dorr type A femurs after total hip arthroplasty using a cementless tapered wedge stem: a matched comparative study with type B femurs, J. Arthroplasty Elsevier. https://www.sciencedirect.com/science/article/pii/S0883540318311173 (accessed November 17, 2020).

[4] R. Gaillard, R. Kenney, J.L. Delalande, C. Batailler, S. Lustig, Ten- to 16-year results of a modern cementless dual-mobility acetabular implant in primary total hip arthroplasty, J. Arthroplasty 34 (2019) 2704–2710, https://doi.org/10.1016/j.arth.2019.06.051.

[5] M.M. Kukreja, S.K. Nishiyama, P.S. Kang, Management of radiation induced, bilateral Primary THA loosening with an aseptic Paprosky Type IIB acetabular defect and a contralateral septic Type IIIB acetabular defect: a case report and review of literature, Int. J. Surg. Case Rep. 42 (2018) 218–223, https://doi.org/10.1016/j.ijscr.2017.12.012.

[6] P. Messer-Hannemann, G.M. Campbell, M.M. Morlock, Deformation of acetabular press-fit cups: influence of design and surgical factors, Clin. Biomech. 69 (2019) 96–103, https://doi.org/10.1016/j.clinbiomech.2019.07.014.

[7] J. Schilcher, I. Ivarsson, R. Perlbach, L. Palm, No difference in periprosthetic bone loss and fixation between a standard-length stem and a shorter version in cementless total hip arthroplasty. A randomized controlled trial, J. Arthroplasty 32 (2017) 1220–1226, https://doi.org/10.1016/j.arth.2016.11.015.

[8] H. Stigbrand, O. Gustafsson, G. Ullmark, A 2- to 16-year clinical follow-up of revision total hip arthroplasty using a new acetabular implant combined with impacted bone allografts and a cemented cup, J. Arthroplasty 33 (2018) 815–822, https://doi.org/10.1016/j.arth.2017.10.006.

[9] F.E. Rowan, A.J. Salvatore, J.K. Lange, G.H. Westrich, Dual-mobility vs fixed-bearing total hip arthroplasty in patients under 55 years of age: a single-institution, matched-cohort analysis, n.d. https://www.sciencedirect.com/science/article/pii/S0883540317304059.
[10] N. Lebeau, M. Bayle, R. Belhaouane, M. Chelli, E. Havet, B. Brunschweiler, P. Mertl, Total hip arthroplasty revision by dual-mobility acetabular cup cemented in a metal reinforcement: a 62 case series at a minimum 5 years' follow-up, Orthop. Traumatol. Surg. Res. 103 (2017) 679–684, https://doi.org/10.1016/j.otsr.2017.04.009.
[11] A. Grazioli, E.T.H. Ek, H.A. Rüdiger, Biomechanical concept and clinical outcome of dual mobility cups, Int. Orthop. 36 (2012) 2411–2418, https://doi.org/10.1007/s00264-012-1678-3.
[12] G. Dikmen, V.E. Ozden, K. Karaytug, R. Tozun, Dual-mobility cups in revision acetabular reconstructions: short-term outcomes in high-risk patients for instability, Acta Orthop. Traumatol. Turcica 53 (2019) 329–333, https://doi.org/10.1016/j.aott.2019.05.002.
[13] P. Maisongrosse, B. Lepage, E. Cavaignac, R. Pailhe, N. Reina, P. Chiron, J.M. Laffosse, Obesity is no longer a risk factor for dislocation after total hip arthroplasty with a double-mobility cup, Int. Orthop. 39 (2015) 1251–1258, https://doi.org/10.1007/s00264-014-2612-7.
[14] P.D. Sunilkumar, K.J. Oh, H.W. Cho, S.M. Kim, Monolithic dual mobility cup total hip arthroplasty has high complication rates with surgical fixation in elderly with femur neck fracture, J. Arthroplasty (2020), https://doi.org/10.1016/j.arth.2020.06.091.
[15] B. Darrith, P.M. Courtney, C.J. Della Valle, Outcomes of dual mobility components in total hip arthroplasty: a systematic review of the literature, Bone Jt. J. 100B (2018) 11–19, https://doi.org/10.1302/0301-620X.100B1.BJJ-2017-0462.R1.
[16] G. Sayac, T. Neri, L. Schneider, R. Philippot, F. Farizon, B. Boyer, Low revision rates at more than 10 Years for dual-mobility cups cemented into cages in complex revision total hip arthroplasty, J. Arthroplasty 35 (2020) 513–519, https://doi.org/10.1016/j.arth.2019.08.058.
[17] D. Hoornenborg, J. van Loon, S. de Waard, I.N. Sierevelt, K.T.M. Opdam, G.M.M.J. Kerkhoffs, D. Haverkamp, Dynamic trial fitting by an expanding trial cup does not jeopardize primary acetabular component stability, Clin. Biomech. 78 (2020) 105077, https://doi.org/10.1016/j.clinbiomech.2020.105077.
[18] M.A. Malahias, L. Kostretzis, A. Greenberg, V.S. Nikolaou, A. Atrey, P.K. Sculco, Highly porous titanium acetabular components in primary and revision total hip arthroplasty: a systematic review, J. Arthroplasty 35 (2020) 1737–1749, https://doi.org/10.1016/j.arth.2020.01.052.
[19] I.R. Wilson, T.R. Turgeon, T.C. Gascoyne, C.J. Della Valle, R.W. McCalden, Midterm results of a contemporary, porous-coated acetabular system in patients undergoing primary total hip replacement for degenerative hip disease: a prospective, multicenter study, J. Arthroplasty 35 (2020) 1862–1867, https://doi.org/10.1016/j.arth.2020.02.068.
[20] V.K. Alamanda, I. Demartino, H.G. Potter, M.F. Koff, B. Lin, A. Muskat, G.H. Westrich, Multiacquisition variable-resonance image combination magnetic resonance imaging used to study detailed bone apposition and fixation of an additively manufactured cementless acetabular shell, Arthroplast. Today 6 (2020) 694–698, https://doi.org/10.1016/j.artd.2020.07.019.
[21] N.A. Beckmann, R.G. Bitsch, M.B. Janoszka, M.C. Klotz, T. Bruckner, S. Jaeger, Treatment of high-grade acetabular defects: do porous titanium cups provide better stability than traditional titanium cups when combined with an augment? J. Arthroplasty 33 (2018) 1838–1843, https://doi.org/10.1016/j.arth.2018.01.068.

[22] H.A.H. Hosny, A. El-Bakoury, S.C.M. Srinivasan, R. Yarlagadda, J. Keenan, Tritanium acetabular cup in revision hip replacement: a six to ten years of follow-up study, J. Arthroplasty 33 (2018) 2566–2570, https://doi.org/10.1016/j.arth.2018.03.040.
[23] W.J. Long, S. Nayyar, K.K. Chen, D. Novikov, R.I. Davidovitch, J.M. Vigdorchik, Early aseptic loosening of the Tritanium primary acetabular component with screw fixation, Arthroplast. Today 4 (2018) 169–174, https://doi.org/10.1016/j.artd.2017.11.009.
[24] S. Agarwala, C. Jadia, M. Vijayvargiya, Incorporation of screwless press-fit acetabular cups and disappearance of polar gaps, J. Clin. Orthop. Trauma. 11 (2020) 85–90, https://doi.org/10.1016/j.jcot.2018.11.012.
[25] A. Fraile Suari, F. Marqués López, M. Cuenca Llavall, M. Tey Pons, A. León García, Reconstruction for pelvic discontinuity and massive acetabular defects, Rev. Esp. Cir. Ortop. Traumatol. 64 (2020) 64–73, https://doi.org/10.1016/j.recot.2019.06.004.
[26] P. Messer-Hannemann, H. Weyer, G.M. Campbell, M.M. Morlock, Time-dependent viscoelastic response of acetabular bone and implant seating during dynamic implantation of press-fit cups, Med. Eng. Phys. 81 (2020) 68–76, https://doi.org/10.1016/j.medengphy.2020.05.012.
[27] A. Jahnke, C.C. Bott, C.A. Fonseca Ulloa, G.W. Jahnke, M. Rickert, B.A. Ishaque, G.A. Ahmed, In vitro examination of the primary stability of three press-fit acetabular cups under consideration of two different bearing couples, Med. Eng. Phys. 67 (2019) 49–54, https://doi.org/10.1016/j.medengphy.2019.03.008.
[28] F. Morosato, L. Cristofolini, F. Castagnini, F. Traina, Effect of cup medialization on primary stability of press-fit acetabular cups, Clin. Biomech. 80 (2020) 105172, https://doi.org/10.1016/j.clinbiomech.2020.105172.
[29] M.A. Ritter, P.M. Faris, E.M. Keating, G. Brugo, Influential factors in cemented acetabular cup loosening, J. Arthroplasty 7 (1992) 365–367, https://doi.org/10.1016/S0883-5403(07)80025-8.
[30] C. Hipfl, V. Janz, J. Löchel, C. Perka, G.I. Wassilew, Cup-cage reconstruction for severe acetabular bone loss and pelvic discontinuity: mid-term results of a consecutive series of 35 cases, Bone Jt. J. 100B (2018) 1442–1448, https://doi.org/10.1302/0301-620X.100B11.BJJ-2018-0481.R1.
[31] D. Arvinte, M. Kiran, M. Sood, Cup-cage construct for massive acetabular defect in revision hip arthroplasty- A case series with medium to long-term follow-up, J. Clin. Orthop. Trauma. 11 (2020) 62–66, https://doi.org/10.1016/j.jcot.2019.04.021.
[32] C.Y. Chang, C.T. Wu, H. Numan, F.C. Kuo, J.W. Wang, M.S. Lee, Survival analysis of allografting and antiprotrusio cage in treating massive acetabular bone defects, J. Arthroplasty (2020), https://doi.org/10.1016/j.arth.2020.08.050.
[33] M. Acharya, W.A. Elnahal, Strategies of management of traumatic periprosthetic acetabular fractures around a pre-existing total hip arthroplasty, J. Clin. Orthop. Trauma. 11 (2020) 1053–1060, https://doi.org/10.1016/j.jcot.2020.10.033.
[34] T.Z.S. Benko", A. Santiago-Martin, C. Ruddlesdin, G. Selzer, Aseptic loosening of 2 rim-fix, hydroxyapatite-coated acetabular cups, J. Arthroplasty 17 (2002) 519–523, https://doi.org/10.1054/arth.2002.32188.
[35] S.P. Issa, D. Biau, P. Leclerc, A. Babinet, M. Hamadouche, P. Anract, Stemmed acetabular cup as a salvage implant for revision total hip arthroplasty with Paprosky type IIIA and IIIB acetabular bone loss, Orthop. Traumatol. Surg. Res. 106 (2020) 589–596, https://doi.org/10.1016/j.otsr.2020.01.012.
[36] J.E. Webb, R.J. McGill, B.T. Palumbo, W.E. Moschetti, D.M. Estok, The double-cup construct: a novel treatment strategy for the management of Paprosky IIIA and IIIB acetabular defects, J. Arthroplasty 32 (2017) S225–S231, https://doi.org/10.1016/j.arth.2017.04.017.

[37] G. Kimura, E. Onishi, K. Kawanabe, … S.K.-J. of O., undefined 2014, Characteristic loosening of the distally straight cylindrical femoral stem, Springer. (n.d.). https://link.springer.com/content/pdf/10.1007/s00776-014-0538-y.pdf.
[38] Y. Okutani, K. Goto, Y. Kuroda, T. Kawai, … Y.O.-J. of O., undefined 2019, Long-term outcome of cemented total hip arthroplasty with the charnley-type femoral stem made of titanium alloy, Elsevier. (n.d.). https://www.sciencedirect.com/science/article/pii/S0949265819302428.
[39] K. Goto, Y. Furuya, K. Oda, R. Minami, … K.S.-J. of O., undefined 2018, Long-term results of total hip arthroplasty using charnley elite-plus stem and the effect of stem geometry on radiographic distal femoral cortical hypertrophy, Elsevier. (n.d.). https://www.sciencedirect.com/science/article/pii/S0949265817303457.
[40] Y. Kim, J. Park, Y.J.-T.J. of Arthroplasty, Long-term survival (up to 34 years) of retained cementless anatomic femoral stem in patients< 50 years old, Arthroplastyjournal.Org. (n.d.). www.arthroplastyjournal.org/article/S0883-5403(20)31154-2/abstract.
[41] M. Innmann, M. Streit, T. Bruckner, … C.M.-T.J. of, undefined 2018, Comparable cumulative incidence of late periprosthetic femoral fracture and aseptic stem loosening in uncemented total hip arthroplasty—a concise follow-up, Elsevier. (n.d.). https://www.sciencedirect.com/science/article/pii/S0883540317310276.
[42] I. Ferreira, J. Mariano, F. Lucas, F.J.-I. journal of, undefined 2019, Osteosynthesis of bilateral vancouver B2 periprosthetic femoral fracture after a bilateral RM total hip arthroplasty at 24 and 21-years follow-up: a case report, Elsevier. (n.d.). https://www.sciencedirect.com/science/article/pii/S221026121930255X.
[43] J. Zeng, Q. Qiu, H. Lan, Z.W.-I.J. of S. Case, undefined 2020, Preprosthesis fracture of femoral and loosening of the femoral prosthesis during closed reduction of hip dislocation: a case report, Elsevier. (n.d.). https://www.sciencedirect.com/science/article/pii/S2210261220304740.
[44] J. Griffiths, L. Roumeliotis, D. Elson, … Z.B.-T.J. of, undefined 2020, Long term performance of an uncemented, proximally hydroxyapatite coated, double tapered, titanium-alloy femoral stem: results from 1465 hips at 10 years, Elsevier. (n.d.). https://www.sciencedirect.com/science/article/pii/S0883540320308925.
[45] P. Panichkul, R. McCalden, … S.M.-T.J. of, undefined 2019, Minimum 15-year results of a dual-offset uncemented femoral stem in total hip arthroplasty, Elsevier. (n.d.). https://www.sciencedirect.com/science/article/pii/S0883540319306576 (accessed November 17, 2020).
[46] J. Gabor, V. Singh, J. Padilla, … R.S.-J. of, undefined 2020, Outcomes with two tapered wedge femoral stems in total hip arthroplasty using an anterior approach, Elsevier. (n.d.). https://www.sciencedirect.com/science/article/pii/S0972978X20302579 (accessed November 17, 2020).
[47] T. Hofstaedter, M. Najfeld, G. Fessel, L.O.-A. today, undefined 2020, Discrepancy of trial rasp and femoral stem relative position within the femoral canal of a coated tapered system: an intraoperative, intrapatient, Elsevier. (n.d.). https://www.sciencedirect.com/science/article/pii/S2352344120301515 (accessed November 17, 2020).
[48] L. Warth, T. Grant, N. Naveen, … E.D.-T.J. of, undefined 2020, Inadequate metadiaphyseal fill of a modern taper-wedge stem increases subsidence and risk of aseptic loosening: technique and distal canal fill matter!, Elsevier. (n.d.). https://www.sciencedirect.com/science/article/pii/S0883540320301674 (accessed November 17, 2020).
[49] Y. Kim, J.P.-T.J. of arthroplasty, undefined 2019, Long-term outcomes of ultra-short metaphyseal-fitting anatomic cementless femoral stem in total hip arthroplasty with ceramic-on-ceramic articulation for, Elsevier. (n.d.). https://www.sciencedirect.com/science/article/pii/S0883540319304048 (accessed November 17, 2020).

[50] S. Tian, K. Goswami, J. Manrique, ... K.B.-T.J. of, undefined 2019, Direct anterior approach total hip arthroplasty using a morphometrically optimized femoral stem, a conventional operating table, without fluoroscopy, Elsevier. (n.d.). https://www.sciencedirect.com/science/article/pii/S0883540318310805 (accessed November 17, 2020).

[51] G. Ullmark, Occult hip prosthetic loosening diagnosed by [18F] fluoride-PET/CT, Elsevier. (n.d.). https://www.sciencedirect.com/science/article/pii/S2352344120301072.

[52] E. Tyrpenou, A. Khoshbin, ... S.M.-T.J. of, undefined 2020, A large-scale fifteen year minimum survivorship of a cementless triple tapered femoral stem, Elsevier. (n.d.). https://www.sciencedirect.com/science/article/pii/S088354032030276X.

[53] L. Martins, F. Garcia, C.P.-T.J. of arthroplasty, undefined 2014, Aseptic Loosening Rate of the Mayo Femoral Stem with Medium-Term Follow up, Elsevier. (n.d.). https://www.sciencedirect.com/science/article/pii/S0883540314004318 (accessed November 17, 2020).

[54] S. Sivaloganathan, C. Maillot, C. Harman, ... L.V.-... & T.S., undefined 2020, Neck-sparing short femoral stems: a meta-analysis, Elsevier. (n.d.). https://www.sciencedirect.com/science/article/pii/S1877056820301900 (accessed November 17, 2020).

[55] J. Schilcher, I. Ivarsson, R. Perlbach, L.P.-T.J. of arthroplasty, undefined 2017, No difference in periprosthetic bone loss and fixation between a standard-length stem and a shorter version in cementless total hip arthroplasty, A randomized, Elsevier. (n.d.). https://www.sciencedirect.com/science/article/pii/S0883540316308154 (accessed November 17, 2020).

[56] B. Thorat, A. Singh, R.V.-J. of C.O. and Trauma, undefined 2020, role of a bone conserving short stem femoral component in revision total hip arthroplasty: a case report, Elsevier. (n.d.). https://www.sciencedirect.com/science/article/pii/S0976566220304677 (accessed November 17, 2020).

[57] P. Laboudie, F. El Masri, ... L.K.-T.J. of, undefined 2020, Short vs standard-length femoral stems cemented according to the "French Paradox": a matched paired prospective study using ein bild roentgen analyze, Elsevier. (n.d.). https://www.sciencedirect.com/science/article/pii/S0883540320309803 (accessed November 17, 2020).

[58] T. Shishido, T. Tateiwa, Y. Takahashi, ... T.M.-J. of, undefined 2018, Effect of stem alignment on long-term outcomes of total hip arthroplasty with cementless bi-metric femoral components, Elsevier. (n.d.). https://www.sciencedirect.com/science/article/pii/S0972978X17303215 (accessed November 17, 2020).

[59] C. Wallace, J. Chang, B. Kayani, ... P.M.-T.J. of, undefined 2020, Long-term results of revision total hip arthroplasty using a modern extensively porous-coated femoral stem, Elsevier. (n.d.). https://www.sciencedirect.com/science/article/pii/S0883540320307178.

[60] Y. Naito, M. Hasegawa, S. Tone, ... H.W.-T.J. of, undefined 2020, Minimum 10-year follow-up of cementless total hip arthroplasty with a 32-mm cobalt-chromium head on highly cross-linked polyethylene and a tapered, fiber, Elsevier. (n.d.). https://www.sciencedirect.com/science/article/pii/S0883540320309645.

CHAPTER FIVE

Interactions of bone with trauma implants and possible failures

5.1 Introduction

Clinical studies have reported the loosening and failure of trauma implants. Trauma implants are fixed through bone tissue using of various types of screws. Integration between the screw and bone and the bending strength of the implant (plate, nail, or rod) are key factors for the long-term effectiveness of trauma implants in managing fractures in bone trauma injuries. The effective purchase and integration of the screw and bone tissue, particularly trabecular bone, increases the stability of the bone—implant construct. However, this increases subjected stress to the trauma plate, nail, or rod implant. Thus, the design of both the screw and the main trauma implant is carried out using experimental testing and computational analysis to examine the biomechanical response of the bone—implant construct under simulated static and dynamic physiologic loading conditions. Although technical concepts and various biomechanical and clinical evaluation methods are employed for the performance and function of trauma fracture fixation during healing, failure of bone—implant constructs is observed. Traditionally, rigidity of fracture fixation was greatly emphasized by implant developers to maintain the achieved fracture reduction interoperatively during healing. However, the significant rate of delayed union and nonunion motivated biomechanical engineers to study the effect of the mechanical performance of the bone—implant on the biological process of bone formation at fracture gap surfaces between bone fragments. The concept of micromotion at the fracture gap was highlighted by biologists as a key factor in mechanical stimulation of ontogenesis cells and other biological activity in the formation of fibrocartilaginous, callus, and ultimately bone tissue formation at the fracture site. This was resulted in the development of the bone—implant construct with lower stiffness and rigidity to increase strain at the microscale between bone fragments. This advantage accelerates cyclic loading conditions that constantly generate mechanical signals for the biological induction of fracture healing. Along with the optimized design of

Interactions of Bone with Orthopedic Implants and Possible Failures
ISBN: 978-0-323-95410-5
https://doi.org/10.1016/B978-0-323-95410-5.00003-7

the plate for sufficient strength and flexibility, specifically designing of trauma implant has been a key solution to enhance bone fragment stability at the fracture site. One type of fixation implant may not provide effective buttressing and capture all bone fragments at the fracture site. Thus, various types of fixation implants have been developed to manage one fracture type with different degrees of severity and fragmentation at the fracture. Fracture fixation has been managed for most skeletal bones. Reasons for fixation failure might be different for various parts of the body. In some sections, body weight loading conditions have the prominent effect, whereas in some sections, the high range of rotational or translation motion causes failure of the bone–implant construct.

In this chapter, the interaction of bone with trauma implants and possible failures are reviewed. Interactions between the bone and screws are discussed in various parts of the lower and upper extremities, pelvis, and clavicle. Possible failures of the trauma bone–implant construct in each skeletal part are discussed and illustrated. There is an attempt to highlight main failures in fixing fractures in bone injuries.

5.2 Bone trauma injury

Bone tissue might be injured owing to the trauma. Normally, the introduction of high loading conditions in the form of an impact causes the bone to break or fracture. Such conditions might be observed in severe accidents or a fall from a height in young patients. Bone trauma injuries might occur in elderly patients even from a simple fall caused by imbalance while in walking or sliding on the floor. Fracture of the bone tissue creates bone fragments at the injured site or zone. The severity of the fracture might be diagnosed as a short- or lengthy crack. In a severe condition that needs surgical intervention, the fracture is completely propagated, in which the bone is fragmented into two, three, or more fragments. The fragments might be dislocated, from which adjacent soft tissues (e.g., ligaments, tendon, muscle) are at risk for irritation or rupture. The bone fracture causes a body reaction because of the accumulation of blood at the injury zone, where inflammation and swelling occur. The activity of macrophage cells significantly increases at fracture site, from which the activity of osteoclast cells increases. This causes the resorption of bone tissue at fracture gaps and increases the volume of bone loss at the injury zone. Therefore, a bone trauma injury is considered an emergency and the patient must be sent to a hospital for primary treatment to control internal bleeding and reduce dislocated fragments at the anatomic position as far as possible to avoid soft tissue injuries and promote bone loss volume. The final treatment of bone fractures is to restore and fix bone fragments at their anatomic position to facilitate healing of the bone fracture by forming bone tissue at the fracture gaps (Fig. 5.1) [1]. Through successful bone fracture fixation and healing, there are no longer bone fragments and the integrity of the bone at the fracture zone is retained. Bone fracture management is implemented through internal or external fixation. For each type of fixation method,

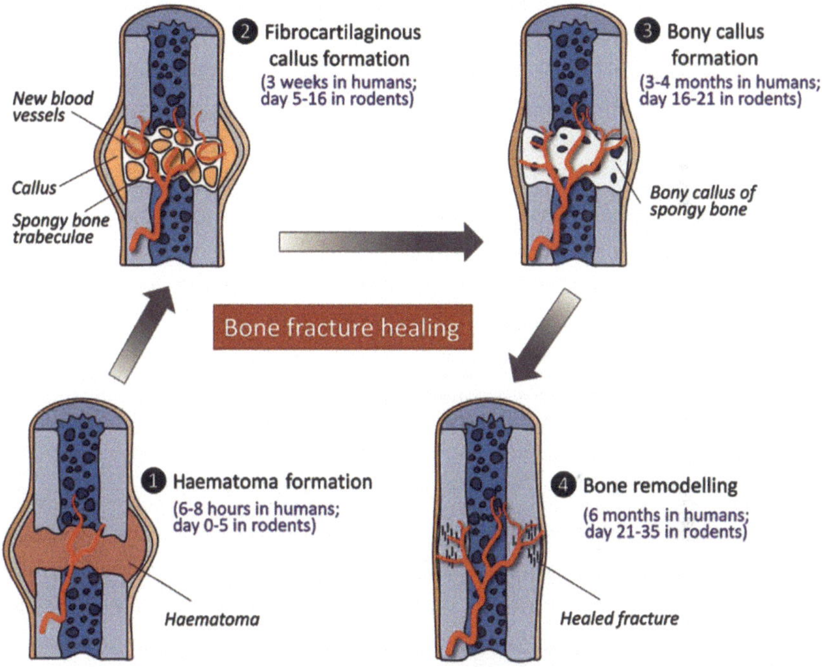

Figure 5.1 Illustration of the fracture healing at the fracture gap. Fracture healing is initiated by blood accumulating at the fracture site immediately after the fracture occurs. Because bone fragments have not been yet reduced in anatomic position, blood might continue to accumulate and cause inflammation and swelling at the injury zone. Through the temporary reduction of bone fragments, inflammation is reduced, and after surgical management of the fracture with trauma implants, the formation of fibrocartilaginous callus is initiated. By successful progress of this process, bony callus formation is achieved at the fracture gap. The formed callus at this stage continues to strengthen with a bone density similar to that of cortical or trabecular bone matrix at the fracture gap. *Through the courtesy of L. Tian, N. Tang, T. Ngai, C. Wu, Y. Ruan, et al., Hybrid Fracture Fixation Systems Developed for Orthopaedic Applications: a General Review, Elsevier, Undefined 2019, (n.d.). https://www.sciencedirect.com/science/article/pii/S2214031X18300548. (Accessed December 8, 2020), and permission from Elsevier.*

various types of implants have been developed to restore bone fragments during healing. External fixation is used for simple close fractures at the midshaft of the long bones or as an initial or supportive fixation device in open fractures involving soft tissue injuries (a closed fracture is when the skin has not been ruptured, whereas in open fractures, the skin is ruptured) [2,3]. On the other hand, internal fixation is advantageous for simple or multifragmentary fractures at the metaphyseal zones of the long bones, multifragmentary fractures at the midshaft of long bones, and any fracture conditions in small bones (e.g., foot and hand bones). After completion and successful fracture healing, external fixation implants are completely removed, whereas internal fixation implants are maintained fixed to the bone if complications of discomfort (e.g., soft tissue irritation) do not bother the patient. Fixation of the implant to bone fractures creates a bone—implant construct that tolerate or bears physiologic loading conditions at the injured site. The integrity of the bone—implant construct might fail through fracture of the implant, refracture of the bone, loosening of implant—bone integrity, suppression of the bone through the implant, or dislocation of bone fragments. Such failures occur because of ineffective interactions between the bone and implant, which are influenced by the design parameters and features of the implant. In the next sections, the varieties of trauma implants and their design concepts along with possible failures are reviewed and discussed.

5.3 Interactions of bone with trauma implants

Most trauma injuries are treated with internal fixation implants. Thus, in this chapter, these categories of trauma implants are evaluated in view of their design concepts, interactions with bone, and failures [1,4]. Internal fixation implants may be categorized into two main groups of trauma plating and intramedullary nailing systems. (The lead author has written a comprehensive reference book, *Trauma Plating Systems, Elsevier, 2017*, for the group of trauma plating systems.) Internal trauma implants are known as plates and nails, which are fixed to the bone by adequate screws. The component that is purchased through bone tissue, either cortical or trabecular bone tissue, is screws. Therefore, the integrity of the bone—implant construct is achieved or fails at the interface of the bone and screws. However, all components of the bone—implant construct composed of bone fragments, screw, and plate and nail have an effect on the performance of the fracture fixation construct. Trauma implants are normally manufactured by machining from stainless steel and titanium alloys. In some cases, the implant might be formed through forging in complex areas. Furthermore, trauma implants have been fabricated from carbon fiber—reinforced polyetheretherketone (CFRPEEK, 30% weight-based carbon fiber). The effectiveness of CFRPEEK plate was compared with titanium in an animal implantation study [5] (Fig. 5.2). A trauma plate is designed as two groups of straight and anatomic plates. The straight plates are used for fracture fixation at the midshaft of the long bones, whereas the anatomic plates are used to fix fractures at the proximal or distal third of the long bones. Anatomic plates are composed of two parts: one is positioned on the diaphyseal portion of the distal or proximal section of the long bones, and the second part is fitted on the metaphyseal portion of the long bones, which has anatomically complex morphology or geometry. The plates feature multiple screws on the whole part to facilitate inserting the screws through various parts of the bone, which is positioned underneath the plate. The plate thickness might be different in various parts in straight or anatomic plates. In anatomic plates, the diversity of thickness is required; the thickness of the diaphyseal part is higher than that of the metaphyseal part. In straight plates, the thickness of the plate is traditionally maintained constant. Newer designs feature thickness at the two ends as lower than that of the plate's center portion to enhance the low-profile advantage of the plate (the thickness starts at the lowest range at the two ends and gradually increases until the highest range at the center part). The plates are designed without sharp edges to eliminate soft tissue irritation or rupture. The provided screw holes on the plates come with or without threads for the insertion of locking and compression or cancellous screws, respectively. The bottom surface of the plates has the lowest contact surfaces with the bone for the highest level of vascularization at the interface of the plate and bone. Because of the compression of the plate to the bone, the bone tissue tends to remodeled to decrease the level of interactive compressive stress at the plate-bone interface, from which the bone morphology might be altered. As the result, the thickness of the underneath cortices layer (cortical bone layer) might be decreased which increase the risk of bone refracture, possibly when the implant is removed from the body after completion of the fracture healing. The combined threaded and nonthreaded screw holes are designed to facilitate inserting locking or compression/cancellous screws at a screw hole, for effective alternative solutions to screw purchase (such screw holes are provided in the straight part of the plate).

Interaction of the bone with the plate could be expressed from various facets. Plate fixation is accomplished with screws. However, its design

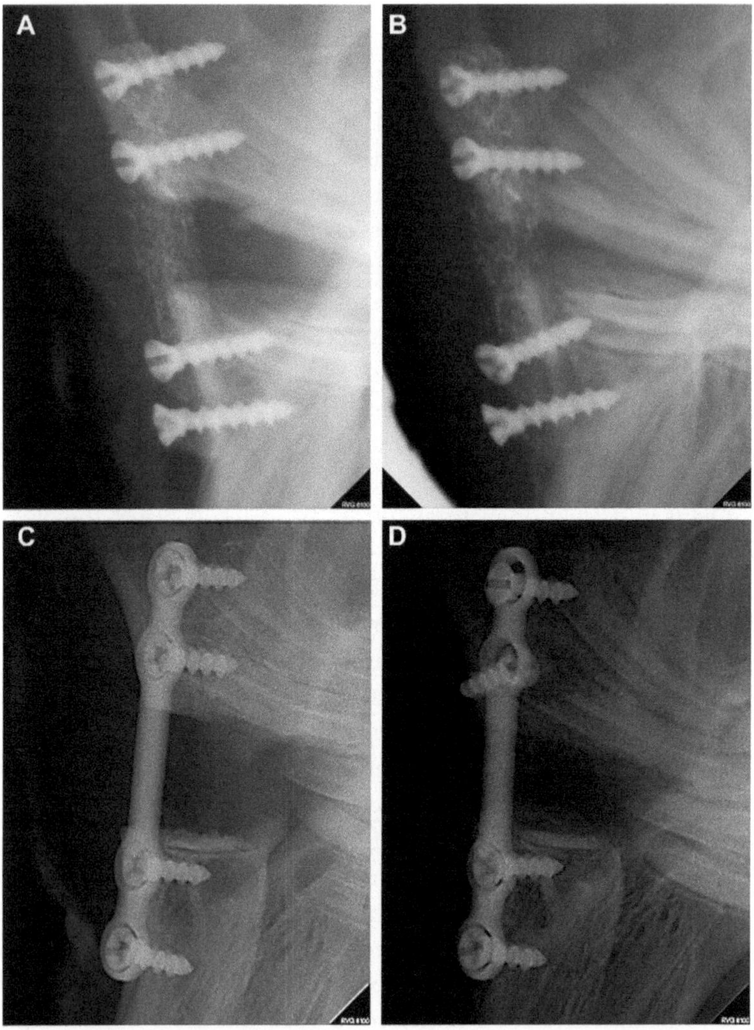

Figure 5.2 Radiographic illustration of plating fixation using (A, B) carbon fiber–reinforced polyetheretherketone (CFRPEEK), and (C, D) titanium (Ti) plates, whereas screws are all made of Ti material. Plating systems are implanted in a rabbit to study the function of fracture fixation under loading conditions. Polyetheretherketone polymer is radiolucent. Thus, in the radiograph image, compounded carbon fiber particles are roughly detectable in CFRPEEK plating fixation. The fracture gap is maintained in the far distance to show the superior function of CFRPEEK compared with the Ti plate in the mechanical induction of fracture healing. Owing to the much lower mechanical modulus of CFRPEEK compared with the Ti material, the flexibility of the CFRPEEK plate–bone construct induces bone remodeling and the formation of bone tissue at the gap space (compare images [B] and [D], which were taken 6 weeks after implantation). Likewise, because of the rigid fixation of Ti plating, localized stress at the screws is

parameters, features, and benefits have a direct influence on its performance when it is stabilized on bone. Plate fixation is highly rigid or semirigid. The plate thickness, length, width, number of screw holes, lateral dents, and so on are considered in the mechanical design. In conjunction with these geometrical parameters, the materials and manufacturing processing of the plate have a significant effect on the mechanical performance of the bone–implant construct. As highlighted in Chapter 6 of *Trauma Plating Systems* (Elsevier, 2017), the working length of the plate has a considerable effect on the concentration of stress at the center section of the working zone (the working zone is the distance or length between locking screws, which described the straight part of the plate, where the optimum number of locking screws are required to reduce stress concentration on the one hand, and excessive bone formation at the fracture gap on the other hand (Fig. 5.6)). A semirigid function plate under physiologic loading is advantageous in most trauma plating fixations. In such a condition, the plate is slightly flexible and has elastic deformation in at the microscale to prevent increased localized stress concentration at critical zones of the plate. Semirigidity or microflexibility of the plate generates micromotion or microelastic strain at the fracture gaps, which increases mechanical stimulation signals for the activity of osteogenic cells and ultimately promotes the formation and maturation of bone tissue at the fracture gaps. Conversely, in highly rigid fixation, the risk for plate fracture and nonunion of bone fragments increases. The behavior of the plate and screws depends on their type of connection. The mechanical and biological integrity of screws at the cortical and trabecular bone tissues increases the rigidity of the bone–implant construct at the screw sections, from which most loading stresses are transferred and borne by the plate. Conversely, with bone and screws that have poor integrity, localized shear stress at the interface of the bone and screw causes the dissociation of screws and thus their loosening at the primary position. For the effective implantation of a plating system to fix fractures of bone fragments, the

increased, and the risk for screw loosening is accelerated (D). *Through the courtesy of S. Lee, S. György, J. Choi, J. Choi, S.K.-J. of Cranio, Carbon Plate Shows Even Distribution of Stress, Decreases Screw Loosening, and Increases Recovery of Preoperative Daily Feed Intake Amount in a Rabbit Model of mandibular continuity defects, Elsevier, Undefined 2014, (n.d.). https://www.sciencedirect.com/science/article/pii/S101051821300262X. (Accessed December 8, 2020), and permission from Elsevier.*

preoperative evaluation of bone quality at the injured zone is required for the adequate selection of an implant and for surgical planning to insert the screws.

With respect to the failure of nailing systems, loosening of screws and dislocation of the nail are highlighted. These occur as a result of the ineffective integrity of the screws and bone. The nail component might not be under as much bending stress as what is borne by the trauma plate, and thus its fracture might not be observed. Screws at the nail—bone construct are under a high level of compressive stress to avoid dislocation or movement of the nail through the intramedullary canal of the long bone at the diaphyseal zone. The nail is supported mechanically by cortices of the diaphyseal zone, against which the highly stiff bone—implant construct achieved resists bending stress. (Bending stress might be increased in some fixations such as fracture management of the femur.) In such a case, the nail might be deformed, the anatomic alignment of the bone might thus be altered and as a result, the physiologic loading and motion conditions at adjacent joints might be significantly affected. One reasons for nail deformation is weakening of the cortices in long-term implantation owing to stress shielding, in which most bending loading is borne by the metallic nail (manufactured from stainless steel for either a lengthy or short nail and titanium alloy for a short nail), and the bone cortices weaken in accordance with Wolff's law.

Biological integration of the screws with trabecular bone is advantageous for promoting bone—screw integrity. It results in a significant reduction in screw loosening. The biological integration of screws can be achieved by coating bioactive materials such as hydroxyapatite (through plasma spraying or biomimetic coating) or creating columnar porous titanium-hydroxyapatite (through plasma electrolyte oxidation or vapor deposition). Bone on-growth and ingrowth through the surface of screw threads strengthens of periprosthetic bone tissue at the interface and at a distance from the bone—screw interface. Strengthening of the bone at the bone—screw interface needs to be initiated at early implantation up to the completion of fracture healing. If fracture union is delayed, bioactive coated materials completely degrade, and thus the advantages of osteoconductivity and osteoinductivity in the bioactive layer no longer exist. (The concepts of osteoconductivity and osteoinductivity were been completely expressed by the lead author in the book *Osseointegration of Orthopedic Implants* [Elsevier, 2019], and the osteoconductive surface engineering method for enhancing orthopedic implant osseointegration was reviewed in-depth in the book *Osseoconductive Surface Engineering for Orthopedic Implants* [Elsevier, 2021] by the lead author.) Even the bone matrix might be formed at the bone—screw interface in early implantation; however, weakening of the bone matrix at the interface might progress in cases of delayed union. The effect of bioactive layering with a high adhesion strength and coated material composition, crystallinity, and mechanical properties is beneficial for the enhanced healing of fracture gaps. The main issue with using bioactive layering on screws is the difficulty of removing screws after fracture healing is completed when the implant has clinical complications such as soft tissue irritation. The high extent of torsional torque for screw removal may cause a portion of the bone which has been strongly integrated to the screw to be removed, or it may even cause a refracture of the bone, particularly in patients with low—mineral density bone.

The next sections review interactions and failures for trauma fracture fixation in various bone injuries based on published clinical studies. The failures and interactions are more serious in some parts of the musculoskeletal system; this is examined more in-depth in this chapter.

5.4 Interactions of lower extremities bones with trauma implants

5.4.1 Interactions of proximal portion of femur bone with trauma implants and possible failures

Fracture fixation of the lower extremities bones is managed by trauma implants. Femur, tibia, patella, and foot bones are treated through internal plating, nailing, and external fixation. Lower extremity bones are under high body weight loading conditions that increase the complexity of bone—implant construct failure. The fixation of fractures at the proximal femur at the capsular and intracapsular sections is one of the most challenging cases in musculoskeletal bones [6—13]. In this zone, there is a high level of bending stress and its localization or concentration at the femoral neck. Although abductor and abductor muscle forces provide loading equilibrium, through a deficiency in the function of these muscles, fracture and instability of the fracture fixation construct might result [7—12]. In particular, in a patient with low—mineral density bone and a high level of BMI (body mass index), the risk for fracture fixation failure increases. Implants such as a dynamic hip screw, trochanteric plate, cannulated screws, short metadiaphyseal nail, and proximal femur locking plate might be used to reduce the fracture at the proximal femur zone, particularly intracapsular and capsular sections. The main concept of fracture fixation using these implants is to stabilize one to three screws from the lateral aspect of the trochanteric through the femoral head zones. The screws are normally cannulated for precise and effective insertion through the narrow cross-section of the femoral neck; their diameter is more than 5 mm. Because of the long length of these screws, stabilized bone fragments at the proximal femoral head may not be stable under physiologic loading conditions at the hip joint. Because of rotational motion of the femoral head in various axes or planes at the hip joint, multiple rotational moments and stresses tend to move the bone fragment at the femoral head and the integrity of inserted screws with trabecular bone may not be sufficiently high to maintain stability until fracture healing is complete. Therefore, the bone fragment might be dislocated, as illustrated in Figs. 5.3—5.5. The threads of screws may enhance their purchase through the intended alignment while establishing mechanical integrity at the bone—screw interface [14]. However, the sharp edges of the threads may cause stress concentration zones under compressive loading conditions, from which bone suppression at the interface of the screw and bone is induced. Such an issue could be reduced by using multiple screws with a lower diameter, by which compressive stress is distributed through multiple contact surfaces and the risk for bone suppression at the bone—screw interface is reduced. However, such an issue may be inevitable, and fixation of femoral head bone fragments needs to be managed clinically considering patients with low—mineral density bones and a high range of BMI.

5.4.2 Interactions of midshaft and distal portions of femur bone with trauma implants and possible failures

Lower than the proximal femur, fixation of the fracture at the diaphyseal portion of the femur is also challenging. Both plating and nailing fixations might be used at this zone to stabilize bone fragments in anatomic position. The anatomy of the femur at the diaphyseal zone is not completely straight from the lateral view or in the sagittal plane (Fig. 5.6). Replacement of the curved nail might not be possible; thus, the use of plating fixation with an anatomic design might be more needed compared with nailing fixation in which the nail is designed and manufactured to be straight. Midshaft femur plating fixation is managed from a lateral aspect of the femur to provide buttressing to be in equilibrium with adductor muscles at the medical aspect of the femur. The midshaft femur plate might be designed to be straight or anatomic. Straight fixation of the midshaft femur fracture in the sagittal plane may cause misalignment of contact surface conditions at the hip or knee joint (or both), from which localized stress at some parts of articulation

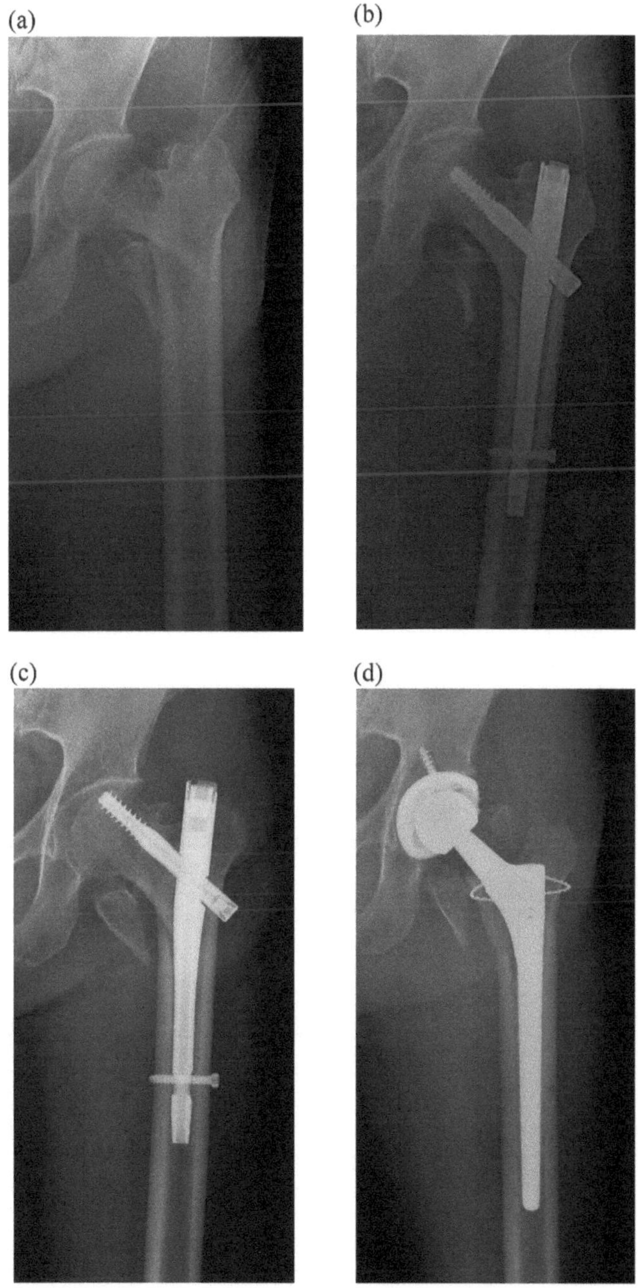

Figure 5.3 Radiographic illustration of failure in fracture fixation of femoral intertrochanteric (A–C). Fixation failed as the result of lag screw cutout and loosening of the implant system (C). Revision surgery was managed by total hip replacement. In this

surfaces at these joint might increase and cause cartilage to degenerate and wear of and ultimately progress to osteoarthritis.

A fracture in the distal third of the femur is managed with a distal femur plate [15]. This plate is implanted at the lateral aspect of the femur. The plate is designed to have a buttressing effect from the diaphyseal zone up to the metaphyseal zone. However, because of the attachment of the muscle-tendon at the lateral aspect of the metaphyseal zone, the head of the distal femur plate might not dedicate the profile of the lateral condyle of the femur in the sagittal plane. This would challenge the effective and secure buttressing of bone fragments at this zone. Thus, bone fragments close to the knee joint or those involving the articular surface (intra-articular fracture) might not be manageable with the distal femur plate. Another issue is that the anatomy of distal condyles is specific, and purchase of the screw from the posterior zone makes it anatomically impossible to extend the plate profile to capture the posterior zone. Owing to the long length of the inserted screws through the screw holes of the distal femur plate from the lateral to medical aspects, dislocation of bone fragments, particularly those at the medial side, might be a major clinical challenge in the success rate of distal femur fracture fixation. If there are bone fragments at the posterior side of the distal femur, the distal femur plate might not be a secure solution for treating fracture fixation at this zone, and other implantation methods, such as headless screw fixation, might be used as a second fixation implant to stabilize fragments that could not be captured by distal femur plating (Fig. 5.7) [16].

Failures in fixing femur bone fractures depend on the anatomy of the bone at the proximal, midshaft, and distal zones. In the proximal and distal zones, interactions between the screws and bone fragments are at risk for failure because of the long length of the screw and the ineffective load

patient (aged 74 years, female), effective integration of the lag screw and femoral head was not achieved. The sharp threads of the lag screw at the interface with trabecular bone may induce stress concentration at the sharp edge of the thread tips and destroy the bone after cyclic motions at the hip joint in short-term or long-term implantation. It is important to study the bone mineral density, hip joint muscle functionality, and the bone mineral density of patients before implantation. The interface of the bone and screw at the thread zone might need to be monitored regularly after implantation to prevent catastrophic failure of the bone–implant construct integrity. *Through the courtesy of courtesy of N. Moon, W. Shin, J. Kim, S. Woo, S. Son, K.S.- Injury, Cementless Total Hip Arthroplasty Following Failed Internal Fixation for Femoral Neck and Intertrochanteric Fractures: a Comparative Study with 3–13 Years' Follow-Up of 96, Elsevier, Undefined 2019, (n.d.). https://www.sciencedirect.com/science/article/pii/S0020138319300221. (Accessed December 8, 2020) and permission from Elsevier.*

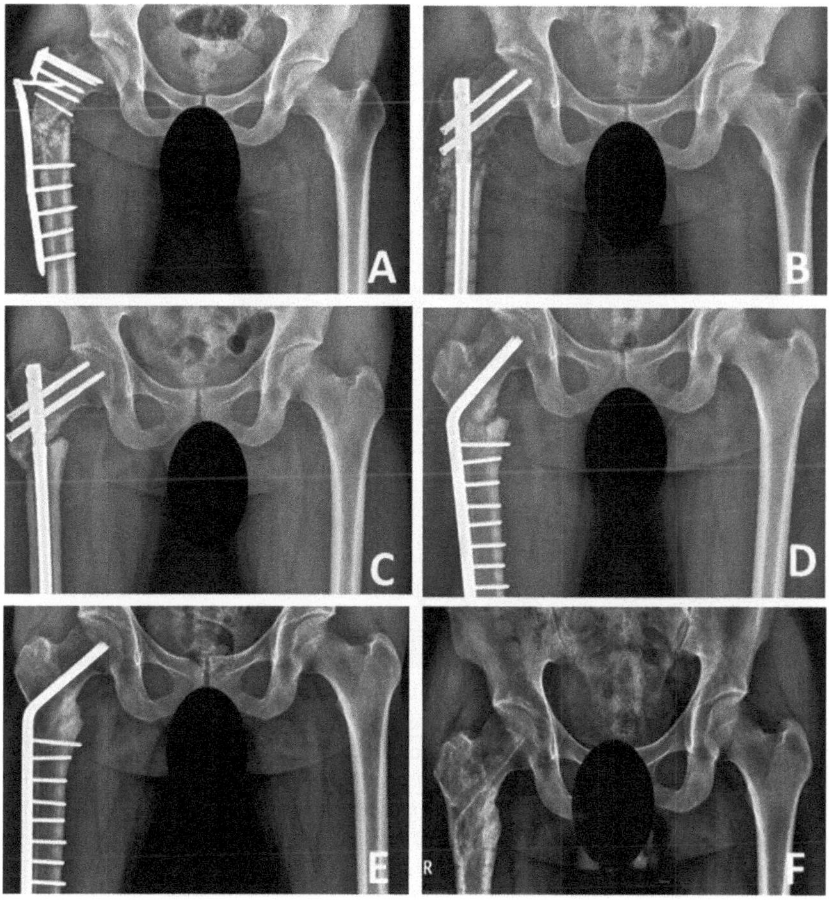

Figure 5.4 Radiographic illustration of intertrochanteric fracture failure managed with fixation implants. (A) Anatomic plate. (B, C) Intramedullary nail. (D, E) Angled blade plate. The fracture finally healed after 2 years. Valgus osteotomy treatment was managed in a 27-year-old male patient with preoperative varus deformity in the femoral head zone. The anatomic plate fractured (A) at the proximal portion where there were five screw holes and thus concentrated bending stress at the failed cross-section. Nonunion was observed using the intramedullary nail at the fracture site (C). Although the nailing system stability was achieved, interfragmentary compression at the fracture site was not obtained, and thus the biological process of fracture healing was not initiated. The solution of a combination of plating and nailing fixations was effective for the progression of fracture healing and bony fusion at the fracture gap (E). Incongruent bony fusion is observed at the medial aspect of the fracture site. *Through the courtesy of B. Min, K. Lee, J. Oh, C. Cho, J. Cho, B.K.- Injury, The Treatment Strategies for Failed Fixation of Intertrochanteric Fractures, Elsevier, Undefined 2019, (n.d.). https://www.sciencedirect.com/science/article/pii/S0020138319302888. (Accessed December 8, 2020), and permission from Elsevier.*

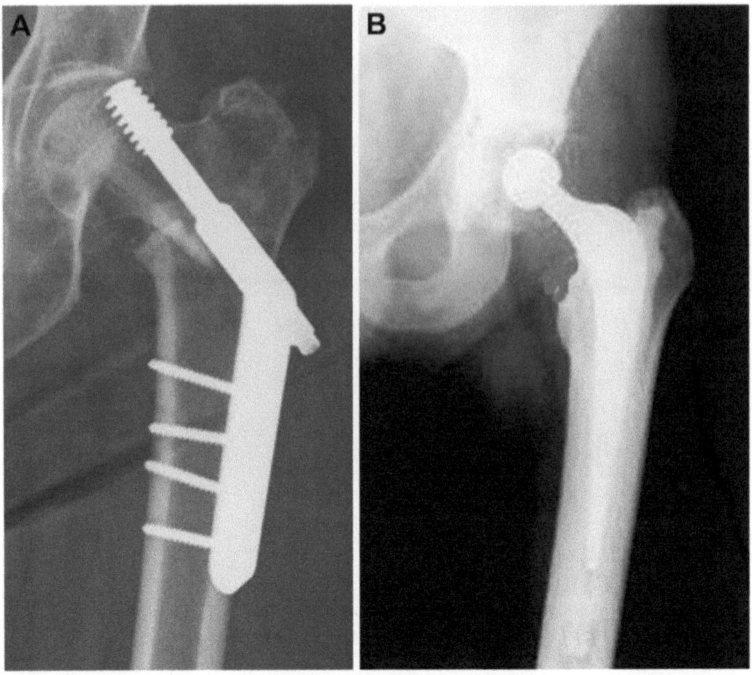

Figure 5.5 Radiographic illustration of the lag screw cut-out in a trochanteric femoral fracture in a meal 67 years old patient. The failure would be observed at the interface of the threaded portion of the lag screw and bone. *With courtesy of E. Morsi, A. Drwish, A. Saber, I.N.-T.J. of Arthroplasty, The Use of Standard Cemented Femoral Stems in Total Hip Replacement after Failed Internal Fixation of Intertrochanteric Femoral Fractures., Elsevier, Undefined 2020, (n.d.). https://www.sciencedirect.com/science/article/pii/S088354032030351X. (Accessed December 9, 2020), and permission from Elsevier.*

transferring between bone fragments and screws in multirotational motion of the bone at these zones. It is essential to maintain movement of the hip and knee joints at the lowest level during healing of fractures at the proximal and distal zones of the femur, respectively. Screws at their end do not have effective buttressing on the captured fragments far from the plate. Thus, the level of movement between gaps might be exceeded under physiologic load-bearing conditions in normal activities such as walking. Strain exceeded at fracture gaps generates higher mechanical signals than the optimal level required to form bone tissue. Separate insertion of other screws to capture the fragment far from the plate is an alternative method to reduce undesirable motion at fracture gaps. The main strategy for preventing failure of the bone—implant construct in trauma fracture fixation is to increase the rigidity

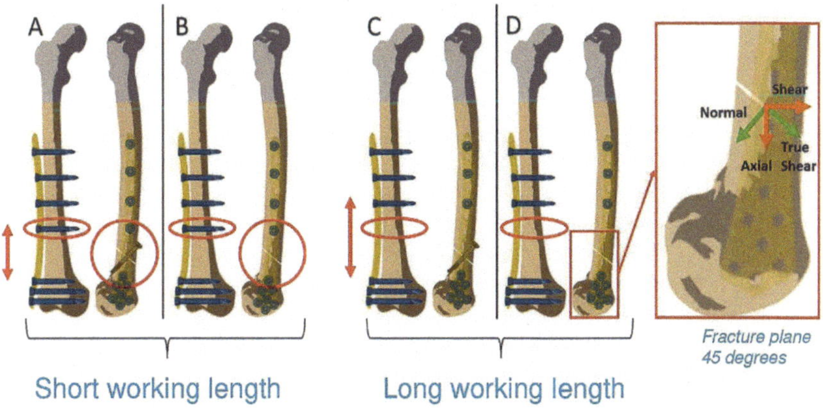

Figure 5.6 Schematic illustration of the working length concept in the plating fixation of bone fractures. The optimal working length would prevent increased stress at the midsection of the nonscrewed zone and promote the range of micromotion and strain at the fracture gap for the better induction of bone formation. Furthermore, interfragmentary screw fixation at the fracture site, particularly in the oblique fracture plane, would enhance the stability of the fracture and reduce the rate of mechanical shear stress at the fracture gap. Such a screw fixation would improve the compression of the bone fragments at the fracture section to accelerate bony fusion. *Through the courtesy of S. Märdian, W. Schmölz, K. Schaser, et al., Locking Plate Constructs Benefit from Interfragmentary Lag Screw Fixation with Decreased Shear Movements and More Predictable Fracture Gap Motion in Simple Fracture, Elsevier, Undefined 2019, (n.d.). https://www.sciencedirect.com/science/article/pii/S0268003319303420. (Accessed December 9, 2020), and permission from Elsevier.*

of bone fragments relative to each other to facilitate healing of the fracture at the gap lines. The challenge of using a high number of screws is to reduce the bone stock at the bone fragments and increase the risk for bone refracture initiating from interfaces of the screw and bone. In contrast to joint prostheses, osteoconductive surface engineering of the screws might not be effective in enhancing biological integration between the bone and implant because there is a lack of contact interfaces between the screws and fracture gaps. The release of bioactive agents through the fracture gaps, such as calcium phosphate with a low dissolution rate (e.g., tetracalcium phosphate), growth factors, morphogenetic proteins, osteoblastic expression genes, and other biological factors that have a direct effect on bone formation, would accelerate biological interaction between the bone fragments to shorten the healing period and avoid the progression of mechanical deficiency at the interface of the bone fragments and screws in long-term implantation. However, this idea has not been yet investigated,

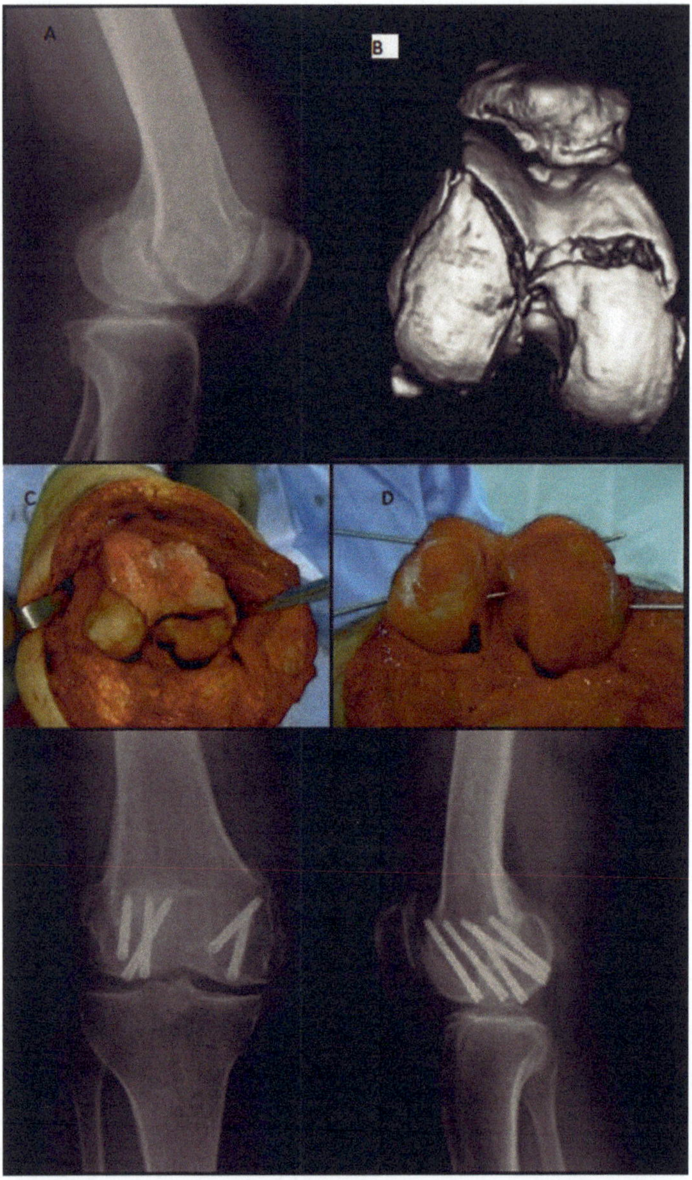

Figure 5.7 Radiographic illustration of distal femur fracture fixation with a posterior bone fragmentary fracture pattern (A-D). The use of a headless screw for this fracture management would be beneficial to avoid impingement of the screw head with the articular surface of the knee joint (E). *Through the courtesy of V. Maheshwari, S. Sharma, D. Goyal, et al., Clinical Experience with Management of Hoffa Fractures Using Headless Compression Screw and Headed Screw, Elsevier, Undefined 2019, (n.d.). https://www.sciencedirect.com/science/article/pii/S0976566218304338. (Accessed February 16, 2021), and permission from Elsevier.*

and it might be valuable for consideration as a large project by well-established global institutes. The lead author has provided the concept of screws made of CFRPEEK and magnesium alloy (Mg). The Mg part could be embedded with bioactive agents to be released through the fracture gaps. With porous structuring of the Mg layer and flow of the blood through the pores, the degradation rate of the Mg material at the fracture gap would be higher than other parts of the screw in contact with the trabecular or cortical bone. The degradation rate of the magnesium alloy would be increased where the flow rate of the blood is greater.

5.4.3 Interactions of proximal portion of tibia bone with trauma implants and possible failures

The fixation of tibia fractures is another type of fracture management in the lower extremities [17—19]. The proximal third zone is the most affected area of the tibia for fracture management. Similar to the distal femur zone, the proximal tibial plate is fixed from the lateral aspect where the length of inserted screws are long from the lateral to the medial side of the metaphyseal zone of the proximal third. Bone fragment dislocation or suppression through the threads of the screws are reported in fracture fixation at the metaphyseal zone of the proximal tibia. Converse to the distal femur plate, the proximal tibial plate could be designed anatomically to dedicate the profile of the lateral aspect of the proximal tibia and have a more effective buttressing effect than the distal femur plate. Therefore, the screws' trajectories could be managed to cover most of the volume of the metaphyseal zone of the proximal tibia, from which bone fragments could be captured by the screws. As discussed for distal and proximal femur plating fixation, the most medial bone fragments might not be effectively stabilized by the purchase of one or multiple screws from the lateral side because of the lack of a buttressing effect of the screws to such fragments far from the lateral plate. However, there is a chance of plating fixation from the medial side in managing proximal tibia fractures from which buttressing of bone fragments at the medial side is facilitated. The use of headless screws is alternative fixation implant strategy to stabilize medial fragments at the metaphyseal zone of the proximal tibia. Both medial plating and screw fixation could be used as an auxiliary implantation solution in conjunction with lateral fixation for multiply fragmented bone fractures at the proximal tibia. Another issue that might be serious in fracture management of the tibia at the proximal third is fracture at the periprosthetic bone after total knee replacement. As illustrated in Fig. 5.8, the progression of bone resorption at the periprosthetic

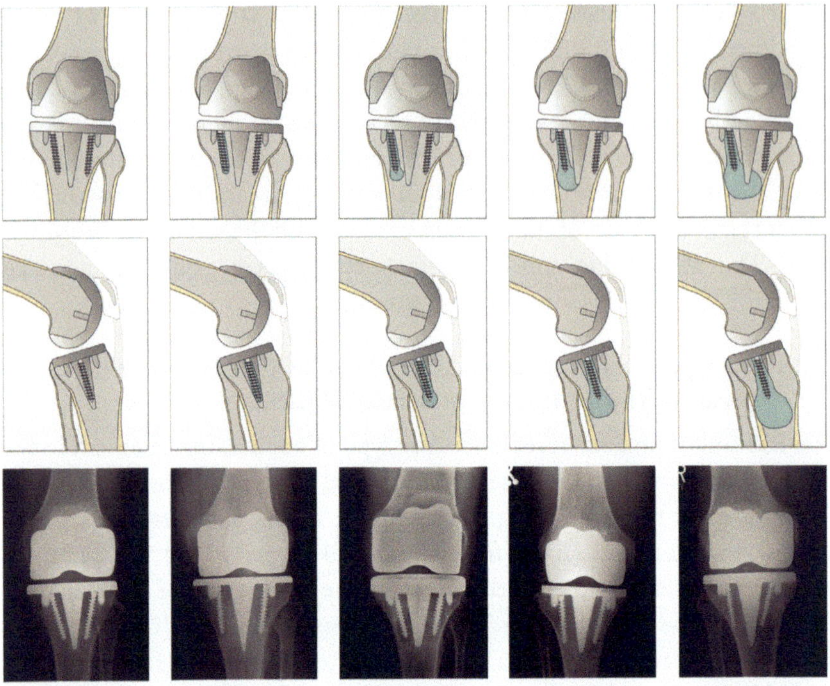

Figure 5.8 Schematic and radiographic illustration of resorption at the periprosthetic bone of screws used to replace the tibial tray prosthesis. The zone between the tray stem and screws were not be loaded effectively. The bone has begun to lose mineral density and is fractured at the affected zone. *Through the courtesy of M. Klutzny, G. Singh, R. Hameister, et al., Undefined 2019, Screw Track Osteolysis in the Cementless Total Knee Replacement Design, Elsevier, Undefined 2019, (n.d.). https://www.sciencedirect.com/science/article/pii/S088354031930004X. (Accessed December 8, 2020), and permission from Elsevier.*

bone near the stem portion of the tibial tray prosthesis causes the bone matrix to weaken at this zone. The risk for bone fracture at the affected zone is thus increased and lateral, medial, or combined plating fixation is required to stabilize the periprosthetic fracture (Fig. 5.9).

5.4.4 Interactions of midshaft and distal portions of tibial bone with trauma implants and possible failures

The tibia might be fractured in the midshaft zone where the bone is structured with a thick cortical layer. Plating and nailing fixation are alternative methods for fracture management at this zone. Compared with the femur, tibia is a straight bone in either the frontal or sagittal planes. Thus, a

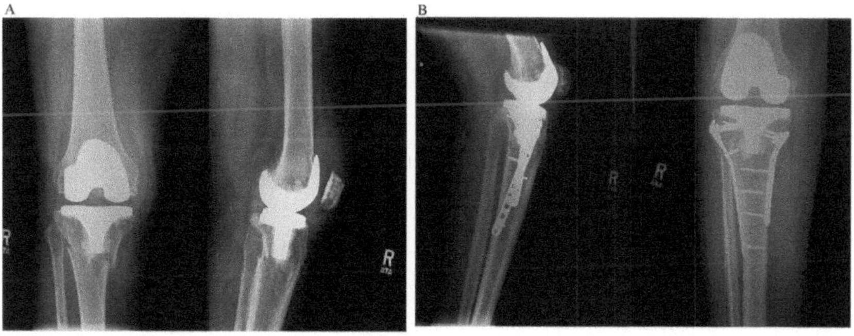

Figure 5.9 Radiographic illustration of fracture of the periprosthetic bone of the tibial tray prosthesis. When a short tibial tray is used in total knee arthroplasty, the stem portion of the prosthesis engages with the metaphyseal zone of the proximal tibia. Bone tissue surrounding the stem is not loaded owing to stress shielding at the lateral interface of the metallic stem with the periprosthetic bone. As a result, bone resorption occurs at the interface and may propagate through the diaphyseal zone and thus cause bone fracture (A). This case needs to be treated by plating fixation. To buttress the bone fragments better in an oblique fracture line, hybrid lateromedial plating fixation is beneficial (B). *Through the courtesy of M. Morwood, S. Gebhart, N. Zamith, H.M.- Injury, Outcomes of Fixation for Periprosthetic Tibia Fractures Around and below Total Knee Arthroplasty, Elsevier, Undefined 2019, (n.d.). https://www.sciencedirect.com/science/article/pii/S0020138319301111. (Accessed December 8, 2020), and permission from Elsevier.*

intramedullary nail could be used through the diaphyseal canal of the tibia. In plating fixation, the adequate plate length, optimum use of locking screws, optimal working length, and plate thickness need to be considered to manage fractures at the midshaft of the tibia. Ineffective technical forethought may cause the bone–implant construct to fail, as illustrated in Fig. 5.10. In the distal part of the tibia, various plating fixations might be used for fracture fixation at this zone. Conventional plating fixation is managed from the medial aspect of the tibia at the distal zone. However, in a fracture with articular involvement or a fracture line close to the joint, anterolateral plating fixation is used as a more effective plating method. Fixation of a fibula fracture at the distal zone is managed by plating to preserve the physiologic role of talofibular ligaments. Using a syndesmotic screw to enhance the stability of the distal fibula fracture fixation is effective. In this technique, one or two long screws are purchased from the lateral aspect of the fibula through the medial aspect of the tibia [20,21]. Syndesmotic screws enhance the stability of the distal fibula fracture fixation, but increased localized strain at this screw may cause resorption at the

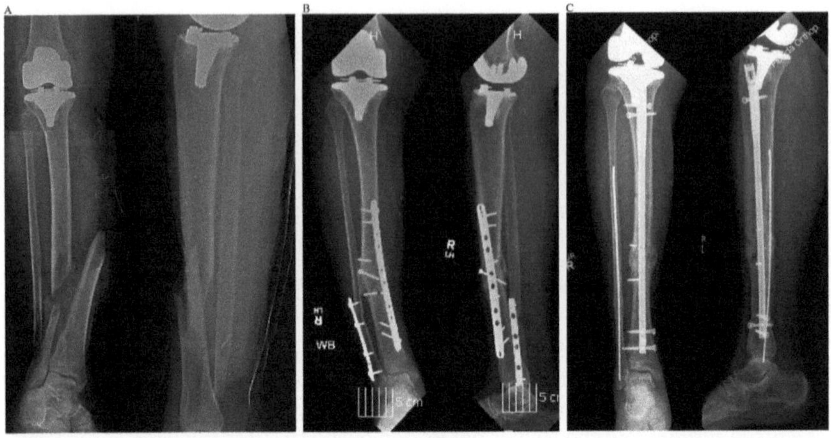

Figure 5.10 Radiographic illustration of bone fracture fixation at the diaphyseal zone of the tibia and fibula. The fracture occurred at the midshaft zone (A), which was treated with plating fixation (B). The failed plating fixation was then treated with intramedullary nailing fixation (C). Failure of plating fixation occurred owing to the ineffective rigidity of the bone–implant construct. An optimal number of locking screws would enhance the bending strength of the bone–implant construct against bending stress. *Through the courtesy of M. Morwood, S. Gebhart, N. Zamith, H.M.- Injury, Outcomes of Fixation for Periprosthetic Tibia Fractures Around and below Total Knee Arthroplasty, Elsevier, Undefined 2019, (n.d.). https://www.sciencedirect.com/science/article/pii/S0020138319301111. (Accessed December 8, 2020), and permission from Elsevier.*

periprosthetic bone to progress at its interface with the tibia (Fig. 5.11) [21], or even screw fracture (Fig. 5.12) [22]. An Mg alloy screw is effective for fixing a lateral malleolus fracture [23] (the effectiveness of the Mg alloy screw in managing bone fractures is discussed in Section 5.4.5).

5.4.5 Interactions of small bones with trauma implants and possible failures

Small foot bones can be fractured, and plating and screw fixation are applicable for fracture management of these bones. The calcaneus, metatarsal, phalanges, talus, navicular, cuboid, and cuneiform structure the foot; fractures might be occurred in all of these. Owing to the use of short screws in fixing fractures of the small bones of the foot, dislocation of bone fragments does not carry a high risk compared with that observed in fixing bone fracture in long bones and using long screws in the metaphyseal zone. Screw fixation in small bones is managed as interfragmentary fixation, in which the screw is obliquely purchased through the bone [24]. The screw that is purchased through the bone fragments in small bones has adequate compression to fix the bone fragments in anatomic position. Thus, excessive

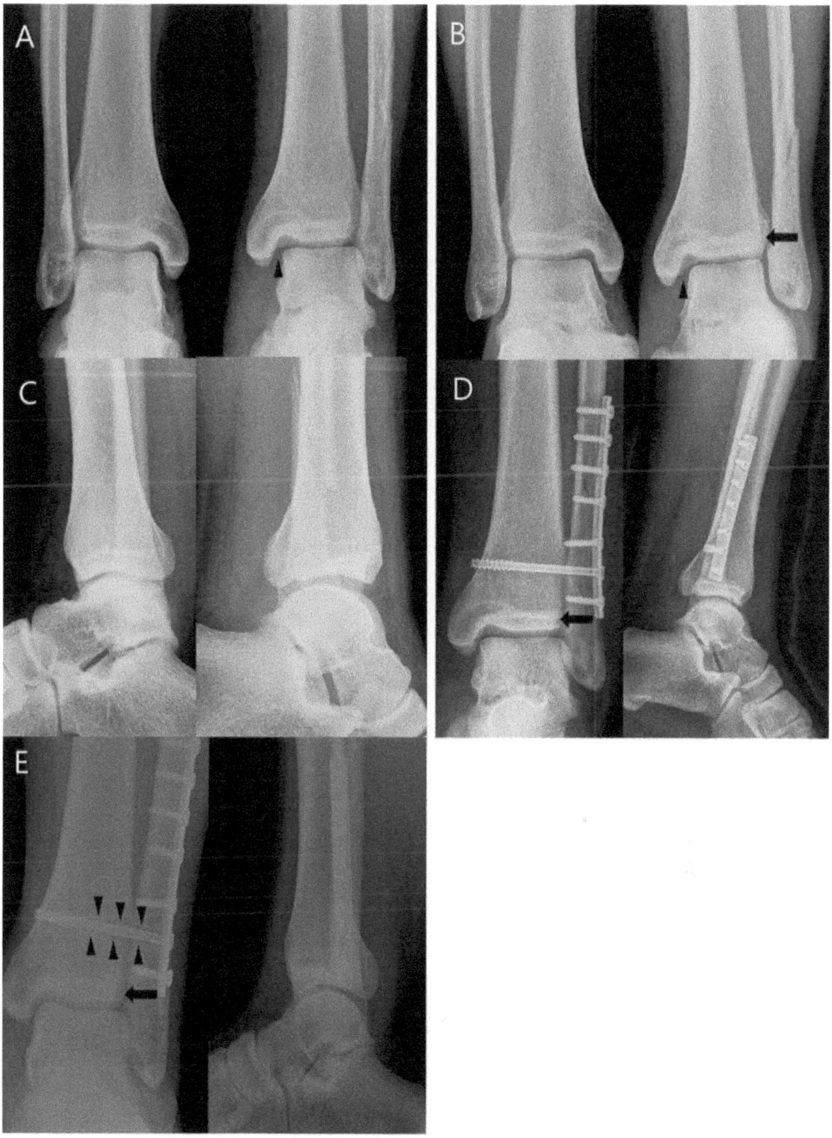

Figure 5.11 Radiographic illustration of fixation of the fibula fracture (A–E). (E) Loosening of the unthreaded portion of a partially threaded syndesmotic screw (*arrowheads*). Relative motion between the tibia and fibula causes stress to concentrate at the interface of the syndesmotic screw and bone. Because of the lack of a smooth surface at the unthreaded screw (shank) portion, bone–implant integration is not achieved and bone resorption at this zone occurs. A fully threaded screw is advantageous in syndesmotic fixation of the tibial and fibula. *Through the courtesy of S. Lee, S. Moon, M. Park, B. Jo, et al., Syndesmosis Fixation in Unstable Ankle Fractures Using a Partially Threaded 5.0-mm Cannulated Screw, Elsevier, Undefined 2018, (n.d.). https://www.sciencedirect.com/science/article/pii/S1067251617307330. (Accessed December 9, 2020), and permission from Elsevier.*

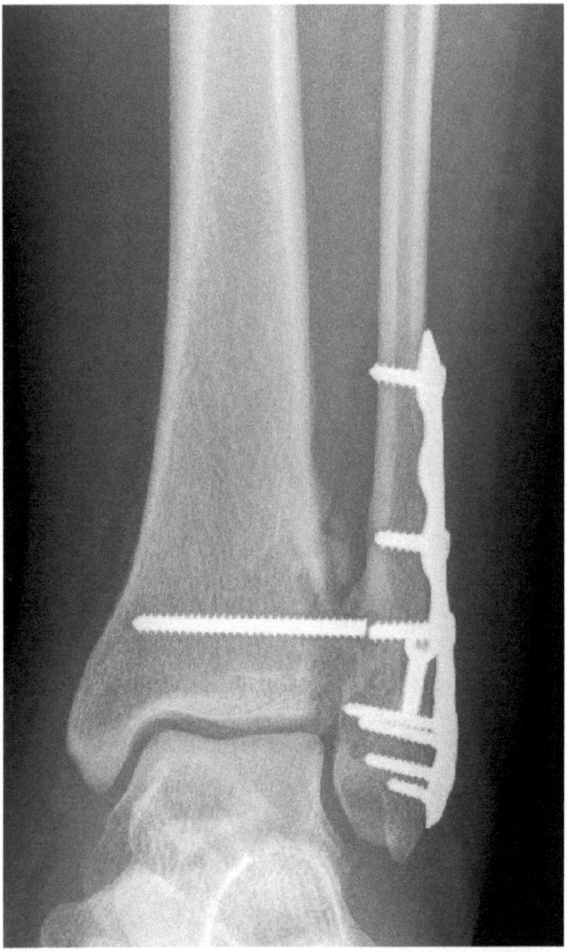

Figure 5.12 Radiographic illustration of fixation of the fibula fracture. In this case, a fully threaded syndesmotic screw was used, but it was fractured. The fracture occurred as the result of concentrated compressive stress at the inlet section of the screw to the tibia. The body weight compressive load was transferred at the interface of the screw and tibia, where the bone tends to bend the screw. Thus, the screw acts as a single cantilever beam where maximum stress concentrates at the screw inlet cross-section where the fracture occurred. Using two syndesmotic screws is beneficial to distribute stress through two cross-sections and reduce stress. Likewise, it is advantageous to design the short inlet portion of the screw smoothly without a thread to reduce concentrated stress, because threads have sharp edges. Furthermore, a syndesmotic screw with a higher diameter has greater bending strength. *Through the courtesy of M.D. Riedel, J. Briceno, C.P. Miller, J.Y. Kwon, Technical tip: removal of a broken tri-cortical syndesmotic screw using a "perfect circle" technique, Injury 49 (2018) 877–880. https://doi.org/10.1016/j.injury.2018.02.022, and permission from Elsevier.*

strain on bone fragments is prevented to maintain relative compressive force among the small fragments. In order to enhance the fracture management of the small bone, using of biodegradable screws made of magnesium (Mg) alloy (e.g., magnesium alloy composed of magnesium, yttrium, rare-earth, zirconium elements) have been introduced and commercialized since 2013 [23,25]. The use of this type of screw is interesting and important because of its mechanical and biological effects in healing bone fractures at the fracture gap (Fig. 5.13). Because of the osteoconductive advantages of

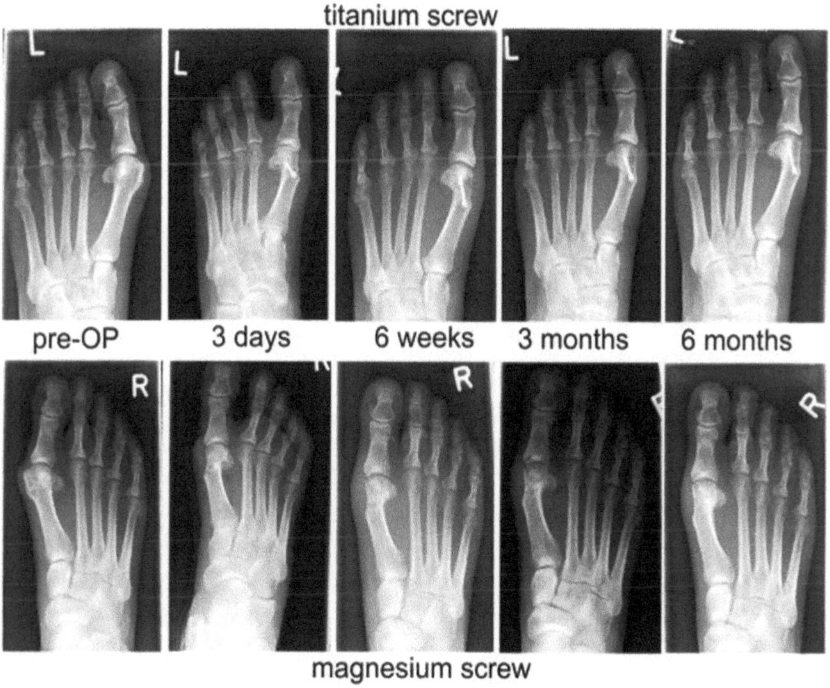

Figure 5.13 Radiographic illustration of treatment of bone deformity at the first metatarsal using *(top images)* titanium alloy; (*bottom images*) magnesium alloy. Better bone formation at the osteotomy zone is observed in radiograph images of fragment fixation using the magnesium alloy screw. The radiolucent zone at the right aspect of the titanium screw result from stress shielding at the periprosthetic bone. The screw at the treated zone causes consistent stress shielding, whereas degradation of the magnesium alloy causes physiologic loading at the fixation site and strengthens the bone tissue. *Through the courtesy of H. Windhagen, K. Radtke, A. Weizbauer, J. Diekmann, Y. Noll, U. Kreimeyer, R. Schavan, C. Stukenborg-Colsman, H. Waizy, Biodegradable magnesium-based screw clinically equivalent to titanium screw in hallux valgus surgery: short term results of the first prospective, randomized, controlled clinical pilot study, Biomed. Eng. Online 12 (2013). https://doi.org/10.1186/1475-925X-12-62, and permission from BioMed Central Ltd.*

Mg, the activity of osteogenic cells on its surface is promoted, and gap healing is accelerated adjacent to the Mg alloy screw, which progresses through the gap surface. This commercialized Mg alloy screw was reported to degrade completely in animal studies. Use of the Mg alloy screw was reported in bone loss at the periprosthetic bone [23]. This occurred because of the release of hydrogen gas upon the degradation of Mg alloy, which interfered with bone remodeling. This effect was gradually removed by reforming the bone at the affected zone. Plating fixation in small bones is crucial to provide buttressing from an adequate aspect of the bone. For instance, to fix a fracture or correct the first metatarsal, plantar or medial plating fixation was introduced. Biomechanically, medial plating has greater mechanical stiffness, and thus a lower range of strain at the fracture gap is achievable, which is a better fixation solution for treating of bone deformity. In bone deformity correction, bone osteotomy is created to realign bone fragments in anatomic position [26].

5.5 Interactions of upper extremities bones with trauma implants

5.5.1 Interactions proximal portion of humerus bone with trauma implants and possible failures

Failure of the bone in interaction with trauma implants in the proximal third of the humerus might be serious [27,28]. Plating and nailing fixation are used in this zone. Plating fixation is managed from the lateral aspect at the tuberosity section and screws are purchased through the humeral head to stabilize bone fragments. The nail may be chosen to be long or short for implantation through the intramedullary canal in the whole or proximal third of the humerus, respectively. In the shoulder joint, the activity of rotator cuff and deltoid muscles has a significant effect on the bone–implant construct, as reviewed in Chapter 3. Fracture and loosening at the bone–implant construct occur at the shoulder joint eventually after implantation. The risk for screw cutout and perforation to the glenohumeral joint is a serious complications occurs in plating fixation of proximal humerus fractures [29,30]. The integrity of bone fragments and inserted screws might not high enough to prevent overloading from occurring at the interface of the bone and screws. Abduction force applied by the deltoid muscles and the capture of greater tuberosity by the rotator cuff muscles generates a bending coupling moment at the fracture gap section and dissociates bone fragments at the humeral head. Because of the role of resistance of screws in preventing the dissociation of bone fragments, the localized increase in shear stress is generated at the interface of the screw and bone. The generation of such excessive loading conditions in the form of cyclic loading in long-term implantation gradual suppresses the bone at the bone–screw interface. Bone resorption occurs and the bone matrix degenerates, as illustrated in Figs. 5.14 and 5.15 (the technical concept of a screw cutout is shown in Fig. 5.16 [31]). In this condition, screws perforate the glenohumeral joint and bone fragments might dislocate while the plate–screw construct remains intact and acts as a stiff component that damages the less stiff trabecular bone matrix at the humeral head. Even after the fracture has healed, because of the existence of the proximal humerus plating system in the proximal third of the humerus, the formed bone tissue at the fracture gap may not be strong enough and the failure at the interface of the screw and bone may occur. Failure of proximal humerus fracture fixation needs to be studied for alternative solutions to manage fractures at the humeral head zone. Although the final solution is to use anatomic or reverse total shoulder arthroplasty (aTHA and rTHA, respectively), the best solution is to maintain bone stock at the humeral head without degeneration or damage through

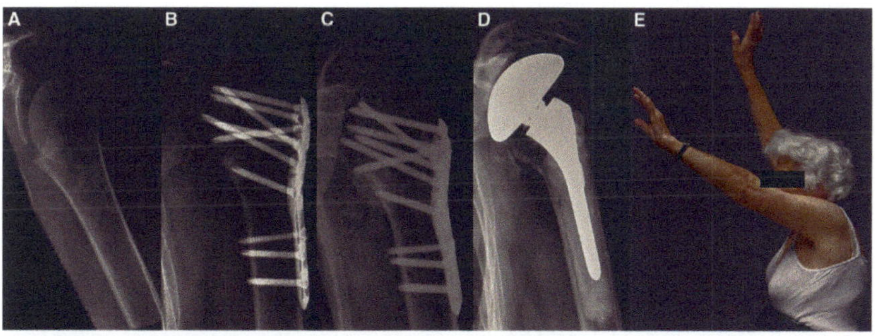

Figure 5.14 Radiographic illustration of bone–implant construct at proximal humerus plating fracture fixation (A–C). Because of the progression of bone resorption at the interfaces of the screws and trabecular bone tissue at the humeral head, the integrity of the screws and bone is affected. Thus, the volume of bone stock is reduced and the screws have shifted through the glenohumeral joint. As a result of significant bone loss at the humeral head, the affected bone was removed and reverse total shoulder replacement was used for treatment in this failed case (D, E). *Through the courtesy of B. Jost, C. Spross, H. Grehn, C.G.-J. of shoulder and elbow, Locking Plate Fixation of Fractures of the Proximal Humerus: Analysis of Complications, Revision Strategies and Outcome, Elsevier, Undefined 2013, (n.d.). https://www.sciencedirect.com/science/article/pii/S1058274612002716. (Accessed December 22, 2020), and permission from Elsevier.*

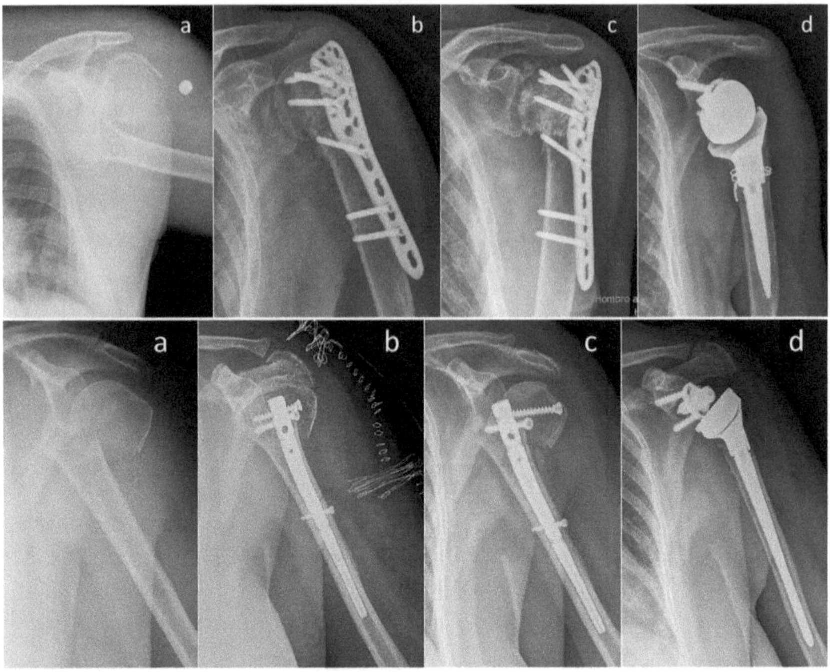

Figure 5.15 Radiographic illustration of bone–implant construct at proximal humerus plating and nailing fracture fixation (A–C). Arm limb motions at the shoulder joint generate relative motion between bone fragments at the proximal humerus. The fragments tend to dissociate from each other and create excessive shear stress at the interface of the bone and screw. In plating fixation (top images, A–C), trabecular bone at the humeral head is greatly affected. It was necessary to remove bone loss at this zone, and reverse total shoulder replacement (rTSR) was used to treat this failed case (top images). In nailing fixation (bottom images, A–C), relative motion between fragments cause the screws to loosen on the nail at the humeral head. Thus, the bone stock is not as affected as that in plating fixation, where the distal fragment and nail are dislocated. Nailing fixation was treated with rTSR (bottom image, D). *Through the courtesy of C. García-Fernández, Y. Lopiz, B. Rizo, L. Serrano-Mateo, B. Alcobía-Díaz, A. Rodríguez-González, F. Marco, Reverse Total Shoulder Arhroplasty for the Treatment of Failed Fixation in Proximal Humeral Fractures, Elsevier, (2018). https://doi.org/10.1016/j.injury.2018.06.042, and permission from Elsevier.*

fracture management techniques. (The interaction of bone with prostheses in aTHA and rTHA implantation methods are reviewed and discussed in depth in Chapter 3) Purchase using multiple screws (four to seven screws) with a diameter greater than 4 mm stabilizes bone fragments at the humeral head zone, but this occupies a considerable volume of trabecular bone, and thus overall stiffness of the bone to bear increased shear stress at its surfaces

Figure 5.16 Screw loosening and cutout inside trabecular bone tissue (A, B) under simultaneous compressive axial force and rotational torque around the axial axis. An experimental evaluation was carried out on the low-density (12.5-pcf to 0.2-g/cm^3) femoral head as the artificial bone to simulate osteoporotic bone. The loading conditions were loaded cyclically to assess the effect of physiologic motion and loading conditions at the hip joint. Because of the great mechanical mismatch between the metallic screw and bone, stress shielding at the interface of the screw and bone progressed and the bone integrity of the screw and bone was affected. Ultimately, suppression of the bone by the screw occurred. The cutout effect initiated from threads of the lag screw and extended to the unthreaded (shank) portion of the screw. *Through the courtesy of G.K. Kouvidis, M.B. Sommers, P.V. Giannoudis, P.G. Katonis, M. Bottlang, Comparison of migration behavior between single and dual lag screw implants for intertrochanteric fracture fixation, J. Orthop. Surg. Res. 4 (2009). https://doi.org/10.1186/1749-799X-4-16 and, permission from BioMed Central Ltd.*

with screws is significantly reduced. A hypothesized solution is the development of a type of screw with a high level of strength and stiffness as well as gradual degradability to return the occupied bone stock at the humeral head. The lead author proposed a concept in Chapter 16 of *Trauma Plating Systems* (Elsevier, 2017) as the Advance Healing Plating System for this purpose. In that concept, the core part of the screw is fabricated from CFRPEEK composite material and the shell part of the screw is manufactured from Mg alloy compounded with bioactive nanoparticles.

5.5.2 Interactions of proximal portion of humerus bone with trauma expandable cage and possible failures

One fixation implant developed to prevent bone resorption and destruction by the purchase of multiple screws through the humeral head is an expandable cage manufactured from nitinol alloy, a shape memory alloy with superelastic properties that can be deformed elastically without a destructive effect on the material [32,33]. As illustrated in Fig. 5.17, this implant can occupy the metaphyseal zone of the humeral head with a scaffold through which one or two screws can pass. By grafting the internal space of the cage with bioactive materials including a calcium phosphate composition, bone formation can take place through the intervals of the cage and create

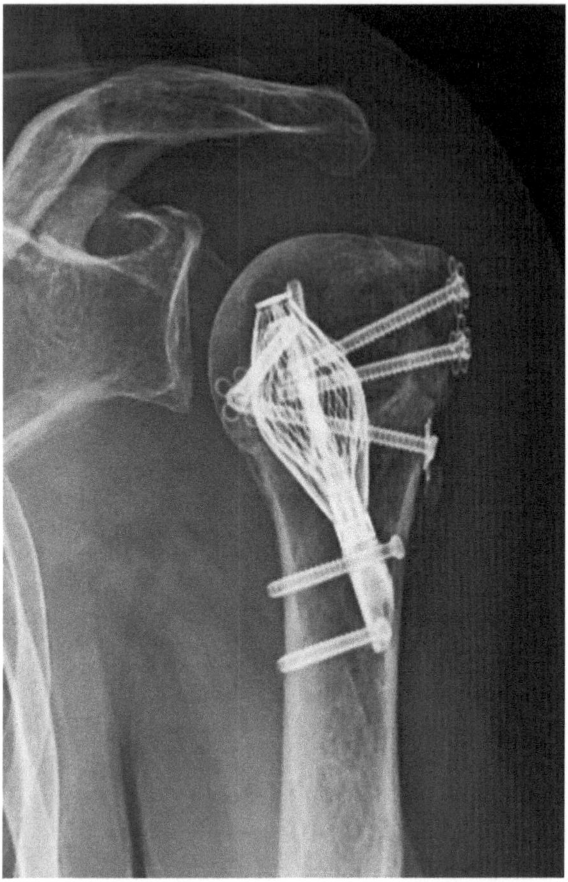

Figure 5.17 Radiographic illustration of using an expandable cage made from shape memory nitinol alloy. Because of the high risk for bone resorption in the humeral head as a result of plating or nailing fixation, employing a cage to conduct bone ingrowth through its volume would be beneficial for the higher integrity of the bone and implant at the humeral head. This would enhance the strength of the bone–implant construct from dissociation under physiologic loading and motional conditions at the shoulder joint. *Through the courtesy of J. Hudgens, J. Jang, K. Aziz, et al., Three-and 4-part Proximal Humeral Fracture Fixation with an Intramedullary Cage: 1-year Clinical and Radiographic Outcomes, Elsevier, Undefined 2019, (n.d.). https://www.sciencedirect.com/science/article/pii/S1058274619303027. (Accessed December 22, 2020), and permission from Elsevier.*

a strong bone matrix. Such a fixation method would be highly beneficial for a multifragmentary fracture pattern with three and more bone fragments at the humeral head zone, particularly those with fragments at the medial column close to the glenohumeral joint. Initially, the expanding function of the cage may destroy the bone. However, during short-term implantation with greatly limited motion at the shoulder joint, the formation of the cage internal bone could induce bone formation at the gap surface of the bone fragments. However, such a solution would be the final answer before using reverse total shoulder arthroplasty as an attempt to preserve bone stock at the humeral head. The concept of using an expandable cage through the internal volume of the humeral head was derived from intramedullary fibular grafting of the medial column in the treatment of humeral fractures. This cage might be used for a case in which the risk for bone destruction or loss at the humeral head zone is high.

5.5.3 Interactions of distal portion of humerus bone with trauma implants and possible failures

The fixation of fractures at the distal third of humerus is normally managed by plating [34–37]. Two plates from the lateral and medical aspects are implanted to stabilize the fracture at the distal humerus. The anatomy of the distal humerus is specific, where the width of the metaphyseal zone suddenly increases from the diaphyseal zone and the height of the metaphyseal zone is low. Such shaping of the humerus at the distal portion necessitates a highly curved anatomic plate at the medial aspect; a dorsolateral plate is designed with lower curving. The insertion of multiple screws from the medial plate might not be possible because of the short height of the distal metaphyseal bone. Even the use a long screw might not achievable from the medial aspect owing to the presence of the olecranon fossa. Thus, it might be necessary to use another plate as the dorsolateral plate to capture the bone fragment at the column. Hybrid buttressing of bone fragments from two sides and capturing of bone fragments are crucial to fix fractures at this zone. The conditions for managing fractures of the distal humerus fractures are at high risk for fragment dislocation or screw loosening owing to increased shear stress at the interface of the bone fragment and screws, particularly when a short screw is used to stabilize plates through the bone (Fig. 5.18) [38], from which a fracture nonunion or delayed union results (Figs. 5.19 and 5.20) [35,36]. It is crucial to avoid the unnecessary purchase of the lengthy screws through the metaphyseal zone to preserve a high range of bone stock and prevent screw cutout and loosening as the result of bone resorption at the interface of the bone and screw. The strong biological integration of the screw and bone increase the success rate of distal humerus fracture management.

5.5.4 Interactions of distal portion of radius bone with trauma implants and possible failures

One of the most common types of bone trauma injury is fracture of the radius at the distal metaphyseal zone [39—45]. Because of the large number of fractures at this zone, various types of plating fixation have been developed. The design of the plate is general, anatomic, or fragment-specific to increase of the surgeon's alternatives for various fracture types and fragmentary conditions. Plates such as the volar rim plate, double-column plate, radial styloid plate, dorsal plate, minidorsal plate, and single-column plate

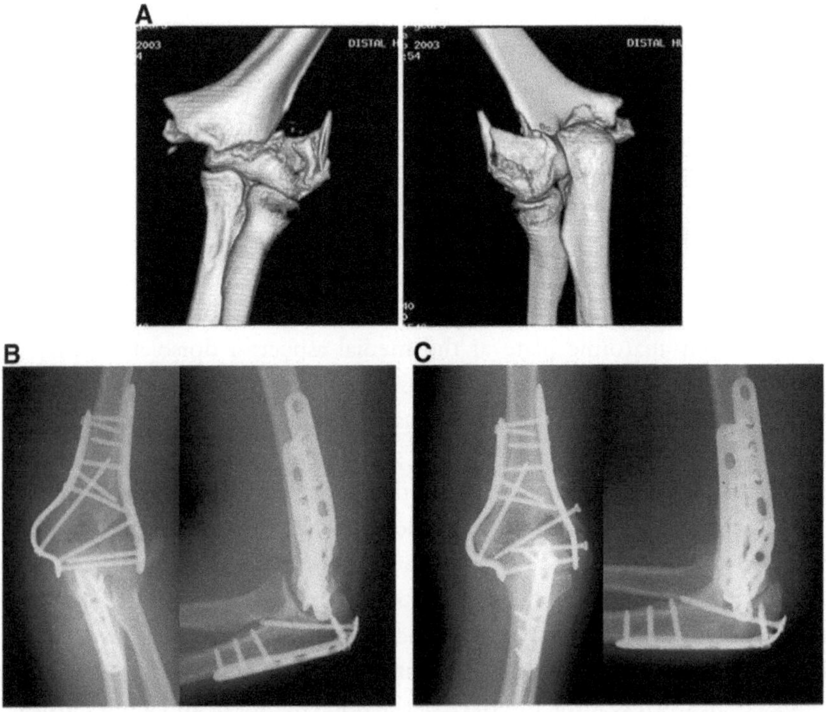

Figure 5.18 Radiographic illustration of distal humerus fracture fixation (A—C). Screw loosening and dislocation are observed at medial plating fixation. Poor integrity of the screw and bone cause the progression of bone resorption at the bone—screw interface. Because of the poor stabilization of bone fragments, there is relative motion at the medial bone—implant plating fixation, dislocation of the screw out of the plate, and irritation of the soft tissue. *Through the courtesy of J. Simone, P. Streubel, et al., Low Transcondylar Fractures of the Distal Humerus: Results of Open Reduction and Internal Fixation, Elsevier, Undefined 2014, (n.d.). https://www.sciencedirect.com/science/article/pii/S1058274613006241. (Accessed December 23, 2020), and permission from Elsevier.*

are some distal radius implants used to fix simple to multifragmentary fractures. The most challenging distal radius fracture fixation is when there is a small fragment at the radial styloid, lunate facet, scaphoid facet, sigmoid notch, or dorsal side zones of the distal radius. The purchase of screws through such a small bone fragment might be insufficient to capture the fragment and prevent its dislocation in soon or long after postimplantation. As illustrated in Fig. 5.21, a small bone fragment at the lunate facet zone was fixed with a double-column distal radius plate but bone healing was not achieved at the fracture gap and it was dislocated. Although the plate provided adequate buttressing to hold the fracture, the screws purchased through this fragment could not reduce it effectively enough to facilitate

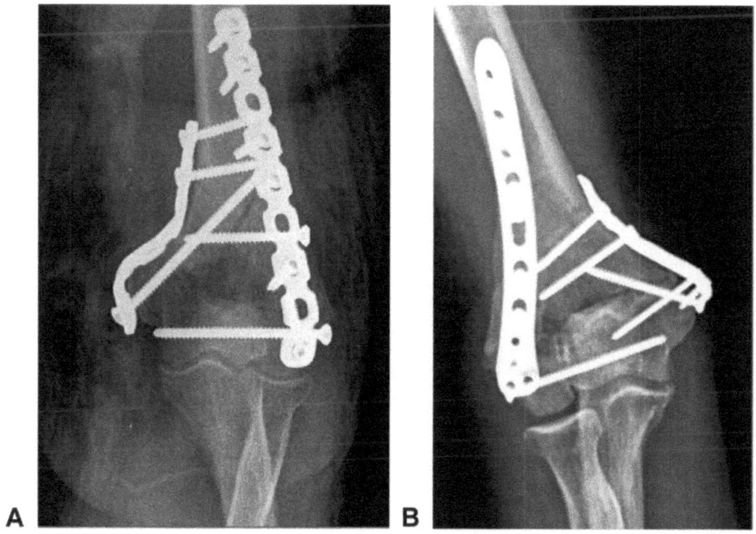

Figure 5.19 Radiographic illustration of failure of distal humerus fracture fixation as (A) loosening of the implant; and (B) nonunion of the fracture. The distal humerus has a complex geometry for placing the plate and screws to capture bone fragments. Thus, the risk for ineffective stabilizing bone fragments is high. The failure of stabilization after surgery causes screws to loosen. Excessive relative motion between bone fragments generates high strain at the fracture gap and creates overloading at the interfaces of the screw and bone and ultimately the occurrence of screw loosening. The failure of stabilization might occur during surgery, when bone fragments are fixed at an excessive gap distance. This would cause nonhealing of the fracture at the gap surfaces. *Through the courtesy of F. Claessen, Y. Braun, et al., Plate and Screw Fixation of Bicolumnar Distal Humerus Fractures: Factors Associated with Loosening or Breakage of Implants or Nonunion, Elsevier, Undefined 2015, (n.d.). https://www.sciencedirect.com/science/article/pii/S0363502315009442. (Accessed December 8, 2020), and permission from Elsevier.*

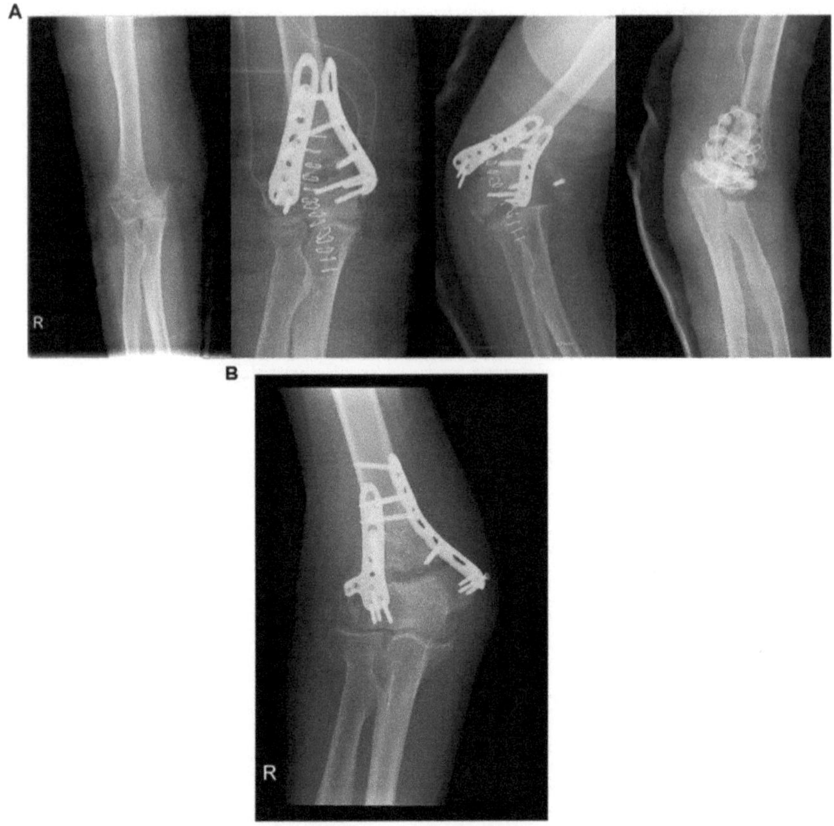

Figure 5.20 (A) Radiographic illustration of distal humerus fracture fixation. (B) Screw loosening at medial plating fixation. In this case, delayed union was reported and union was observed at 8 weeks after surgery. *Through the courtesy of Y. Sur, Y. Kim, H.P.-J. International, Results of Plate Fixation for Transcondylar Fracture of the Distal Humerus: a Rare Pattern of Fractures, Elsevier, Undefined 2020, (n.d.). https://www.sciencedirect.com/science/article/pii/S266663832030044X. (Accessed December 9, 2020), and permission from Elsevier.*

bone formation at the gap surface. Such a condition in distal radius plating fixation is evidence for the development of screws with biological advantages such as what those observed in Mg alloy biodegradable screws, as highlighted in Section 5.4.5 to fix the first metatarsal. Furthermore, a nitinol expandable cage (as used for humeral head fracture fixation) is used for simple fracture management of the distal radius, as illustrated in Fig. 5.22 [46].

5.6 Interactions of pelvic bone with trauma implants and possible failures

Plating fixation is effective for managing pelvic fracture compared with screw or external fixation [47–55]. Plating fixation can be managed posteriorly and anteriorly in the pelvis (various types of pelvic plating fixation are reviewed in Chapter 15 of *Trauma Plating Systems*). With the use of reconstruction plates, pelvic fractures can be stabilized with good to excellent results. Complications of fracture union are nonunion, malunion, and delayed union, particularly at the acetabular wall or sacroiliac joint zone

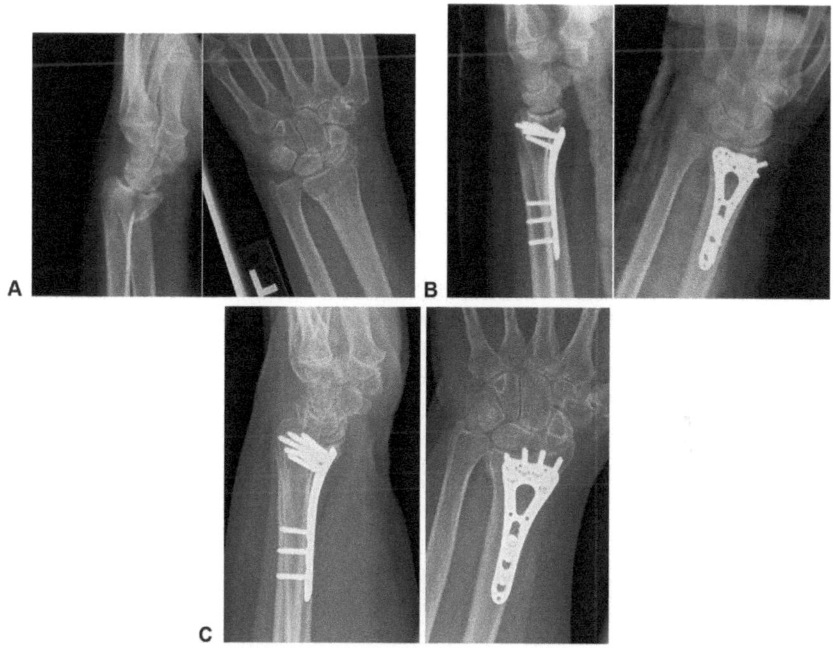

Figure 5.21 Radiographic illustration of distal radius plating fixation. Loosening and dislocation of the small bone fragment at the lunate facet zone are reported in this case. Fixation of small fragments near the articular surface of the wrist joint is challenging. Capture of a small fragment by screws might not be achieved from only the volar aspect. Specific plating implants were designed to manage small bone fragments at the distal radius, such as radial styloid plate and volar rim plate. *Through the courtesy of J. Beck, N. Harness, H.S.-T.J. of hand surgery, Volar Plate Fixation Failure for Volar Shearing Distal Radius Fractures with Small Lunate Facet Fragments, Elsevier, Undefined 2014, (n.d.). https://www.sciencedirect.com/science/article/pii/S0363502314000793. (Accessed December 23, 2020), and permission from Elsevier.*

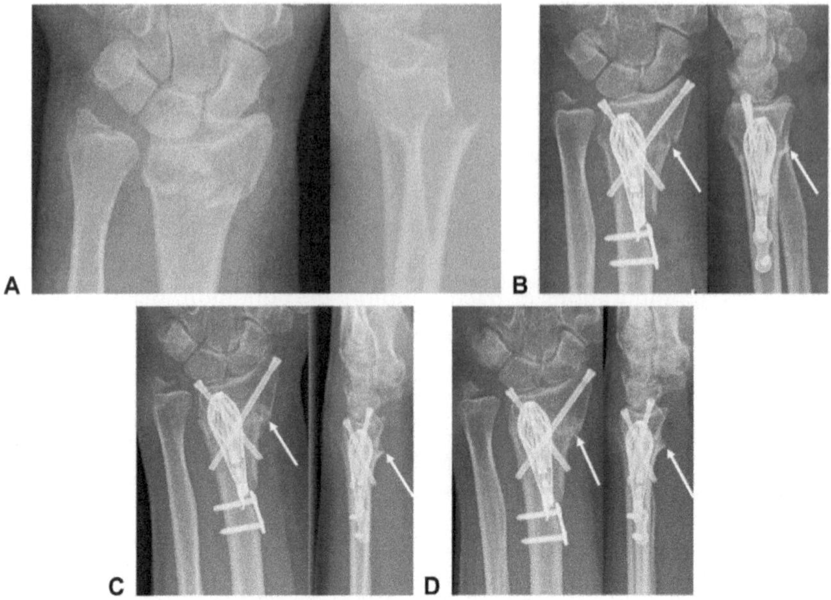

Figure 5.22 Radiographic illustration of distal radius fracture fixation by expandable cage. The cage is made of sharp memory alloy (nitinol) for elastic expansion without undesirable plastic deformation and a reduction of fatigue strength. This cage is intended for a simple fracture pattern at the distal radius zone. Fracture union was achieved, although undesirable bone trophy is observed at the healed section. Use of this implant is advantageous for severely osteoporotic bone where the integrity of screws and bone is not achievable in conventional plating fixation. *Through the courtesy of M. Strassmair, M. Jonas, et al., Distal Radial Fracture Management with an Intramedullary Cage and Fragment Fixation, Elsevier, Undefined 2016, (n.d.). https://www.sciencedirect.com/science/article/pii/S0363502316301356. (Accessed December 23, 2020), and permission from Elsevier.*

(Figs. 5.23 and 5.24) [56,57]. The most challenge type of pelvic fracture fixation is when fracture occurs at the acetabular wall and pubic symphysis (Figs. 5.23 and 5.25) [57,58]. Loosening of the screw or plate fracture is a cause of failure in managing a fracture of the pelvis. Based on the thickness of the bone at the location of the screw insertion, a short or long screw may be used. The risk for short screw loosening is higher; thus, multiple short screws in a specific zone are inadequate. It is best to use a longer plate with more inserted screws to reduce the risk for screw loosening. The shape of the pelvis is complex and various muscles are attached to or originate from the pelvis. Each muscle loads a tensile force to the pelvis, and when a fracture occurs, the equilibrium of loading conditions is affected at the fracture site.

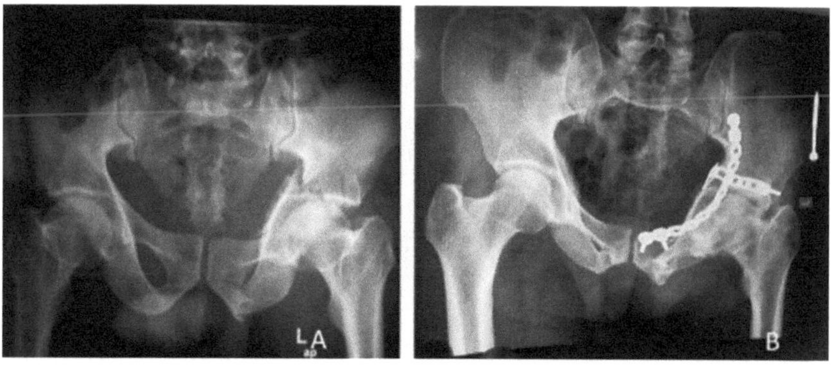

Figure 5.23 Radiographic illustration of malunion in the fixation of a fracture at the acetabular wall and hip arthritis at 6 years after surgery (A, B). It is challenging to maintain the stabilization of bone fragments at the acetabular wall without dislocation. Reduction of the fixation is affected as the result of contact loading conditions at the acetabulum cavity that are created by the body weight reaction force and motion of the femoral head. These conditions increase the risk for fragment dislocation. During excessive motion between fragments, union of the fracture is not achieved and loosening or fracture of the implant results in long-term implantation. During mild relative motion at the fracture gap, delayed union results with incongruence of the bone anatomy at the fracture site, as occurred in this case. *Through the courtesy of U. Meena, S. Tripathy, R. Sen, et al., Predictors of Postoperative Outcome for Acetabular Fractures, Elsevier, Undefined 2013, (n.d.). https://www.sciencedirect.com/science/article/pii/S1877056813002132. (Accessed December 23, 2020), and permission from Elsevier.*

Stabilization of the two sides of the bone fracture gap is managed by implanting a plate as the bridging element to facilitate bone fusion at the fracture gap. However, the length of the plate needs to be sufficiently long to increase the rigidity of the bone–implant construct. Furthermore, the plate needs to be mechanically stiff against fatigue loading conditions that are introduced to the plate as the result of excessive loading conditions at the fracture site. A limitation in managing fractures of the pelvis is that interfragmentary screw insertion might not be manageable to enhance the stability of the bone–implant construct. Using screws with a processed osteoconductive surface is advantageous for bone growth on the surface of the screw and higher shear strength of the interface layer at the contact surfaces of the screw and bone. The development of porous screws might be used in conjunction with normal screws to establish in-growth bone formation through the body of such a screw and increase the bone–implant construct stability in managing fractures of the pelvis.

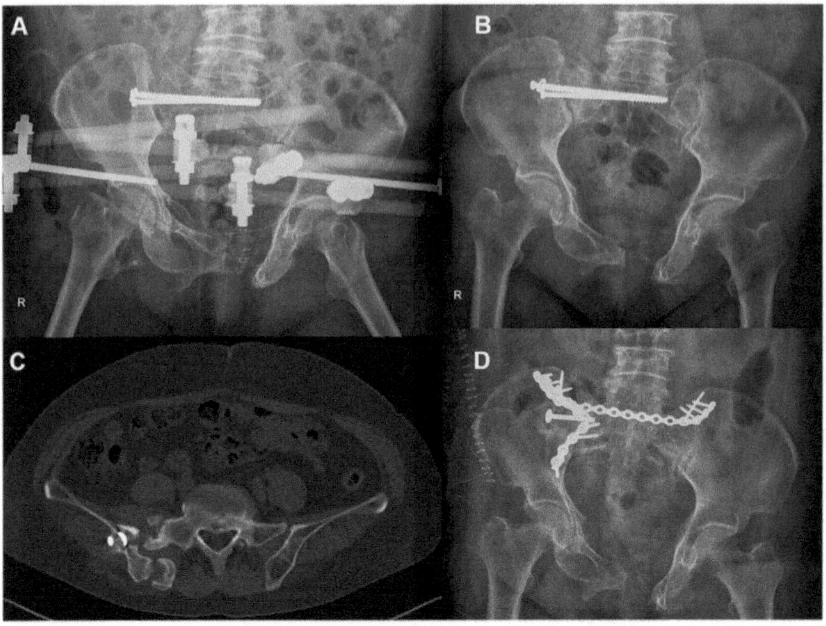

Figure 5.24 Radiographic illustration of screw loosening in the fixation of a fracture at the sacral zone (A—C). Fracture nonunion was observed 6 months after surgery. Revision surgery was carried out using of three plating fixations to fix the fracture from the posterior aspect of the pelvic ring (D). The stability and rigidity of the bone—implant construct at the sacral zone were affected as the result of physiologic loading conditions of the attached or originated muscle at the sacral and ilium zones. Insertion of one or two transplant long screws from the ilium to sacral zones is the initial surgical solution to manage such a fracture. However, the failed integrity between the bone and screw progressed owing to high shear stress generated at the bone—screw interface. Multiple plating fixations are a more reliable solution of rigid stabilization of bone fragments in pelvic and sacral bones. *Through the courtesy of courtesy of Y. Yoon, D. Ma, S. Lee, J. Oh, H.S.-A. journal of surgery, Posterior Pelvic Ring Injury of Straddle Fractures: Incidence, Fixation Methods, and Clinical Outcomes, Elsevier, Undefined 2020, (n.d.). https://www.sciencedirect.com/science/article/pii/S101595842030097X. (Accessed December 23, 2020), and permission from Elsevier.*

5.7 Interactions of clavicle bone with trauma implants and possible failures

Failure might be occurred in fixation of clavicle fractures [59—61]. Challenges similar to those in pelvic fracture fixation are present in managing fractures of the clavicle. Multiple muscles are attached to or originated from the clavicle. A fracture destroys overall mechanical equilibrium at

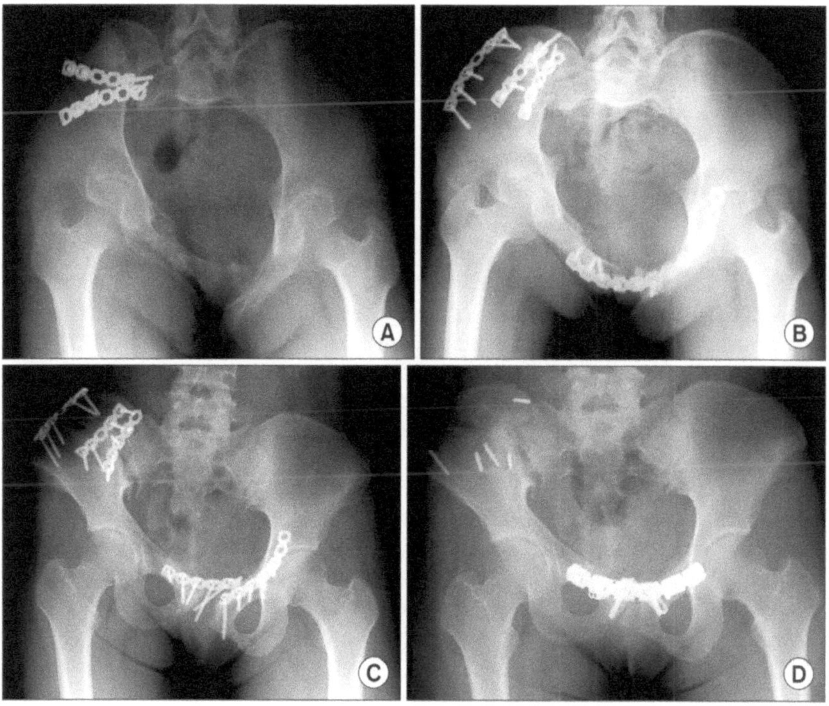

Figure 5.25 Radiographic illustration of plating fixation failure in the treatment of a hemipelvis deformity at the anterior lesion (A–D). The anterior plate was fractured 2 months after surgery (C). A thicker plate was used to stabilize the fracture. *Through the courtesy of K.J. Lee, B.W. Min, G.M. Oh, S.W. Lee, Surgical correction of pelvic malunion and nonunion, CiOS Clin. Orthop. Surg. 7 (2015) 396–401. https://doi.org/10.4055/cios.2015.7.3.396. and The Korean Orthopedic Association.*

the clavicle, which hardens effective clavicle fracture management. Loading conditions at the two sides of the fracture gap tend to increase the instability of the fracture fixation, where the high level of bending stress and shear stress are transferred through the plate and screws, respectively. As the result, screw loosening [59], plate fracture [60], or bone fragment dislocation [62] might occur (Figs. 5.26–5.28).

5.8 Summary and discussion

Based on the reviewed aspects of the interaction between bone and trauma implants and possible failures, the two concepts of bone fragment buttressing and capturing are the key requirements of managing fractures in the treatment of bone trauma injuries. Trauma implants are composed

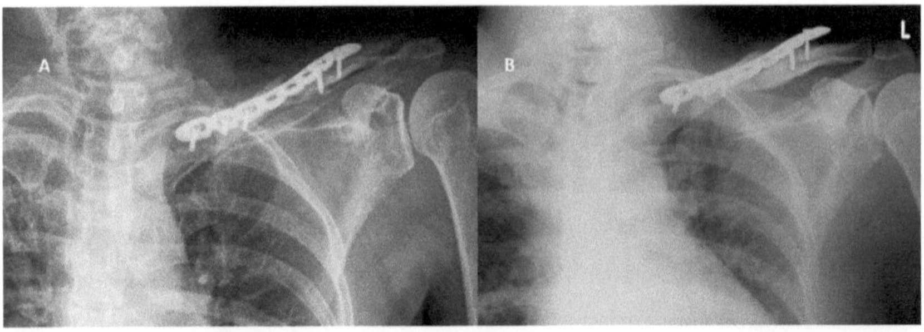

Figure 5.26 Radiographic illustration of screw loosening in fixation management of a clavicle fracture (A, B). Owing to the attachment of multiple muscles to the clavicle, the equilibrium of loading conditions for bone fragments is affected and plating fixation may not maintain relative loading conditions between bone fragments under physiologic conditions. This introduces excessive shear stress at the interface of the bone fragment and screw, and thus bone resorption at the interface and ultimately dissociation and loosening of the screw (B). *Through the courtesy of R. Kundangar, K. Singh, S. Mohanty, K.E.-J. of orthopaedics, Clinical outcome of internal fixation of middle third clavicle fractures using locking compression plate: comparison between open plating and MIPO, Luot.Ch. (n.d.), Undefined 2019. http://luot.ch/wordpress/wp-content/uploads/2019/11/ClavicleFx-open-vs-mipo.pdf. (Accessed December 8, 2020), and permission from Elsevier.*

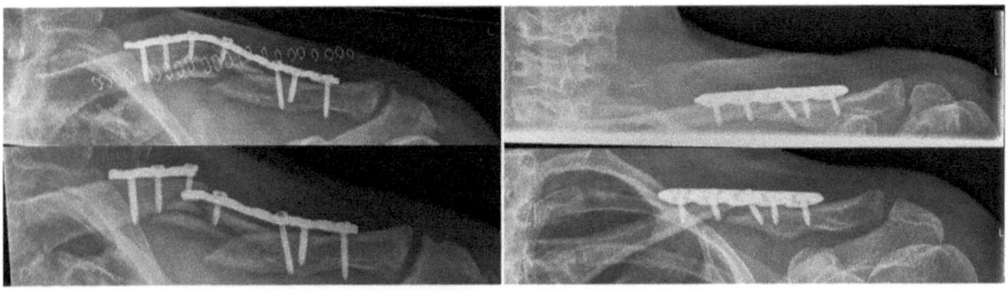

Figure 5.27 Radiographic illustration of failure in fixation of a clavicle fracture. (*Left*) Excessive loading conditions at the bone–implant construct cause a fracture of the plate at the mechanically formed section. The good purchase of screws and dissociation at the interface of the screw and bone concentrate stress at the most critical section of the plate. (*Right*) Dissociation of integrity at the interface of the screw and bone fragment cause dislocation of the bone fragment and failure of the bone–implant construct. Biomechanical evaluation of the clavicle plate experimentally or computationally is useful to strengthen the critical section of the plate. Likewise, the use of an anatomic plate with well-integrated screws and bone are good for loosening the bone–implant construct. *Through the courtesy of M.A. Meeuwis, A.F. Pull Ter Gunne, M.H.J. Verhofstad, F.H.W.M. Van Der Heijden, Construct failure after open reduction and plate fixation of displaced midshaft clavicular fractures, (n.d.). https://www.sciencedirect.com/science/article/pii/S002013831730061X. (Accessed December 8, 2020), and permission from Elsevier.*

of a main implant and screws. The main implant might be a plate, nail, or rod that is fixed through bone tissue using multiple screws. According to one view, the stability and effectiveness of bone fracture fixation are achieved by considering the mechanical aspects of the bone–implant construct. In this view, the greater strength and rigidity of the implant and a higher number of screws used are emphasized. Biological concerns are not as much addressed. Because of the high extent of physiologic loading at the fracture site, the mechanical interaction of trauma fracture fixation with the bone is important. Even the extent of loading conditions is accelerated as a result of the fragmentation of bone tissue, which complicates the load transferring mechanism at the fracture site. Thus, trauma implants are normally manufactured from metallic materials such as stainless steel or titanium alloy for a higher range of mechanical modulus and properties against compressive, bending, tension, and torsion loading conditions in either static or dynamic patterns. A purely mechanical approach to fracture fixation of bone injuries is ineffective after an observation of various types of complications such as fragment dislocation, delayed union, nonunion, implant fractures, bone suppression at the interface with screws, implant loosening, and so forth. Gradually, biological factors involved in apposition and the formation of bone tissue have been highlighted as factors that influence the rigidity and stability of the bone–implant construct. Mechanical-biological investigations showed that bone formation is promoted under loading conditions in which the compactness of the bone is enhanced. Through using of very high rigid implant, the mechanical stimulation signals for bone formation might not be generated. The design of trauma implants tends to reduce the mechanical properties to enhance micromotion or mechanical strain at the fracture gap to induce bone healing. To obtain such an advantage in the fracture fixation of bone injuries, concepts were highlighted for the development of trauma implants. First, bone fragments need to be buttressed by the trauma main implant. Owing to huge use of the trauma plate, this concept was addressed for plating fixation of the bone fracture, particularly in metaphyseal or geometrically complex zones. In fact, it is essential to provide a mechanical barrier for bone fragments to prevent movement in the expected direction. This advantage might not be achievable for all bone fragments unless the use of multiple plates is manageable at the fracture site. Otherwise, the use of single-plating fixation along with other separate screw fixations might be an alternative for cases in which only one plate can be inserted at the fracture site. Buttressing enhances the stability of bone fragments, promoting the mechanical-biological induction of bone

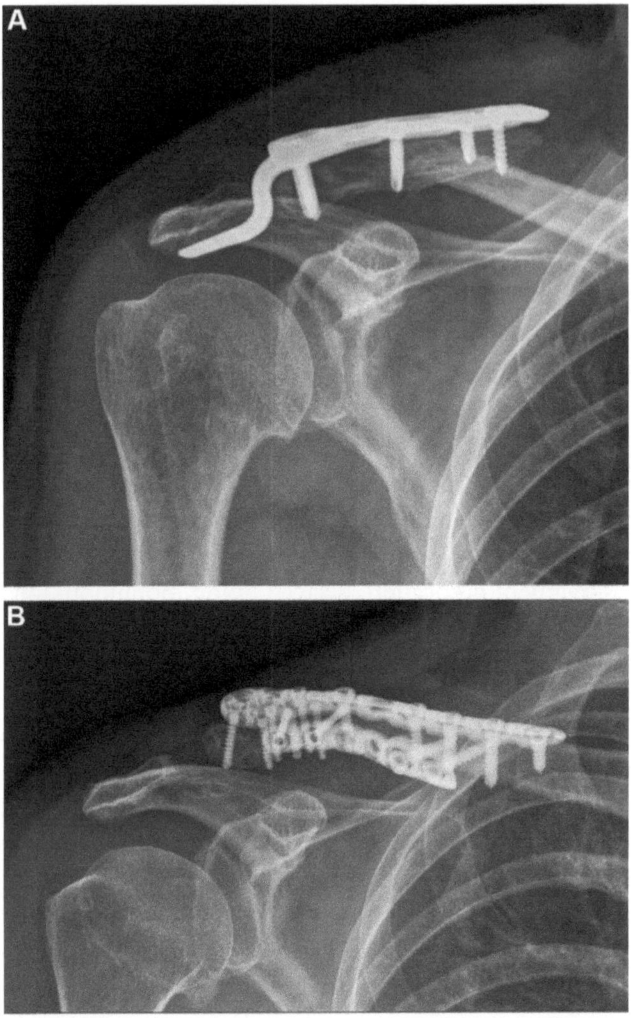

Figure 5.28 Radiographic illustration of bone fragment dislocation in the fixation of a clavicle fracture (A) and revision treatment with double-plating fixation (B). *Through the courtesy of J. Seo, K. Heo, S. Kim, J. Kim, H. Ham, et al., Comparison of a Novel Hybrid Hook Locking Plate Fixation Method with the Conventional AO Hook Plate Fixation Method for Neer Type V Distal Clavicle Fractures, Elsevier, Undefined 2020, (n.d.). https://www.sciencedirect.com/science/article/pii/S1877056819303846. (Accessed December 29, 2020), and permission from Elsevier.*

formation at fracture gaps. Effective buttressing of bone fractures at the distal femur and distal humerus is challenging in a multifragmentary fracture pattern in these zones because of the complex geometry of the bone anatomy, as discussed in Sections 5.4.2 and 5.5.3, respectively. The second concept in trauma fracture fixation is the effective purchase of bone fragments by screws. The size of the bone fragment and the screw diameter are considered for intraoperative planning regarding screw insertion through the bone fragments. The size of the bone fragment needs to be high enough to purchase one or multiple screws to capture the fragment. Multiple screw purchasing occupies a greater volume of bone stock, with a significant reduction in the overall mechanical stiffness of the fragment. As the result, the fragment dissociates from the screws and dislocates. Purchase of the screws from various aspects of the fragment is mechanically better compared with when all screws are purchased to the fragment from one aspect. In a very small bone fragment, stabilization with a wire or using a small-diameter screw is better. Even full buttressing of a small fragment with a plate might be an unreliable solution to hold the fragment. Purchase of the bone fragment at the distal femur, distal humerus, proximal tibia, proximal humerus, and distal humerus is challenging because of the use of long screws for bone fragments far from the plate in a multifragmentary fracture pattern. The third requirement for trauma fracture fixation is the optimal design of the main implant based on the desired interaction with the bone. The concept of a low profile and reduced rigidity are considered by designers to promote micromotion at the fracture gap. Likewise, the design parameters, features, and benefits of the implant are rigorously evaluated through experimental and computational biomechanical methods. For instance, for the plate, design concepts such as a locking mechanism reduce thickness at the metaphyseal zone, and lateral wedges/chamfers/fillets, the number of screw holes, screw trajectories, combined threaded-unthreaded screw holes, a three-dimensional curved profile in multiple planes, and so on, are implemented to fix bone fractures. The optimal use of the number of screws to stabilize the plate or nail is important to reduce the stiffness of the bone—implant construct for decrease concentrated stress and fracture of the implant. Such a requirement is considered in managing fractures of bone injuries at the diaphyseal portion of the long bones. A low number of screws inserted increase the level of strain at the fragment gap and exceed bone formation and occurrence of the cortical hypertrophy. Increase the level of strain could also increase the risk of nonunion, loosening and dislocation of the bone fragments at the diaphyseal portion. The fourth concept is the biological and mechanical integration of screws through bone fragments. The mechanical integration of the screw is achieved by threads, whereas biological integration results from the establishment of a bioactive or osteoconductive layer on the surface of the screw threads. Based on the type (cortical or trabecular) and conditions (osteoporotic, osteopenia, or normal) of bone tissue, the thread design (e.g., crest edge width, pitch, thread root, thread angle, helix angle) might be different. The formation of the osteoconductive layer on the screw has been completely discussed in the book *Osseoconductive Surface Engineering for Orthopedic Implants*, written by the lead author. Enhanced interaction of the bone and screws significantly reduces the rate of bone—implant construct failure and lessens the screw loosening, bone fragment dislocation, malunion, nonunion, and delayed union. The fifth factor that has a significant effect on the interaction between the bone and implant in trauma fracture fixation is the adequate reduction of the bone fragment in anatomic position while fracture gaps are completely filled by overlapping fragments. For bone loss at the fracture site, the use of bone graft material is necessary to maintain bone healing in its anatomic and physiologic structure. Trauma implants from other materials such as CFRPEEK and magnesium alloy are used with good clinical outcomes in the fracture fixation of bone injuries with a low level of loading conditions or combined with metallic implants in a load-bearing bone—implant construct. In this chapter, although bone—implant failures are examined, the success rate of bone fracture fixation is considerable in published clinical studies. However, the clinical demands are getting altered to returning back of the fractured site to the full healthyconditions, thus, it would not be limited to pain elimination and mild functional outcomes at the injured zone. Thus, researchers of bone injury and scientists still seek a better understanding of the interaction of bone with trauma implants and possible failures. In this regard, bone fracture treatment incorporating bone molecular and cell biomechanics, biochemistry, and biology is being investigated combined with conventional treatment methods with trauma implants.

5.9 Remind and learn

In this section, some questions are provided for effective reminders of the reviewed and discussed contents in this chapter. Answering these questions enhance learning about the interactions of bone with trauma implants and possible failures.

Q 5.1: What is bone tissue trauma? What are causes of trauma in young and elderly patients? Discuss the incidence of trauma at the affected zone and the medical treatment plan from the occurrence of trauma until the fractures have healed.

Q 5.2: Discuss the internal and external fixation of bone fractures. What types of fractures might be treated by internal and external fixation methods? What are the possible failures of the bone–implant construct in bone fracture fixation using internal fixation methods? Refer to Fig. 5.1: What are the biological steps of bone fracture healing?

Q 5.3: What types of materials and manufacturing processes might be used to develop trauma implants? Discuss the design characteristics for the straight and anatomic plate in trauma plating system (consider Section 5.3 for a complete discussion on trauma plates).

Q 5.4: Refer to Fig. 5.2: What are the benefits of the CFRPEEK composite polymer compared with metallic materials in the development of trauma implants?

Q 5.5: What is the optimal design of plate stiffness in the fixation of a bone fracture at the diaphyseal zone? Discuss your answer in view of design parameters such as thickness, the overall profile, screw holes, and so on, and fixation parameters such as the working length, semirigidity, and so forth.

Q 5.6: What are the benefits of the optimal design of plate stiffness fixation of a bone fracture at the diaphyseal zone? Discuss your answer in view of micromotion, the biological activity of bone formation at fracture gaps, and so on.

Q 5.7: What are the advantages and disadvantages of integrity between the bone and screws? What failures may occur as a result of ineffective integrity at the interface of the bone and screws?

Q 5.8: Discuss the biomechanical function of the nailing system in bone fractures. What is the behavior of the nail against bending stress? What failure might occur in the nail under increased bending stress? What are the main types of failure of bone–implant constructs in nailing fracture fixation?

Q 5.9: How may the biological integration of the bone and screws be achieved? What are the advantages and disadvantages of the biological integrity of the screws and bone tissue? Discuss how the effective osteoconductive or bioactive layering on the screw surface enhances the clinical outcomes of fracture fixation.

Q 5.10: What is the main disadvantage of biological integrity between the bone and screw? What failures might occur as a result of this disadvantage?

Q 5.11: Discuss the overall characteristics of fracture fixation at the intracapsular and capsular sections of the proximal femur. What types of fixation methods might be used for this part of the femur? Discuss the challenges of fracture fixation at this zone in view of long screws that are used, a high extent of physiologic loading and conditions of motion at the hip joint, and screw threads. Why would the risk for failure in this fracture fixation would be greater in a patient with high range of body mass index and low—mineral density bone?

Q 5.12: Refer to Figs. 5.3 and 5.5: What is the screw cutout in the bone? Discuss why such an issue might have occurred in fracture fixation at the intracapsular and capsular sections of the proximal femur. Refer to Fig. 5.4: Discuss failures that might have occurred in different types of fixation methods at this zone.

Q 5.13: What types of fixation method might be used to manage bone fractures at the midshaft of the femur? What are the advantages and disadvantages of plating fixation compared with nailing fixation? What is the benefit of anatomic fixation of a midshaft femur fracture compared with straight fixation in a lateral aspect?

Q 5.14: Discuss the overall design characteristics of the distal femur plate and why the plate cannot be designed anatomically to dedicate the distal condyles in a sagittal plane. What supplement or substitution method might be used to fix bone fragments at the posterior or medial sides of the distal condyles?

Q 5.15: What is the challenge of using a high number of screws to capture bone fragments? Discuss why osteoconductive surface engineering in trauma screws might not be as effective when it is used for joint replacement prostheses. What design concept might be investigated in future to enhance biological activity at fracture gap areas?

Q 5.16: Refer to Fig. 5.6: Discuss the optimal working length in fixing bone fractures at the diaphyseal zone. What is the benefit of interfragmentary screw fixation to enhance fracture fixation at the diaphyseal zone?

Q 5.17: Refer to Fig. 5.7: What are the benefits of using headless screw fixation to manage a bone fracture at the distal femur condyles?

Q 5.18: Discuss the management of bone fractures at the proximal tibia using lateral plating fixation. What are the benefits of medial plating or screw fixation in this zone? Why might periprosthetic bone near the stem portion of the tibial tray be resorbed? What might failure occur as a result of this phenomenon, and what would be the solution to treat bone failure at the affected zone?

Q 5.19: Discuss the overall conditions of fracture management at the midshaft and distal zones of the tibia. Discuss the effectiveness of anterolateral plating and syndesmotic screw fixation in fracture management at the distal zone of the tibia and fibula, respectively.
Q 5.20: What are the main causes of failure in fracture management of the midshaft tibia using plating fixation? Refer to Fig. 5.11: What are the reasons for bone resorption at the interface of the syndesmotic screw and tibia? Refer to Fig. 5.12: What is cause of syndesmotic screw fracture? Discuss your answers to these questions including possible solutions for enhanced interaction between the bone and implant.
Q 5.21: What are the benefits of using an Mg alloy screw to treat a bone fracture at the lateral malleolus? What is the main challenge of using an Mg alloy screw in bone fracture fixation? Discuss the applicability of using an Mg alloy screw to fix a bone fracture at the zones with a low level of load-bearing conditions.
Q 5.22: Discuss fracture management conditions in small bones. What is interfragmentary screw fixation in the management of small bone fractures? Discuss the benefit of plating fixation from the aspect of fixation in the management of small bone fractures.
Q 5.23: What type of failure might occur in the fixation of bone fractures at the humeral head? What is the main reason for such a failure at this zone in view of physiologic loading and motion conditions generated at the shoulder joint? Discuss why failure at the humeral head is seriously catastrophic. What is the current solution for failed fracture management at this zone? What other possible solution is hypothesized for more effective management of bone fractures at the humeral head?
Q 5.24: Refer to Fig. 5.15: Discuss the effectiveness of nailing fixation in managing a bone fracture at the humeral head. Refer to Fig. 5.16: discuss the cutout phenomenon at the interface of the screw and trabecular bone.
Q 5.25: Discuss the overall characteristics of the expandable cage to treat bone fractures at the humeral head. What is the effect of this cage in enhancing bone fracture fixation at this zone? What cases might be more selected when using such an implantation method?
Q 5.26: Discuss the challenges of fracture fixation management at the distal portion of the humerus in view of the bony anatomy at this zone. What are the benefits of hybrid plating fixation from two sides of the humerus at the distal metaphyseal zone? Discuss the use of long

and short screws in distal humerus fracture fixation. What failures might occur in fixing bone fractures at this zone?

Q 5.27: Refer to Fig. 5.18: Discuss the influence of physiologic motions and intraoperative bone fragment reduction in the effectiveness of bone fracture fixation at the distal zone of the humerus. Refer to Figs. 5.18 and 5.19: Discuss the risk for nonunion or delayed union in this zone.

Q 5.28: What are the main design concepts for the design of a distal radius plate? What types of distal radius plates might be developed? Refer to Fig. 5.21: What is the main challenge of distal radius fracture fixation? Is buttressing sufficient to manage this challenge? Refer to Fig. 5.22: What is the benefit of using an expandable cage for distal radius fracture fixation?

Q 5.29: What are the challenges of fracture fixation in the pelvis in view of bony anatomy and attached or originated muscles to this bone? What type of plate is beneficial for pelvic fracture fixation? What failures might occur in the pelvis—implant construct? What fixation strategies might be beneficial for enhanced fracture management of the pelvic fracture in view of the screw and plate length? What design concept might be hypothesized to enhance bone-screw integrity at this bone?

Q 5.30: Refer to Figs. 5.23—5.25: Discuss loading conditions at the pelvis, particularly at the acetabular zone, and reasons for union complications (on-union, delayed union, and malunion) that might occur in pelvic fracture fixation.

Q 5.31: Refer to Figs. 5.26—5.28: Discuss loading conditions at the clavicle and reasons for screw loosening, plate fracture, or bone fragment dislocation that might occur in clavicle fracture fixation.

5.10 Think and challenge

In this section, some questions are provided for further thinking about the reviewed and discussed contents of this chapter. The aim is to challenge current facets of the interaction of bone with trauma implants and possible failures by answering these questions, which may enhance design concepts for the future development of implants.

Q 5.32: As reviewed in this chapter, fracture fixation in the pelvis and clavicle might fail as the result of applied muscle forces in bone fragments. Discuss how the stability of the bone—implant construct in these bones could be enhanced. Likewise, discuss whether other novel fixation methods could be hypothesized to restore bone fragments in the pelvis and clavicle.

Q 5.33: As highlighted in this chapter, the concept of using an active implant with the ability to remodel effective against physiologic loading conditions may enhance interaction between the implant and bone. Discuss how such a concept could be designed and processed in the development of trauma implants.

Q 5.34: As reviewed in this chapter, fracture fixation using long screws might be challenging for capturing and stabilizing bone fragments rigidly far from the plate. Discuss how the screw design or material could be enhanced to prevent fragment dislocation at such fracture fixations.

References

[1] L. Tian, N. Tang, T. Ngai, C. Wu, Y. Ruan, et al., Hybrid Fracture Fixation Systems Developed for Orthopaedic Applications: a General Review, Elsevier, Undefined 2019, (n.d.). https://www.sciencedirect.com/science/article/pii/S2214031X18300548. (Accessed December 8, 2020).

[2] P. Pankaj, S.X.- Injury, The Risk of Loosening of Extramedullary Fracture Fixation Devices, Elsevier. (n.d.), Undefined 2019. https://doi.org/10.1016/j.injury.2019.03.051.

[3] C. Donnelley, E. von Kaeppler, H.R.- Injury, Monoplanar External Fixation of Comminuted Open Tibial Shaft Fractures Predicts Loss of Alignment by One Year Compared to a Statically Locked Intramedullary SIGN Nail, Elsevier, Undefined 2020, (n.d.). https://www.sciencedirect.com/science/article/pii/S0020138320308986. (Accessed December 9, 2020).

[4] A. MacLeod, P.P.- Injury, Pre-operative Planning for Fracture Fixation Using Locking Plates: Device Configuration and Other Considerations, Elsevier, Undefined 2018, (n.d.). https://www.sciencedirect.com/science/article/pii/S0020138318302961 (Accessed December 8, 2020).

[5] S. Lee, S. György, J. Choi, J. Choi, S.K.-J. of Cranio, Carbon Plate Shows Even Distribution of Stress, Decreases Screw Loosening, and Increases Recovery of Preoperative Daily Feed Intake Amount in a Rabbit Model of mandibular continuity defects, Elsevier, Undefined 2014, (n.d.). https://www.sciencedirect.com/science/article/pii/S101051821300262X. (Accessed December 8, 2020).

[6] Y. Okudera, H. Kijima, S. Yamada, et al, The location of the fracture determines the better solution, osteosynthesis or revision, in Periprosthetic Femoral Fractures, Elsevier, Undefined 2020, (n.d.). https://www.sciencedirect.com/science/article/pii/S0972978X20301768. (Accessed December 8, 2020).

[7] N. Moon, W. Shin, J. Kim, S. Woo, S. Son, K.S.- Injury, Cementless Total Hip Arthroplasty Following Failed Internal Fixation for Femoral Neck and Intertrochanteric Fractures: a Comparative Study with 3–13 Years' Follow-Up of 96, Elsevier, Undefined 2019, (n.d.). https://www.sciencedirect.com/science/article/pii/S0020138319300221. (Accessed December 8, 2020).

[8] B. Min, K. Lee, J. Oh, C. Cho, J. Cho, B.K.- Injury, The Treatment Strategies for Failed Fixation of Intertrochanteric Fractures, Elsevier, Undefined 2019, (n.d.). https://www.sciencedirect.com/science/article/pii/S0020138319302888. (Accessed December 8, 2020).

[9] J. Chen, G. She, S. Luo, W. Wu, T. Zhuang, S. Huan, N.L.- Injury, Hemiarthroplasty Compared with Internal Fixation for Treatment of Nondisplaced Femoral Neck Fractures in Elderly Patients: a Retrospective Study, Elsevier, Undefined 2020, (n.d.). https://www.sciencedirect.com/science/article/pii/S0020138320301716. (Accessed December 8, 2020).
[10] D. De Meo, D. Ferraro, L. Petriello, B. Zucchi, A.C.- Injury, Outcome and Technical Consideration of Conversion Total Hip Arthroplasty after Failed Fixation of Intracapsular and Extracapsular Hip Fractures: are They Really that, Elsevier, Undefined 2020, (n.d.). https://www.sciencedirect.com/science/article/pii/S0020138320307798. (Accessed December 9, 2020).
[11] N. Hernandez, B. Chalmers, K. Perry, et al., Total Hip Arthroplasty after in Situ Fixation of Minimally Displaced Femoral Neck Fractures in Elderly Patients, Elsevier, Undefined 2018, (n.d.). https://www.sciencedirect.com/science/article/pii/S088354031730671X. (Accessed December 9, 2020).
[12] E. Morsi, A. Drwish, A. Saber, I.N.-T.J. of Arthroplasty, The Use of Standard Cemented Femoral Stems in Total Hip Replacement after Failed Internal Fixation of Intertrochanteric Femoral Fractures., Elsevier, Undefined 2020, (n.d.). https://www.sciencedirect.com/science/article/pii/S088354032030351X. (Accessed December 9, 2020).
[13] L. Halonen, H. Vasara, A. Stenroos, J.K.- Injury, Routine Follow-Up is Unnecessary after Intramedullary Fixation of Trochanteric Femoral Fractures—Analysis of 995 Cases, Elsevier, Undefined 2020, (n.d.). https://www.sciencedirect.com/science/article/pii/S0020138320302552. (Accessed December 8, 2020).
[14] E. Panagiotopoulos, E. Panagiotopoulos, A.P. Fortis, Z. Millis, E. Lambiris, V. Kostopoulos, L. Vellios, Pattern of screw loosening in fractures fixed with conventional and functional plates, Injury 25 (1994) 515—517, https://doi.org/10.1016/0020-1383(94)90092-2.
[15] S. Märdian, W. Schmölz, K. Schaser, et al., Locking Plate Constructs Benefit from Interfragmentary Lag Screw Fixation with Decreased Shear Movements and More Predictable Fracture Gap Motion in Simple Fracture, Elsevier, Undefined 2019, (n.d.). https://www.sciencedirect.com/science/article/pii/S0268003319303420. (Accessed December 9, 2020).
[16] V. Maheshwari, S. Sharma, D. Goyal, et al., Clinical Experience with Management of Hoffa Fractures Using Headless Compression Screw and Headed Screw, Elsevier, Undefined 2019, (n.d.). https://www.sciencedirect.com/science/article/pii/S0976566218304338. (Accessed February 16, 2021).
[17] M. Morwood, S. Gebhart, N. Zamith, H.M.- Injury, Outcomes of Fixation for Periprosthetic Tibia Fractures Around and below Total Knee Arthroplasty, Elsevier, Undefined 2019, (n.d.). https://www.sciencedirect.com/science/article/pii/S0020138319301111. (Accessed December 8, 2020).
[18] M. Potgieter, H. Pretorius, G. Du Preez, M.B.- Injury, Complications Associated with Hexapod Circular Fixation for Acute Fractures of the Tibia Diaphysis: a Retrospective Descriptive Study at a High Volume Trauma Centre, Elsevier, Undefined 2020, (n.d.). https://www.sciencedirect.com/science/article/pii/S0020138319307107. (Accessed December 8, 2020).
[19] M. Klutzny, G. Singh, R. Hameister, et al., Undefined 2019, Screw Track Osteolysis in the Cementless Total Knee Replacement Design, Elsevier, Undefined 2019, (n.d.). https://www.sciencedirect.com/science/article/pii/S088354031930004X. (Accessed December 8, 2020).
[20] L. Xie, H. Xie, J. Wang, C. Chen, C. Zhang, et al., Comparison of Suture Button Fixation and Syndesmotic Screw Fixation in the Treatment of Distal Tibiofibular Syndesmosis Injury: a Systematic Review and Meta-Analysis, Elsevier, Undefined 2018,

(n.d.). https://www.sciencedirect.com/science/article/pii/S174391911831700X. (Accessed December 8, 2020).
[21] S. Lee, S. Moon, M. Park, B. Jo, et al., Syndesmosis Fixation in Unstable Ankle Fractures Using a Partially Threaded 5.0-mm Cannulated Screw, Elsevier, Undefined 2018, (n.d.). https://www.sciencedirect.com/science/article/pii/S1067251617307330. (Accessed December 9, 2020).
[22] M.D. Riedel, J. Briceno, C.P. Miller, J.Y. Kwon, Technical tip: removal of a broken tri-cortical syndesmotic screw using a "perfect circle" technique, Injury 49 (2018) 877–880. https://doi.org/10.1016/j.injury.2018.02.022.
[23] H. Waizy, J. Diekmann, I. Bartsch, In vivo study of a biodegradable orthopedic screw (MgYREZr-alloy) in a rabbit model for up to 12 months Implant-associated magnetic drug targeting: antibiotic therapy of periimplant infections view project MAGNEZIX view project, Artic. J. Biomater. Appl. 28 (2013) 667–675. https://doi.org/10.1177/0885328212472215.
[24] L. Rocchi, G. Merendi, G. Cazzato, D.C.- Injury, Scaphoid waist fractures fixation with staple. Retrospective Study of a Not Widespread Procedure., Elsevier, Undefined 2020, (n.d.). https://www.sciencedirect.com/science/article/pii/S0020138319308162. (Accessed December 8, 2020).
[25] H. Windhagen, K. Radtke, A. Weizbauer, J. Diekmann, Y. Noll, U. Kreimeyer, R. Schavan, C. Stukenborg-Colsman, H. Waizy, Biodegradable magnesium-based screw clinically equivalent to titanium screw in hallux valgus surgery: short term results of the first prospective, randomized, controlled clinical pilot study, Biomed. Eng. Online 12 (2013), https://doi.org/10.1186/1475-925X-12-62.
[26] J.M. Cottom, R.B. Rigby, Biomechanical comparison of a locking plate with intraplate compression screw versus locking plate with plantar interfragmentary screw for lapidus arthrodesis: a cadaveric study, J. Foot Ankle Surg. 52 (2013) 339–342, https://doi.org/10.1053/j.jfas.2013.02.012.
[27] L. Nowak, N. Dehghan, M. McKee, E.S.- Injury, Plate Fixation for Management of Humerus Fractures, Elsevier, Undefined 2018, (n.d.). https://www.sciencedirect.com/science/article/pii/S0020138318303000. (Accessed December 8, 2020).
[28] J. Barlow, A. Logli, S. Steinmann, et al., Locking Plate Fixation of Proximal Humerus Fractures in Patients Older than 60 Years Continues to be Associated with a High Complication Rate, Elsevier, Undefined 2020, (n.d.). https://www.sciencedirect.com/science/article/pii/S105827462030001X. (Accessed December 9, 2020).
[29] B. Jost, C. Spross, H. Grehn, C.G.-J. of shoulder and elbow, Locking Plate Fixation of Fractures of the Proximal Humerus: Analysis of Complications, Revision Strategies and Outcome, Elsevier, Undefined 2013, (n.d.). https://www.sciencedirect.com/science/article/pii/S1058274612002716. (Accessed December 22, 2020).
[30] C. García-Fernández, Y. Lopiz, B. Rizo, L. Serrano-Mateo, B. Alcobía-Díaz, A. Rodríguez-González, F. Marco, Reverse Total Shoulder Arhroplasty for the Treatment of Failed Fixation in Proximal Humeral Fractures, Elsevier, 2018, https://doi.org/10.1016/j.injury.2018.06.042.
[31] G.K. Kouvidis, M.B. Sommers, P.V. Giannoudis, P.G. Katonis, M. Bottlang, Comparison of migration behavior between single and dual lag screw implants for intertrochanteric fracture fixation, J. Orthop. Surg. Res. 4 (2009), https://doi.org/10.1186/1749-799X-4-16.
[32] J. Hudgens, J. Jang, K. Aziz, et al., Three-and 4-part Proximal Humeral Fracture Fixation with an Intramedullary Cage: 1-year Clinical and Radiographic Outcomes, Elsevier, Undefined 2019, (n.d.). https://www.sciencedirect.com/science/article/pii/S1058274619303027. (Accessed December 22, 2020).
[33] P. Paterson, et al., Techniques for using a novel intramedullary cage technology for fixation of proximal humeral fractures, Pdfs.Semanticscholar.Org. (n.d.), Undefined

2016, https://pdfs.semanticscholar.org/9b28/da8b0a9c95c879f9aa99b662ee2e372c16a9.pdf. (Accessed February 16, 2021).
[34] N. Shah, W. Hayes, H. Wang, J.H.- Injury, A Pilot Biomechanical Study Comparing a Novel, Intramedullary Nail/Plate Construct to Standard Dual-Plate Fixation of Intra-articular C2. 3 Distal Humerus Fractures, Elsevier, Undefined 2020, (n.d.). https://www.sciencedirect.com/science/article/pii/S0020138320305416. (Accessed December 8, 2020).
[35] F. Claessen, Y. Braun, et al., Plate and Screw Fixation of Bicolumnar Distal Humerus Fractures: Factors Associated with Loosening or Breakage of Implants or Nonunion, Elsevier, Undefined 2015, (n.d.). https://www.sciencedirect.com/science/article/pii/S0363502315009442. (Accessed December 8, 2020).
[36] Y. Sur, Y. Kim, H.P.-J. International, Results of Plate Fixation for Transcondylar Fracture of the Distal Humerus: a Rare Pattern of Fractures, Elsevier, Undefined 2020, (n.d.). https://www.sciencedirect.com/science/article/pii/S266663832030044X. (Accessed December 9, 2020).
[37] F. Claessen, N. Stoop, et al., Interpretation of Post-operative Distal Humerus Radiographs after Internal Fixation: Prediction of Later Loss of Fixation, Elsevier, Undefined 2016, (n.d.). https://www.sciencedirect.com/science/article/pii/S0363502316303768. (Accessed December 9, 2020).
[38] J. Simone, P. Streubel, et al., Low Transcondylar Fractures of the Distal Humerus: Results of Open Reduction and Internal Fixation, Elsevier, Undefined 2014, (n.d.). https://www.sciencedirect.com/science/article/pii/S1058274613006241. (Accessed December 23, 2020).
[39] J. Beck, N. Harness, H.S.-T.J. of hand surgery, Volar Plate Fixation Failure for Volar Shearing Distal Radius Fractures with Small Lunate Facet Fragments, Elsevier, Undefined 2014, (n.d.). https://www.sciencedirect.com/science/article/pii/S0363502314000793. (Accessed December 23, 2020).
[40] Y. Yanagisawa, A. Ito, Y. Hara, H. Mutsuzaki, et al., Initial Clinical Trial of Pins Coated with Fibroblast Growth Factor-2—Apatite Composite Layer in External Fixation of Distal Radius Fractures, Elsevier. (n.d.), Undefined 2019. https://www.sciencedirect.com/science/article/pii/S0972978X18304239. (Accessed December 8, 2020).
[41] K. Bennett, S.K.-T. case reports, Aseptic Loosening of an APTIS Distal Radioulnar Joint Arthroplasty—Case Report, Elsevier, Undefined 2019, (n.d.). https://www.sciencedirect.com/science/article/pii/S2352644019300305. (Accessed December 9, 2020).
[42] N. Johnson, L. Cutler, J. Dias, A. Ullah, C.W.- Injury, Complications after Volar Locking Plate Fixation of Distal Radius Fractures, Elsevier, Undefined 2014, (n.d.). https://www.sciencedirect.com/science/article/pii/S0020138313004518. (Accessed December 23, 2020).
[43] K. Inagaki, K.K.-J. of O. Science, Distal Radius Fractures— Design of Locking Mechanism in Plate System and Recent Surgical Procedures, Elsevier, Undefined 2016, (n.d.). https://www.sciencedirect.com/science/article/pii/S0949265816000075 (accessed December 23, 2020).
[44] L. Reichel, B. Bell, et al., Radial Styloid Fractures, Elsevier. (n.d.), Undefined 2012, https://www.sciencedirect.com/science/article/pii/S0363502312008179. (Accessed December 23, 2020).
[45] A. Turan, Y.A. Kati, B. Acar, O. Kose, Magnesium bioabsorbable screw fixation of radial styloid fractures: case report, J. Wrist Surg. 09 (2020) 150—155, https://doi.org/10.1055/s-0039-1685489.
[46] M. Strassmair, M. Jonas, et al., Distal Radial Fracture Management with an Intramedullary Cage and Fragment Fixation, Elsevier, Undefined 2016, (n.d.). https://www.

sciencedirect.com/science/article/pii/S0363502316301356. (Accessed December 23, 2020).
[47] H. Schmal, M. Larsen, F. Stuby, P. Strohm, K.R.- Injury, Effectiveness and Complications of Primary C-Clamp Stabilization or External Fixation for Unstable Pelvic Fractures, Elsevier, Undefined 2019, (n.d.). https://www.sciencedirect.com/science/article/pii/S0020138319305145. (Accessed December 8, 2020).
[48] C. Kim, J.K.- Injury, Plate versus Sacroiliac Screw Fixation for Treating Posterior Pelvic Ring Fracture: a Systematic Review and Meta-Analysis, Elsevier, Undefined 2020, (n.d.). https://www.sciencedirect.com/science/article/pii/S0020138320305623. (Accessed December 8, 2020).
[49] R. Zhang, Y. Yin, Z. Hou, Sacroiliac screw versus a minimally invasive adjustable plate for zone II sacral fractures: a retrospective study orthopedic surgery and injury epidemiology view project, Injury (2019), https://doi.org/10.1016/j.injury.2019.02.011.
[50] L.H. Rappoport, I.Y. Luna, G. Joshua, Minimally Invasive Sacroiliac Joint Fusion Using a Novel Hydroxyapatite-Coated Screw: Preliminary 1-Year Clinical and Radiographic Results of a 2-Year Prospective Study, Elsevier, 2017, https://doi.org/10.1016/j.wneu.2017.02.046.
[51] Y. Yoon, D. Ma, S. Lee, J. Oh, H.S.-A. journal of surgery, Posterior Pelvic Ring Injury of Straddle Fractures: Incidence, Fixation Methods, and Clinical Outcomes, Elsevier, Undefined 2020, (n.d.). https://www.sciencedirect.com/science/article/pii/S101595842030097X. (Accessed December 9, 2020).
[52] H. Le Quang, W. Schmoelz, R.L.- Injury, Biomechanical Comparison of Fixation Techniques for Transverse Acetabular Fractures—Single-Leg Stance vs. Sit-To-Stand Loading, Elsevier, Undefined 2020, (n.d.). https://www.sciencedirect.com/science/article/pii/S0020138320305672. (Accessed December 8, 2020).
[53] J. Hack, Y. Kranz, T. Knauf, M. Bäumlein, D.M.- Injury, Stability of Internal versus External Fixation in Osteoporotic Pelvic Fractures—a Biomechanical Analysis, Elsevier, Undefined 2020, (n.d.). https://www.sciencedirect.com/science/article/pii/S0020138320306896. (Accessed December 8, 2020).
[54] H. Wu, C. Song, R. Shang, Q. Shao, X. Liu, H.Z.- Injury, Double Column Acetabular Fractures Fixation Using a Novel Dynamic Anterior Plate-Screw System: a Biomechanical Analysis, Elsevier, Undefined 2020, (n.d.). https://www.sciencedirect.com/science/article/pii/S0020138320308858. (Accessed December 9, 2020).
[55] A. Anizar-Faizi, A. Hisam, et al., Outcome of surgical treatment for displaced acetabular fractures, Ncbi. Nlm. Nih. Gov. (n.d.), Undefined 2014. https://www.ncbi.nlm.nih.gov/pmc/articles/PMC4536391/. (Accessed December 23, 2020).
[56] U. Meena, S. Tripathy, R. Sen, et al., Predictors of Postoperative Outcome for Acetabular Fractures, Elsevier, Undefined 2013, (n.d.). https://www.sciencedirect.com/science/article/pii/S1877056813002132. (Accessed December 23, 2020).
[57] Y. Yoon, D. Ma, S. Lee, J. Oh, H.S.-A. Journal of surgery, Posterior Pelvic Ring Injury of Straddle Fractures: Incidence, Fixation Methods, and Clinical Outcomes, Elsevier, Undefined 2020, (n.d.). https://www.sciencedirect.com/science/article/pii/S101595842030097X. (Accessed December 23, 2020).
[58] K.J. Lee, B.W. Min, G.M. Oh, S.W. Lee, Surgical correction of pelvic malunion and nonunion, CiOS Clin. Orthop. Surg. 7 (2015) 396—401, https://doi.org/10.4055/cios.2015.7.3.396.
[59] R. Kundangar, K. Singh, S. Mohanty, K.E.-J. of Orthopaedics, Clinical outcome of internal fixation of middle third clavicle fractures using locking compression plate: comparison between open plating and MIPO, Luot.Ch. (n.d.), Undefined 2019. http://luot.ch/wordpress/wp-content/uploads/2019/11/ClavicleFx-open-vs-mipo.pdf. (Accessed December 8, 2020).

[60] M.A. Meeuwis, A.F. Pull Ter Gunne, M.H.J. Verhofstad, F.H.W.M. Van Der Heijden, Construct failure after open reduction and plate fixation of displaced midshaft clavicular fractures, (n.d.) https://www.sciencedirect.com/science/article/pii/S002013831730061X. (Accessed December 8, 2020).
[61] Y. Hsiao, T. Lin, Y. Wang, C. Chang, H.T.- Injury, Prognostic Factors and Outcomes of Secondary Surgery after Plate Fixation for Midshaft Clavicle Fracture: Comparison of Traditional DCP and Pre-contoured Locking Plate, Elsevier, Undefined 2020, (n.d.). https://www.sciencedirect.com/science/article/pii/S0020138320305283. (Accessed December 8, 2020).
[62] J. Seo, K. Heo, S. Kim, J. Kim, H. Ham, et al., Comparison of a Novel Hybrid Hook Locking Plate Fixation Method with the Conventional AO Hook Plate Fixation Method for Neer Type V Distal Clavicle Fractures, Elsevier, Undefined 2020, (n.d.). https://www.sciencedirect.com/science/article/pii/S1877056819303846. (Accessed December 29, 2020).

CHAPTER SIX

Interactions of bone with spinal implants and possible failures

6.1 Introduction

Spinal implants are rapidly increasing, as are the number of developers and manufacturers in this industry of orthopedic implants. The motivation for spinal implants is its considerable variety of other categories. The concepts of fracture fixation, joint replacement, bony fusion, anatomic reduction, deformity treatment, surface osteoconductivity enhancement, and bone grafting exist in developing spinal implants to treat spinal column injuries. Spinal implants are recognized for the main categories of screw–rod stabilization, intervertebral fusion cages, corpectomy cages, and total disc replacement (TDR). Another interesting point of spinal implants is the extensive use of polyetheretherketone (PEEK) polymer and its composites for the manufacture of these implants, which is growing. Like other types of orthopedic implants, spinal implants might fail in interaction with bone tissue. Bone–implant failures such as stress shielding, overloading, ineffective load transfer, excessive rigidity, subsidence, dislocation, loosening, and bone resorption or osteolysis occur after the implantation of spinal implants in bone tissue. Failure of a bone–implant construct in the management of spinal column injuries may affect the stabilization of the affected zone and recompress the spinal cord, progressing to pain. Thus, the success of spinal injury treatment through spinal implants is clinically crucial. Spinal implants are designed to be anatomically compatible with the anatomy of the vertebrae. The position of the designed implant is required to provide effective biomechanical compatibility with localized physiologic loading conditions. The effects of body weight and muscle loading conditions influence the extent of compressive loading conditions transferred between the vertebrae from the cervical zone to the lumbar and sacral subcolumns. In addition, movement of the spinal column creates three types of moment or torque in various axes and in the form of a cyclic or dynamic mode, which magnifies loading conditions to the implanted zone. The main strategy for treating spinal injuries is rigid stabilization of the injured zone. This strategy is

Interactions of Bone with Orthopedic Implants and Possible Failures
ISBN: 978-0-323-95410-5
https://doi.org/10.1016/B978-0-323-95410-5.00007-4

challenging owing to immobilization of a portion of the dynamic structure, which is inadequate. In fact, the functionality of the spinal column would be reduced mildly or significantly depends on the number of vertebral segments that would be rigidly stabilized. However, with current and extended techniques, rigid or semirigid implantation surgical implantation is managed as salvage to reduce severe pain in most affected patients. Dynamic solutions have been introduced, but their applicability and effectiveness are controversial and are intended only for mild or soft spinal injuries. In this chapter, the interaction of bone with spinal implants used for rigid or dynamic stabilization of the affected zone, and their possible failures, are reviewed, and possible reasons are discussed.

6.2 Spinal column injury

In the musculoskeletal system, the spinal column has a specific and interesting structure and function. It is composed of two hard and soft tissues. The hard tissue is known as the vertebral segment and the soft tissue as the intervertebral disc. The vertebrae are completely composed of bone tissue whereas the intervertebral discs are composed of fibrous and gelatinous tissue. The spinal column is divided into three sections: lumbar, thoracic, and cervical (Fig. 6.1) [1]. The structure of the spinal column in each section is different, and the shape and size of the vertebrae and intervertebral discs vary. In the lumbar section, there are five vertebrae that are formed concave posteriorly. The degree of rotation in the lumbar column is much lower than that of the thoracic and cervical columns in flexion-extension, lateral bending, axial rotation [2]. The effective loading conditions in the lumbar column have the highest range in the whole spinal column. Loading conditions in the spinal column are classified as compressive force, torsional torque, lateral bending torque, and flexion-extension torque. The effects of body weight loading conditions highly affect this section of the spinal column because of the attachment of multiple muscles from the pelvis bone to the lumber vertebrae. The lumbar vertebrae are the largest vertebrae of the spinal column. The thoracic column is composed of 11 vertebrae with intervertebral discs, which facilitate movement of the spinal column at the thoracic zone. The extent of rotation angle in thoracic vertebrae is higher than for the lumbar column, facilitating flexion-extension of the spinal column up to complete stooping. Effective loading conditions for the thoracic vertebrae are than those of the lumber vertebrae. The third spinal subcolumn is the cervical column, with seven vertebrae and a higher rate of motion

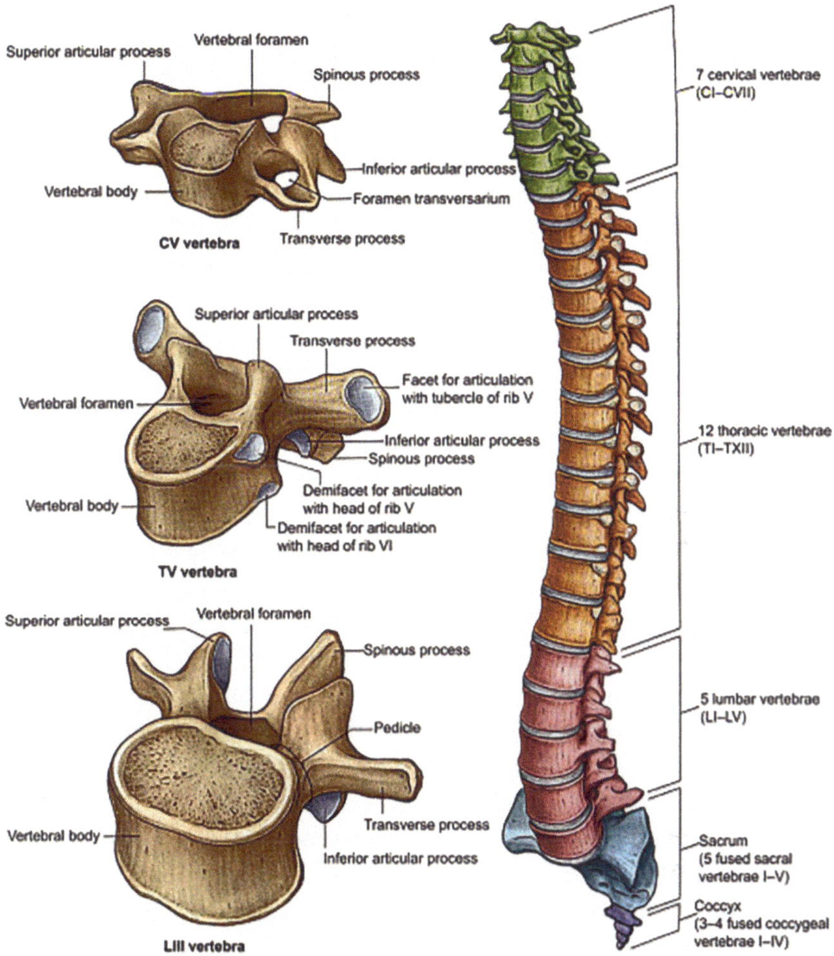

Figure 6.1 Illustration of spinal column anatomy. (Left) Whole column. (Right) Vertebra. The spinal column is divided into three parts: cervical vertebrae (CV), thoracic vertebrae (TV), and lumbar vertebrae (LV). At the most distal portion of the spinal column, the sacrum and coccyx have fused vertebrae. The anatomy of vertebrae in cervical, thoracic, and lumbar parts has some differences; however, all vertebrae have a body portion and two posterior joints connected with adjacent vertebrae to form the spinal column. *Through the courtesy of K. Yang, Modeling the Spine and Upper and Lower Extremities for Impact Scenarios, Elsevier, n.d.. https://www.sciencedirect.com/science/article/pii/B9780128098318000167. (Accessed 5 January 2021) and permission from Elsevier.*

in all three axes of rotation (flexion-extension, lateral bending, and axial rotation). Applied loading conditions through the cervical column are lower than for other spinal subcolumns. The size of cervical vertebrae is smaller than vertebrae in thoracic and lumbar subcolumns.

Vertebrae compose various parts such as the body, spine, and pedicle (Fig. 6.1). The body is the anterior portion of the vertebrae and is connected with intervertebral discs. It is articulated at the facet joint with the posterior portion of the vertebrae including the spinous section. Between the anterior and posterior portions, there is a canal where the spinal cord including a neural network passes. The presence of ligaments at the facet joint and between the spinous portions provides relative stability to the vertebrae in various loading and motion conditions of the spinal column. The other main tissue at the spinal column is the intervertebral disc, which is composed of two parts: nucleus and annulus (Fig. 6.2) [3]. The annulus part is considered the external or shell. It is composed of fibrous tissue with tension flexibility in various axes of translation and rotation. The nucleus part is the center of the disc. It is composed of compressible tissue to absorb relative compressive energy between vertebrae. Nucleus tissue is completely covered with annulus tissue peripherally. Loading conditions between vertebrae are borne by intervertebral discs, facet joints, and the spine. Approximately half of the compressive load is borne by the disc and half by the facet joints.

The spinal column may be injured at the vertebra or intervertebral disc tissues. The injury is diagnosed when the patient feel pains at the spine or other body limbs. Pain is caused by impingement of vertebral or disc tissues with the central neural network passed from the spinal cord. Cortical hypertrophy occurs at the spinal canal, from which the canal narrows. This irritates or suppresses the central neural network system and incurs pain to the patient. The output nerves from the irritated zone through the other body limbs transmit this irritation to the connective limbs, pain is felt. On the other hand, the intervertebral disc may impingement on the spinal cord as well. When the disc tissues degenerate, either annulus or nucleus, disc mechanical strength is reduced; thus, they reshape and compress the central neural system, generating pain at the irritated zone and other connective limbs through output nerves. Cortical hypertrophy of the vertebra at the remodeled zone results from to overloading conditions. It occurs as the result of anterior or central degeneration of the disc and the concentration of mechanical stress at the posterior aspects of the vertebral body and facet joints, and remodeling of bone tissue at this zone to counteract increased loading conditions. This is the mechanical-biological response of bone tissue to excessive loading conditions compared with normal physiologic loading under healthy conditions. Disc degeneration occurs for various reasons. Obesity and osteoporosis cause excessive transference of loading conditions to intervertebral discs, particularly in the lumber subcolumn. Obesity results

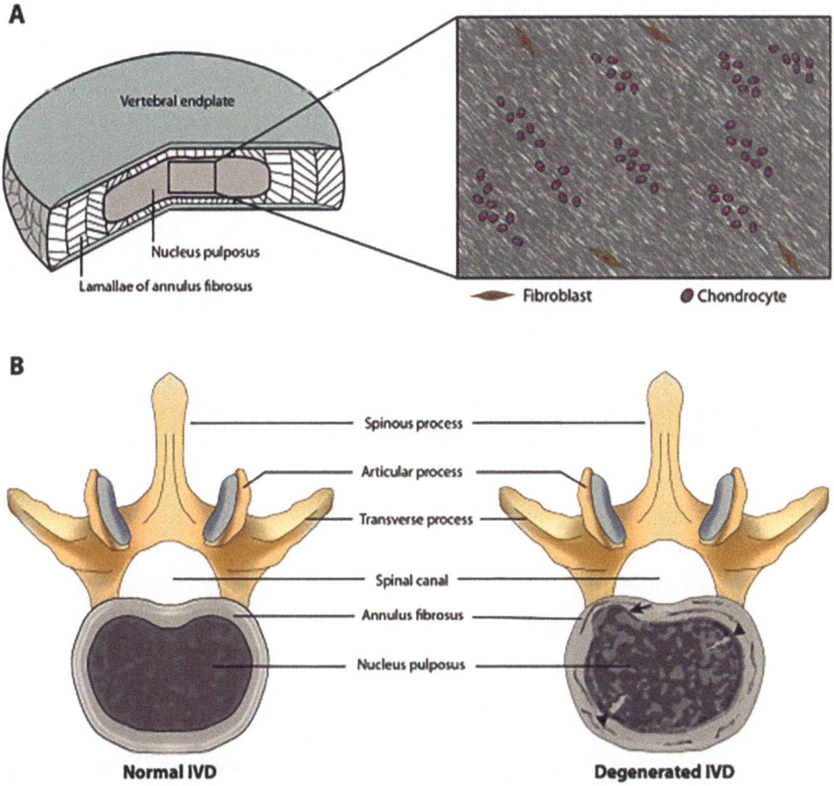

Figure 6.2 Illustration of the intervertebral disc (IVD). (Left) Healthy or normal. (Right) Degenerated. This tissue has multimechanical performance against various physiologic loading conditions. Annulus fibrous tissue tolerates tensile stress in various axes to stabilize the vertebrae in physiologic position while providing limited range of flexibility between vertebrae, from which spinal column motions in axes of flexion-extension, lateral bending, and axial torsion are facilitated. Nucleus tissue bears compressive stress transferred from adjacent vertebrae to the intervertebral disc. These two tissues of the disc highly influence mechanical interactions with the physiologic loading and motion between vertebrae, from which the biological and biochemical reactions of cells in annulus and nucleus tissue are affected and result in degeneration of the intervertebral disc. *Through the courtesy of Y. Feng, B. Egan, J.W.-G. & diseases, undefined 2016, Genetic Factors in Intervertebral Disc Degeneration, Elsevier, n.d. https://www.sciencedirect.com/science/article/pii/S2352304216300149. (Accessed 5 January 2021) and permission from Elsevier.*

in accumulated fat in the stomach, forcing the spinal column to be out of a physiologic *S*-shaped pattern. Through this ineffective physiologic phenomena, loading conditions concentrate through the anterior aspect of the intervertebral zone. Hence, anterior portions of the annulus and nucleus are affected over time. Because of the high flexibility of disc tissues, increased stress at the anterior zone is transmitted to the posterior aspect of the disc and vertebral body. When the disc tissue weakens, it is projected posteriorly and impinges on the spinal cord. When the trabecular bone is degraded or weakened at the vertebral bodies, bone suppression occurs at the central section of the vertebrae and loading equilibrium between vertebrae is affected. Such imbalance results in the concentration of loading conditions anteriorly or posteriorly and acceleration of the intervertebral disc injury and irritation of the spinal cord. The intervertebral disc is a tissue that is influenced by nonphysiologic conditions at the spinal column. A deficiency in function of the intervertebral disc causes reshaping of the spinal column owing to a deficiency of the loading distribution and mechanism between vertebrae. This affects the performance of the attached or originated muscles to the vertebrae. The muscles' contraction and stretching pattern adapts to the deficient conditions occurring in the spinal column. The failure of mechanical function in the intervertebral disc affects the facet joints or pars interarticularis section and may cause vertebral dislocation (Fig. 6.3) [4].

Treatment of spinal injuries in vertebral or disc tissues is extensive and involves of mechanical, biological, chemical, or physical therapy. These methods are categorized as surgical or nonsurgical. Surgical methods are classified as minimally or highly invasive techniques, in which the extent of skin incision and soft tissue involvement is different. Highly invasive surgical treatments are used extensively, particularly for severe injuries (e.g., spinal deformity), whereas minimally invasive techniques are becoming more popular. Surgical treatment is managed with or without using the internal implantation of artificial devices or implants. For severe degeneration or injuries of vertebral or disc tissues, internal implantation of implants is employed as the final solution to reduce severe pain at the affected zone and enhance the functionality of the patient's spinal column. The clinical target of internally implanting spinal implants is to stabilize the vertebrae in a position so that the spinal cord is not irritated, compressed, or impressed. Based on the overall conditions of the patient and specific conditions at the affected site in the spinal column, the clinical strategy is different. It is determined by a specialist neurosurgeon through a preoperative diagnosis and study of the injury. Conditions such as bone quality, conditions at the adjacent vertebrae and discs, the level of obesity, the necessary functionality, the extent of disc height loss, the extent of disc degeneration, the extent of disc herniation, the extent of spinal cord narrowing, and so on need to be studied before surgery. The other clinical target is bony fusion and restoration of the

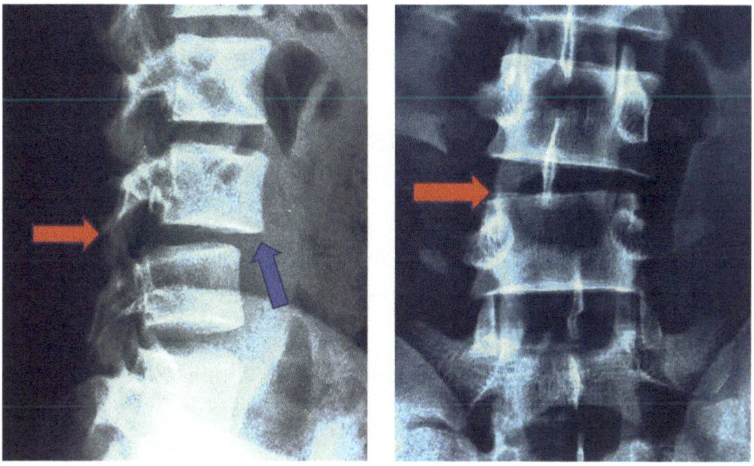

Figure 6.3 Illustration of vertebral dislocation at the lumber column between the L3/L4 (red arrows) in a patient aged 24 years. The cause of vertebral dislocation was reported to be listhesis (blue arrow), which weakened and separated the pars interarticularis. Bone tissue in the section between the superior and inferior articular processes at the facet joints was cracked or fractured and caused forward dislocation or movement of the vertebra. Such conditions might occur as a result of a severe vehicle accident or a fall. The cervical vertebrae are more affected than the lumbar vertebrae for this type of injury. *Through the courtesy of A. Younus, A.K.-I. Neurosurgery, undefined 2020, Lumbar Unifacet Dislocation Without an Associated Fracture-A Case Report and Literature Review, Elsevier, n.d. https://www.sciencedirect.com/science/article/pii/S221475192030534X. (Accessed 5 January 2021) and permission from Elsevier.*

disc height between vertebrae with highly severe disc degeneration, which necessitates removing disc tissue completely or partially. Bony fusion is managed in conjunction with stabilizing the vertebrae at the affected zone to provide stability to the spinal column to facilitate bony fusion between vertebrae. Various types of implants have been developed to manage stabilization or bony fusion at the affected zone of the spinal column in lumbar, thoracic, and cervical subcolumns. Overall, spinal implants may be categorized as screw—rod stabilization, intervertebral fusion cages, corpectomy cages, plating systems, and TDR. A screw—rod stabilization system is used to structure a bone—implant construct at the desired position, whereas intervertebral fusion and corpectomy cages are used for bony fusion and disc height restoration. The plating system is intended for vertebrae stabilization; however, it is adequate for anatomic or normal physiologic stabilization of the affected zone with no deviation or deformity. The next sections discuss interactions of bone with spinal implants and possible failures that might occur as a result of ineffective interactions of the implant and bone.

6.3 Interactions of vertebral bone with screw—rod stabilization system and possible failures

Screw—rod stabilization systems were reviewed in extensive detail in the Chapter 4 of *Osseointegration of Orthopedic Implants*, by the lead author. This system is composed of multiple components such as screws, rods, cross-links, lateral offsets, hooks, and dominos. Each component is designed and developed in different models with similar technical concepts. The aim of the screw—rod stabilization system is to fix vertebrae in the optimal position where the spinal cord is not compressed. First, screws are inserted through the pedicle aspects of the vertebra and purchased until the central zone of the vertebral body. The screw is designed in to be monoaxial, polyaxial, or uniaxial with different functions and features for vertebrae stabilization [5]. Then, the precontoured rod is inserted and fixed to the screw heads to stabilize spinal column alignment in the desired profile. Other components of the system are used for biomechanical enhancement of vertebrae stabilization in accordance with the type and extent of the spinal injury at the affected zone. Adequate purchase of screws to the vertebral bodies with enhanced integrity with bone tissue facilitates the transfer of loading conditions through the rods. Normally, the screws and rods are manufactured from metallic materials (e.g., titanium [Ti] alloy, cobalt-chromium [CoCr] alloy). Load transference through the metallic rod establishes a mechanically rigid structure at the implanted zone with no further movement of the spinal column at that zone. This rigid stabilization is the traditional target for treating severe spinal injuries with inadequate intervertebral disc physiologic function. In such conditions, ligaments at the facet joints and spinous zones might not be physiologically loaded, which may accelerate stress shielding conditions at these zones. The screw—rod system is first developed by monoaxial screws with rigidly connected rod and screws. In the case of the involvement of two vertebrae in a bone—implant construct of spinal stabilization, loading conditions are transferred from the vertebral bodies to the screws and rod. Various types of pedicle screws are illustrated in Fig. 6.4.

Because of the rigid fixation of the screw—rod construct to the vertebrae, loading conditions are concentrated at the linkage of the screw—rod and screw-bone structure. Considering the rigid attachment of the screw and

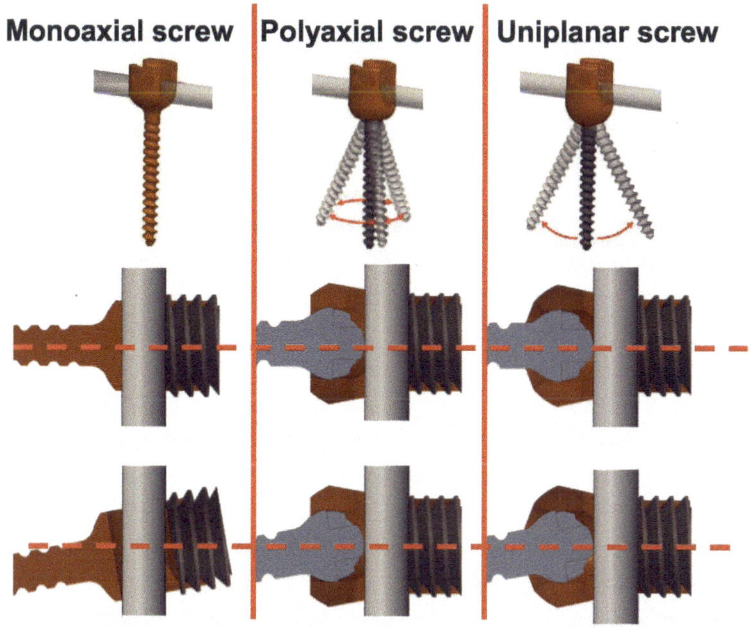

Figure 6.4 Illustration of monoaxial, polyaxial, and uniaxial pedicle screws. Each type of pedicle screw has a specific function and advantages in stabilizing the vertebrae at the affected zone. The monoaxial screw is adequate for enhancing the rigidity of the vertebrae–implant construct. It is beneficial to use this screw at the top and bottom vertebrae to enhance the rigidity of the whole construct. A polyaxial screw is useful for preserving the spinal anatomy at the affected zone, from which rod alignment could be effectively maintained physiologically. It is advantageous to use a polyaxial screw at the medial vertebrae. A uniaxial screw is used to provide rotational freedom in the craniocaudal axis, which is effective in reducing the vertebrae in anatomic position in spinal deformity or scoliosis surgeries. *Through the courtesy of P. Liu, P. Lai, C.L.-C. Biomechanics, undefined 2019, A Biomechanical Investigation of the Retentive Force of Pedicle Screw Structures for Different Screw Tulip Designs, Elsevier, n.d. https://www.sciencedirect.com/science/article/pii/S026800331831009X. (Accessed 6 January 2021) and permission from Elsevier.*

rod, the screw acts as a single cantilever beam. Thus, loading stresses are localized at the interface of the vertebral bone and screw. This condition affects both vertebrae (particularly, the trabecula at the internal zone of the vertebral body) and adjacent healthy intervertebral discs. Cyclic overloading at the bone–screw interface in long-term implantation causes progression of fibrous tissue formation and affects the integrity of the bone and screw [6]. This increases bone resorption at the interface of the bone and screw; it gradually progressed through the surrounding bone tissue far from the screw.

Ultimately, loosening of the screw inside the vertebral body results, from which instability of the bone–implant construct is observed. On the other hand, high rigidity of the bone–implant construct from spinal stabilization between vertebrae introduces a high level of reaction force to adjacent intervertebral discs. Energy transferred through the bone–implant construct is borne by the metallic screw–rod system with negligible strain, from which energy cannot be absorbed by the bone–implant construct and reacts through adjacent discs. This increased level of energy introduces greater loading conditions than that under physiologic conditions. Cyclic actions and reactions of energy and loading conditions at adjacent discs result in gradual failure at disc tissues, either annulus or nucleus.

Another condition is when more than two vertebrae are stabilized by a screw–rod system (Fig. 6.5). The two top and bottom vertebra have similar interactions and possible failures, as discussed for two vertebra stabilization constructs. However, the median vertebra may have a different response or reaction to the transferred loading condition through the bone–implant construct. Because much energy is borne by the metallic screw–rod structure, loading conditions are not transferred through the median disc and vertebrae. They are transferred through the rod and then transmitted through the median screws as physiologic, such as when there is no screw–rod stabilization. By increasing the number of the vertebrae and screws involved, the reaction force from the screw–rod structure to adjacent intervertebral discs increases, and thus the risk for failure in adjacent discs increases. In addition, an increase in the number of screws or of vertebral involvement in spinal stabilization increases the accumulated energy and load at the bone–implant construct. Owing to the attachment of muscles to the vertebrae, the tension load applied to each vertebra cannot be transferred to the lower disc and vertebral segment and instead is transferred to the screw–rod structure and gradually increases from the upper level of the screw–rod structure to the bottom level. Therefore, the risk for rod breakage or screw fracture in the lower screws increases at the close section of the rod to the screws or at the neck of the screw (Fig. 6.5). The scenario is different when polyaxial screws are used to stabilize the spinal column at the affected zone. With polyaxial screws, the main or shaft portion of the screw has a limited range of rotation in all directions relative to the screw tulip and rod (Fig. 6.6), which reduces the level of rigidity at the joint between the screw and rod. This feature of the polyaxial screw is used for easier implantation and to position the precontoured rod in the desired profile. The angular freedom of the polyaxial screw is completely fixed in the final stage

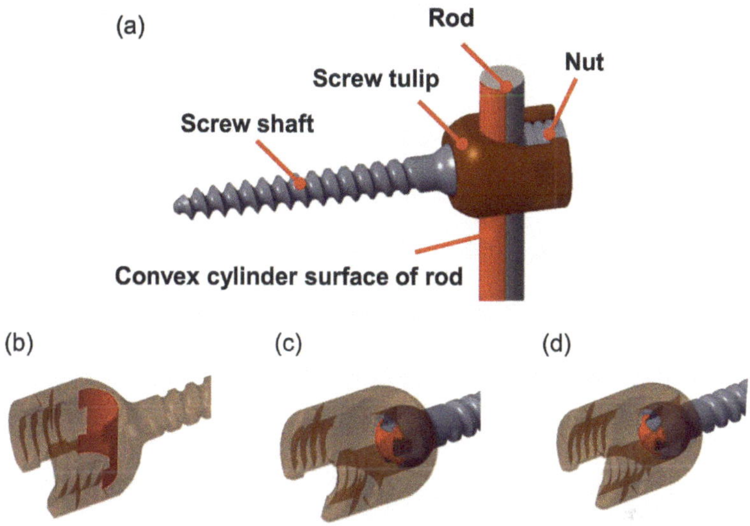

Figure 6.5 (A–D) Illustration of various portions of the polyaxial pedicle screw-. The screw shaft portion is inserted through the vertebral body through the pedicle aspect. In some techniques, the screw is inserted through another aspect of the vertebra. The tulip portion (also known as a king or rod connector) facilitates the connection of the screw shaft and rod. The nut fixes the rod to the head of the screw through a locking washer. *Through the courtesy of P. Liu, P. Lai, C.L.-C. Biomechanics, undefined 2019, A Biomechanical Investigation of the Retentive Force of Pedicle Screw Structures for Different Screw Tulip Designs, Elsevier, n.d. https://www.sciencedirect.com/science/article/pii/S026800331831009X. (Accessed 6 January 2021) and permission from Elsevier.*

of screw–rod stabilization. Thus, the flexibility of the polyaxial screw is converted into rigid fixation similar to a monoaxial screw. However, when the load is increased to a certain range, slippage occurs between the screw tulip and screw shaft. This feature facilitates a lower concentration of stress at the lowest level of screws and thus reduces rod or screw fracture.

Other components of the screw–rod stabilization system are used to enhance load transference or the stability of the bone–implant construct. The cross-link component is establishes a mechanical link between the two columns of screw–rod fixation. Cross-linking enhances the overall integrity of the screw–rod structure to bear the transferred loading conditions to the bone–implant construct. This component increases the stiffness of the construct against compressive axial torsion and lateral bending of the spinal column at the affected zone. As a result, the level of loading conditions is maintained in a more rigid condition in both columns of the screw–rod structures. The other component is a lateral offset connector, which is used

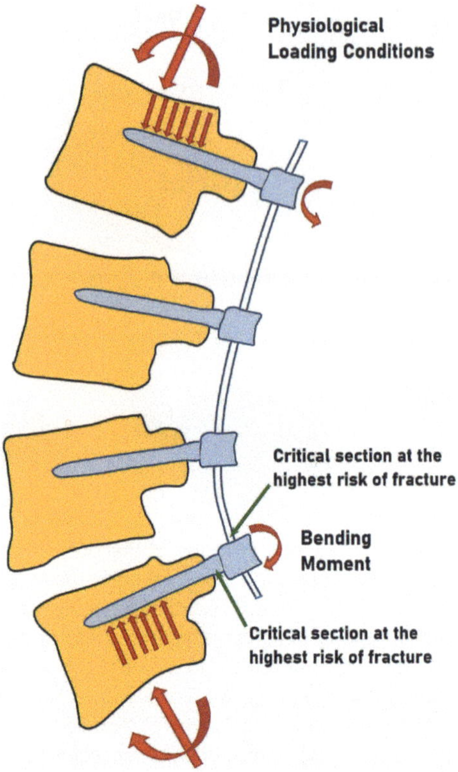

Figure 6.6 Illustration of loading conditions (red arrows) of bone–implant construct in screw–rod stabilization system at the lumbar subcolumn. Loading conditions in the sagittal plane are shown. These are the main physiologic loading conditions applied to the bone–implant construct. Effective action and reaction forces of the body weight load are loaded from proximal and distal portions of the construct in conjunction with the effective bending moment of flexion-extension. (Effective loading of lateral bending and axial torsion motions are not illustrated in this figure.). Physiologic loading conditions are transferred at the interface of the vertebral bone and screw shaft at the top and bottom levels of vertebrae. Adequate integrity of the screw and bone generates a bending moment to the rod in the sagittal plane, from which the rod should bend anteriorly. Median levels of the screws increase the stiffness of the construct; thus, the construct has increased bending strength to the generated bending moment in the sagittal plane. Likewise, the transferred compressive stress is distributed to all screws (mostly at the neck portions), but because of the high rigidity of the construct, the concentration of stress at the neck of the bottom screws or near the rod with the nut (*green arrows*), is expected, which increases the fracture of the screw or rod at these section, respectively.

when alignment of the rod does not coincide with the screw head in one or multiple vertebrae. The lateral offset connector acta as a cross-link between the screw heads and rod alignment in spinal stabilization with more than two involved vertebrae. To provide integrity in screw—rod stabilization between lumbar-thoracic or thoraco-cervical vertebrae, a domino connector facilitates the connection of two rods with different diameters (e.g., 6—3.5 mm). Figs. 6.7—6.10 show possible failures of the screw—rod stabilization system and their reasons [7—13].

6.4 Interactions of vertebral bone with intervertebral fusion cage

As highlighted in Section 6.2, intervertebral fusion cages were developed to facilitate bony fusion between vertebrae. An intervertebral fusion cage has two functions. First, it acts as a spacer between vertebrae to maintain disc height at a certain range; second, it provides a biological bridge between vertebrae that are intended to be fused. This type of spinal implant is developed from various types of materials, mainly titanium alloy, PEEK, hydroxyapatite-polyetheretherketone (HAPEEK), and rarely, bioactive glass [14], biodegradable polymer [15], bone graft, and so on. In this section, intervertebral cages made of common materials are reviewed and discussed.

6.4.1 Interactions of vertebral bone with titanium intervertebral fusion cage and possible failures

Traditionally, intervertebral cages were developed from titanium or titanium alloys to have sufficient mechanical strength under dynamic loading conditions that would be subjected to the cage. High levels of loads between lumbar vertebrae necessitate the use of a high-strength cage to maintain the disc height until bony fusion is complete. In cervical vertebrae, a lower-strength cage might be used compared with that in lumbar vertebrae; however, a greater rate of motion in this part of the spinal column generates a considerable amount of rotation torque, which may affect the implanted cages at the intervertebral space. Although titanium alloy is secure in maintaining disc height, in long-term implantation, vertebral bone subsidence through the titanium cages was observed. Therefore, in the next step, the design modeling of titanium cages was modified to obtain a lower range of stiffness and reduce subsidence. The great mechanical mismatch between the bone and titanium generates a large amount of reaction force to the bone at the contact surfaces of the cage and end plate portion of the vertebral

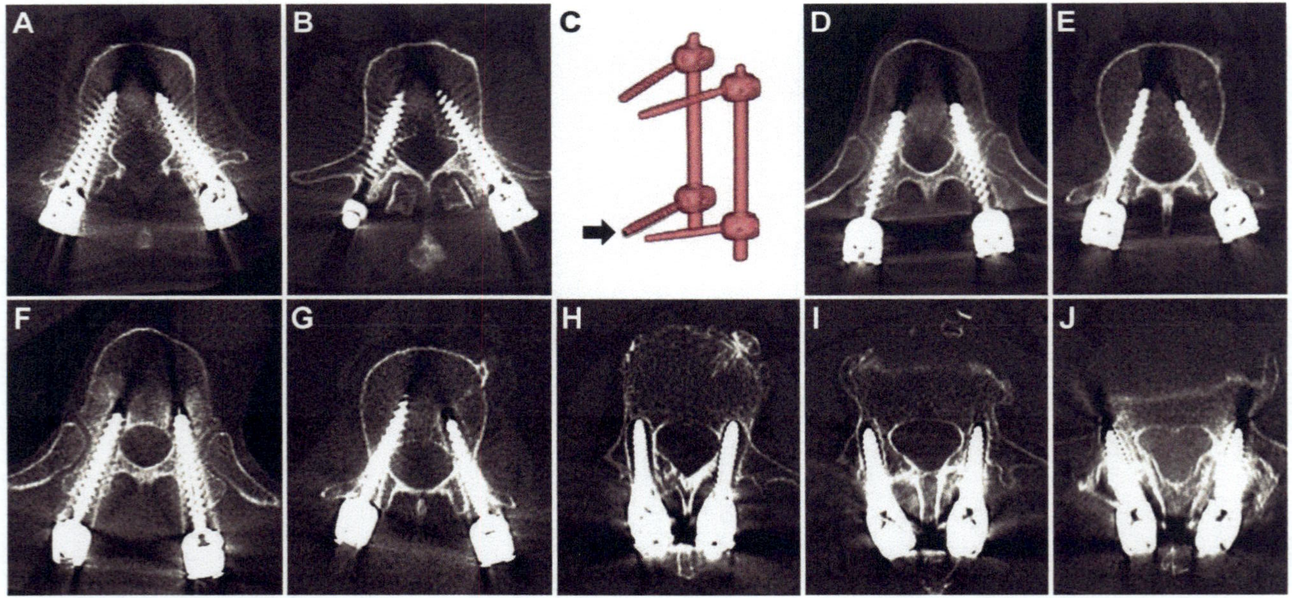

Figure 6.7 Radiographic illustration of screw lucent zones and loosening in three cases. (A–B) Case 1. (C–G) Case 2. (H–J) Case 3. In case 1, no radiographic lucent zone is observed 18 months after the operation (B). In case 2, the T11 and L1 vertebrae are stabilized by a screw–rod stabilization system. Screw reposition at the shaft portion is detected (*black arrow*) at the right screw of the L1 after 1 month (C). The positions of the screws at 1 day and 1 month after the operation were evaluated through superposed stereolithography for the accurate detection of changes in the position of the screw. Axial radiographic computed tomography (CT) at 1 month after the operation for T11 (D) and L1 (E) show no lucent zone at the interface of the screw and bone. At 14 monthsoperation, axial CT scans show radiolucent zones at both screws in T11 (F) and the left screw in L1 (G). Deterioration of screw and bone integrity cause loosening and posterior dislocation of the screw from its initial position and generate bone loss zones at the tip of the loosened screws. The length of the unloosened screw at the right side of L1

vertebra seems to be longer and has a relatively greater bone–implant contact surface. In case 3, radiolucent zones at the interface of the screw shaft and pedicle section of the vertebrae in L3 (H), L4 (I), and L5 (J) are detected on axial CT scans. In this case, the screws were purchased through the section with highest cortical compactness for increased integrity with the vertebrae. Cortical placement of the screw in lumbar vertebrae necessitates the use of a shorter screw and increases interfaces with cortical bone compared with conventional posterolateral pedicle placement. However, such a condition may result in the localization of stress at the shorter zone and overloading conditions, with a reduction in osteogenic cell activity, progression of bone resorption, and an increase in fibrous tissue formation at the bone–screw interface. Although cortical placement might be biomechanically beneficial shortly after implantation [7,8] (particularly for osteoporotic bones), its long-term efficacy might not be good, as seen in case 3. *Through the courtesy of S. Tanioka, F. Ishida, K. Kuraishi, K.T.-W. Neurosurgery, undefined 2019, A Novel Radiologic Assessment of Screw Loosening Focusing on Spatial Position Change of Screws Using an Iterative Closest Point Algorithm with Stereolithography, Elsevier, n.d. https://www.sciencedirect.com/science/article/pii/S1878875019300889. (Accessed 3 January 2021) and permission from Elsevier.*

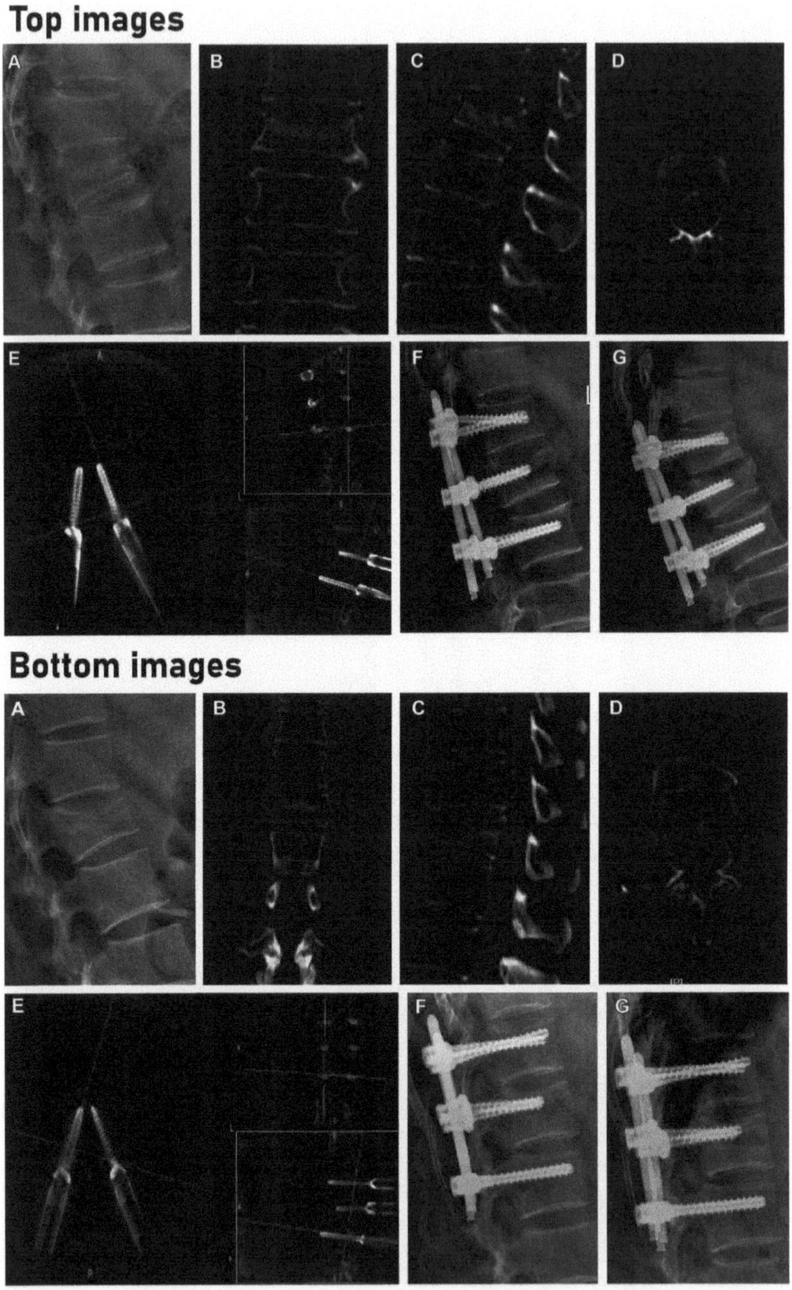

Figure 6.8 Radiographic illustration of screw–rod stabilization system to treat vertebral fracture in (Top: A–G) L1 vertebra using polyaxial screws at the T12 to L2 vertebrae (polyaxial construct); (Bottom: A–G) T12 vertebra using monoaxial screws at the T11

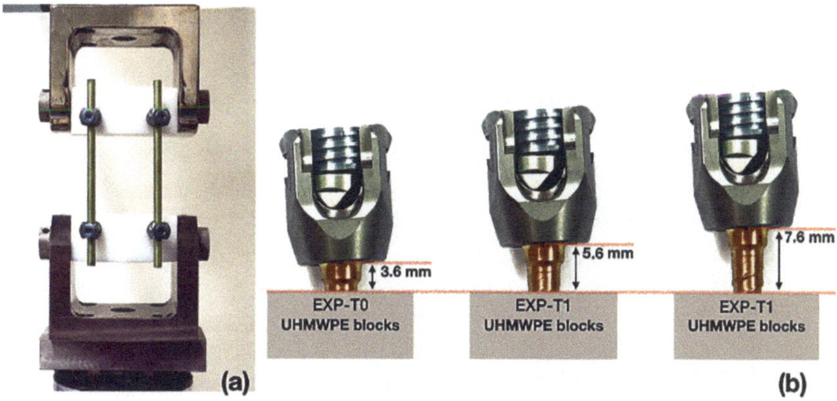

Figure 6.9 Illustration of pedicle screw fixation to ultrahigh molecular weight polyethylene (UHMWPE) blocks with various unsupported lengths (A, B). This biomechanical evaluation test was carried out in accordance with ASTM F1717, in which UHMWPE blocks stimulated vertebra. The results showed that the fatigue strength of the bone–implant construct was reduced when the screw shaft was not purchased completely through the vertebra. Allowing the screw shaft thread (T2 group) to remain out of the vertebra reduced fatigue strength significantly. This caused posterior soft tissue irritation and reduced the stability of distraction at the anterior aspect of the vertebra. *Through the courtesy of Y.L. Chu, C.H. Chen, F.Y. Tsuang, C.J. Chiang, Y. Wu, Y.J. Kuo, Incomplete insertion of pedicle screws in a standard construct reduces the fatigue life: a biomechanical analysis, PLoS One. 14 (2019). https://doi.org/10.1371/journal.pone.0224699 and permission from PLoS One.*

body. Such localized increased stress create overloading conditions to the bone matrix and reduces osteogenic cell activity, generating fibrous tissue with much lower mechanical properties (<2 MPa for fibrous tissue vs. >4000 MPa for bone tissue at the vertebral end plate). As a result, the titanium cage penetrated the affected area of the end plate. Ultimately, the disc

and L1 vertebrae and polyaxial screw at T12 vertebra (mono-poly-mono axial construct). The loss of anterior vertebral body height and Cobb angle was detected in the polyaxial construct at 12 months postoperative follow-up. The collapse of the T12 vertebral body and screw shaft was observed, which was due to sliding of the screw shaft over the tulip component of the polyaxial screw. This shows the weaker ability of the polyaxial screw to maintain distraction stability or rigidity of the top vertebra at the anterior side [10]. *Through the courtesy of W. Qin, K. Chen, H. Chen, P. Yang, H. Yang, H.M.-W. Neurosurgery, undefined 2020, Comparison of Poly-Axial or Poly-/Mono-Axial Mixed Screw Fixation for Treatment of Thoracolumbar Fractures with O-Arm Navigation: A Case Control Study, Elsevier, n.d. https://www.sciencedirect.com/science/article/pii/S1878875020301418. (Accessed 3 January 2021) and permission from Elsevier.*

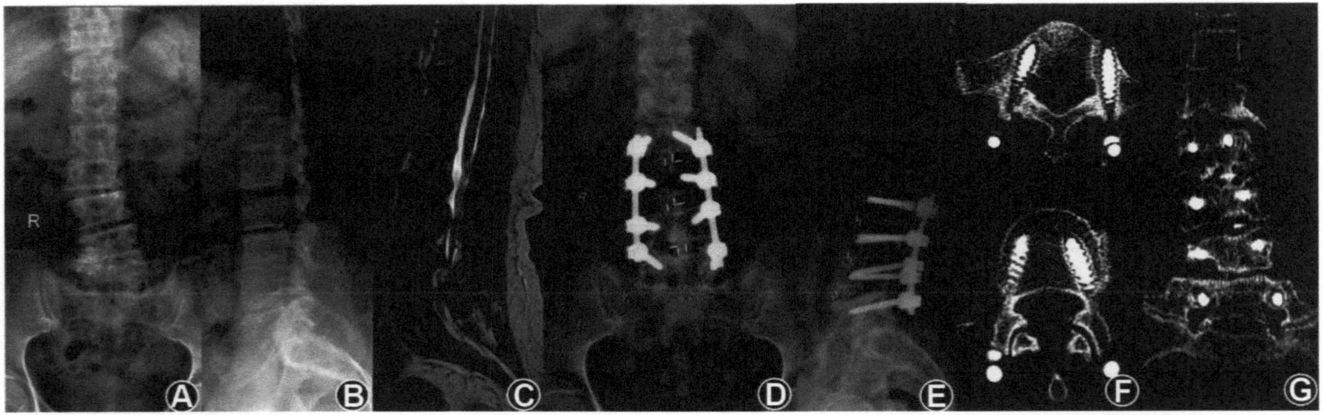

Figure 6.10 Radiographic illustration of screw loosening and vertebra nonfusion (D–G) in treating disc degeneration and scoliosis between L2 and L5 vertebrae (A–B) when disc herniation and canal stenosis were diagnosed in the magnetic resonance image (C). Screw loosening was detected in the 6-months post operation follow-up (D–E) at the L5 vertebra, which progressed with significant radiolucent zones at the interface of the screw and bone on 3-years post operation scans (F–G). At 3 years operation, screw loosening and a radiolucent zone were observed in the left screw of the L2 vertebra. Likewise, fusion was recognized in the intervertebral space of L2/3 and no progressed callus formation was apparent in the interspace of L3/4 and L4/5 through the implanted intervertebral fusion cages. *Through the courtesy of H. Guo, Y. Tang, D. Guo, Y. Ma, K. Yuan, Y. Li, J.P.-W. Neurosurgery, undefined 2020, Pedicle Screw Fixation in Single-Level, Double-Level or Multilevel Posterior Lumbar Fusion for Osteoporotic Spine: A Retrospective Study with a Minimum 2-year Follow-Up, Elsevier, n.d. https://www.sciencedirect.com/science/article/pii/S1878875020309086. (Accessed 3 January 2021) and permission from Elsevier.*

height might be reduced, which is detected as the subsidence of the cage through the bone (Fig. 6.11). The titanium cage is modified by reducing the cage walls to reduce the overall stiffness of the cage while providing greater empty space to insert the bone graft and a thicker bone structure between vertebrae. The optimal range of cage wall needs to be evaluated through experimental and computational methods, when physiologic loading conditions (preferably in cyclic form) are applied to the bone–implant construct. The anatomic design of the cage is another modification strategy for reducing subsidence at the cage wall. In this strategy, the superior surface of the cage is designed to be convex in either the sagittal or frontal planes, whereas the inferior surface is shaped to be slightly concave. The curved profile of the cage at the superior and inferior surfaces dedicates the end plate geometry or anatomy of the vertebral body (Fig. 6.11). An anatomic cage enhances the physiologic loading distribution through the cage and reduces stress localization or concentration at the contact areas of the cage walls. An enhanced load transferring mechanism or pattern at the interface of the vertebra and cage provides a greater compressive load

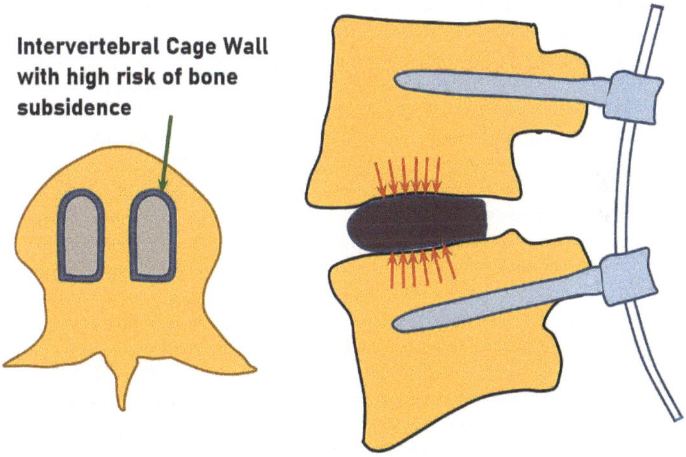

Figure 6.11 Illustration of intervertebral fusion cage replacement at the interspace between vertebrae (A, B). The cage is best designed anatomically with a concave profile at the vertebra end plate. The superior and inferior surfaces of the cage are curved for maximal contact interfaces with the vertebra end plates. This allows load transference at the central area of the cage (B, *red arrows*), where the bone graft is compacted at the internal empty space of the cage. Therefore, preload conditions are created for the bone graft, which accelerate bony fusion. Likewise, the anatomic design reduces loading to the cage walls (A) and thus reduces the risk for subsidence in long-term implantation of the cage.

through the replaced bone graft at the internal space of the cage. The presence of loading conditions on the bone graft is considered a preload for the early mechanical stimulation of the osteogenic cells to initiate bone formation. An anatomic design and bone graft preload accelerates bony fusion between vertebrae, and thus the risk for nonfusion or cage migration is reduced.

These design features reduce undesired effects of the titanium cage for treating degenerated discs. The titanium cage is still being modified for enhanced efficacy in bony fusion treatment. In addition to this anatomic design, the lateral walls of titanium cages are designed with a slot. Such a design feature reduces the overall stiffness of the titanium cage and increases elastic strain at the cage walls, reducing the reaction force or energy to the end plate, and ultimately a lower rate of cage subsidence. The presence of a slot in the cage wall enables the diagnosis and follow-up of the progression or conditions of bony fusion during healing, from which the surgeon can adjust postoperative treatment plans and therapies to enhance the clinical outcomes.

Normally, titanium cages are manufactured by machining, specifically milling. Milling may not have a significant effect on the overall bulk of the machined component if the process parameters (e.g., amount of milling loading, milling tool speed, quality of the milling tool, milling cooling) are optimized. However, milling can generate microcracks and residual stress on the surface of the component, which can be removed or relieved through polishing or surface treatment processes (e.g., electropolishing, stress relief [heat treatment] processes, fine polishing, magnetic polishing, anodizing, hard anodizing, nanostructuring). Additive manufacturing processes (also known as three-dimensional (3D) printing or rapid prototyping processes) are used to fabricate titanium intervertebral cages. This category of manufacturing has various benefits of functionality and performance for intervertebral fusion cages in bony fusion treatment. First, the mechanical properties of the processed component that are achieved with additive manufacturing methods are not be as good as for component that is machined from a material sheet or a rod that was produced through rolling and other hardening processes on a large scale. In additive manufacturing, components are made through localized melting of microsize powder and structuring into a solid form. Therefore, the overall enthalpy of the component is in higher range, which results in weaker molecular bonding and lower mechanical properties. This is a positive feature in the development of titanium intervertebral cages with a lower mechanical modulus and greater elastic strain under physiologic loading conditions between vertebrae. On the other hand, through 3D printing, the structure of the component is organized in a porous form as whole or partial. Porous structuring in the whole is another strategy to reduce the overall mechanical properties of the cage; thus, there is greater elastic strain at the contact surfaces with the vertebra. Porous structuring is designed at the contact surface of the cage with the vertebra at a few micrometers to millimeters deep to establish an osteoconductive layer on the superior and inferior surfaces of the titanium cage. Additive manufacturing has processing challenges that reduce its effectiveness in the homogeneous integrity of the processed component through a laser or electron beam source of energy. The use of these processes is necessary combined with rigorous mechanical validation testing to optimize the process parameters. Also, precise quality control planning is needed to inspect the process parameters during mass production. Fig. 6.12 illustrates the failure of titanium intervertebral cages [16] its feasible reasons. The use of a 3D printed titanium fusion cage [17] is illustrated in Fig. 6.13.

6.4.2 Interactions of vertebral bone with polyetheretherketone (PEEK) intervertebral fusion cage and possible failures

PEEK polymer was introduced as an alternative material for use in the orthopedic implant industry as a substitute for metallic materials such as titanium and stainless steel. Although the mechanical properties of the PEEK polymer are significantly lower than those of titanium (the mechanical modulus is 4 GPa for PEEK vs. 113 GPa for titanium and titanium alloys), its close mechanical modulus to that of cortical bone is a motivation to use this material for implants with a lower risk for mechanical failure as a result of increased loading conditions. One of the first group of implants commercially developed with PEEK polymer was intervertebral fusion cages [18]. This was implemented after implant-based complications occurred from the first generation of titanium intervertebral cages. The product developers were searching for a substitution material to reduce the rate of subsidence to prevent or eliminate significant reduction in disc height at the implanted zone. No failure of the titanium cage was reported; thus, it was concluded that the level of force might not be high at the vertebrae. Thus, the first generation of PEEK polymer cages was developed as a spacer with a low amount of internal bone graft space for implantation between vertebrae. The PEEK cage was created with a thick wall for adequate mechanical strength against loading conditions at the implanted zone. Owing to poor

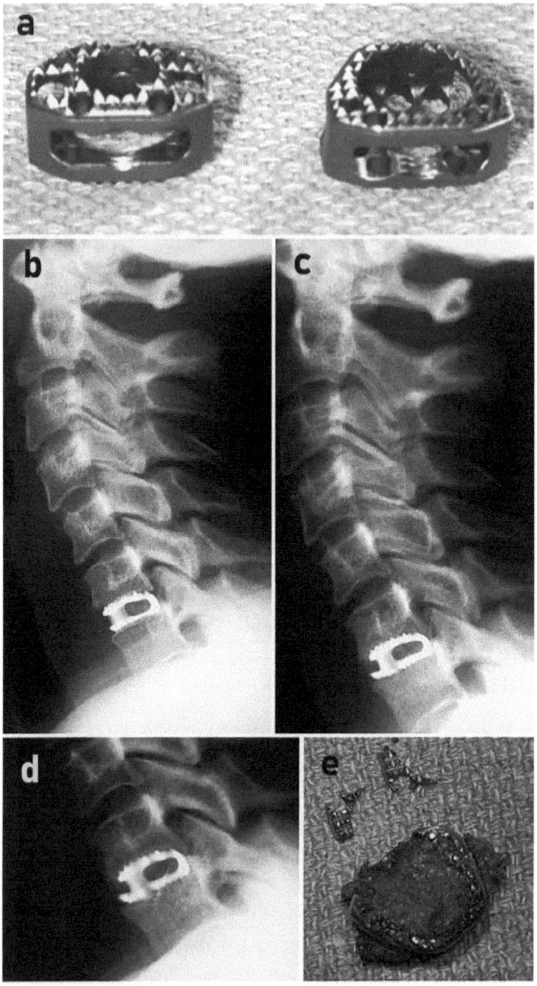

Figure 6.12 (A) Illustration of disc injury treatment at the intervertebral space of C6/7 using titanium anatomic cage in a 39-year-old man. (B). Cage subsidence was detected on a radiographic scan 6 months after the operation, with significant deformation of the cage at the anteroinferior zone (C). Subsidence was progressed and cage fracture at the anteroinferior zone was diagnosed at the 15-months postoperation follow-up (D, E). This cage was implanted as stand-alone with no supportive stabilization system (e.g., plating fixation) or fixator element (e.g., blade, screw, plate). This could be the reason for significant subsidence through the titanium cage. In general, physiologic motions in the cervical subcolumn is high, and thus may introduce increased of rotation torque in three axes, particularly at the lower levels of C5–7. Significant mechanical mismatch between the cage and bone causes overloading at the interface of the metallic cage and bone, and thus bone suppression and the acceleration of cage subsidence. Cage fracture occurred owing to the narrow design of the cage at the anterior zone to reduce cage stiffness. *Through the courtesy of E. Gercek, V. Arlet, J. Delisle, D. Marchesi, Subsidence of stand-alone cervical cages in anterior interbody fusion: warning, European Spine Journal 12 (2003) 513–516. https://doi.org/10.1007/s00586-003-0539-6 and permission from Springer.*

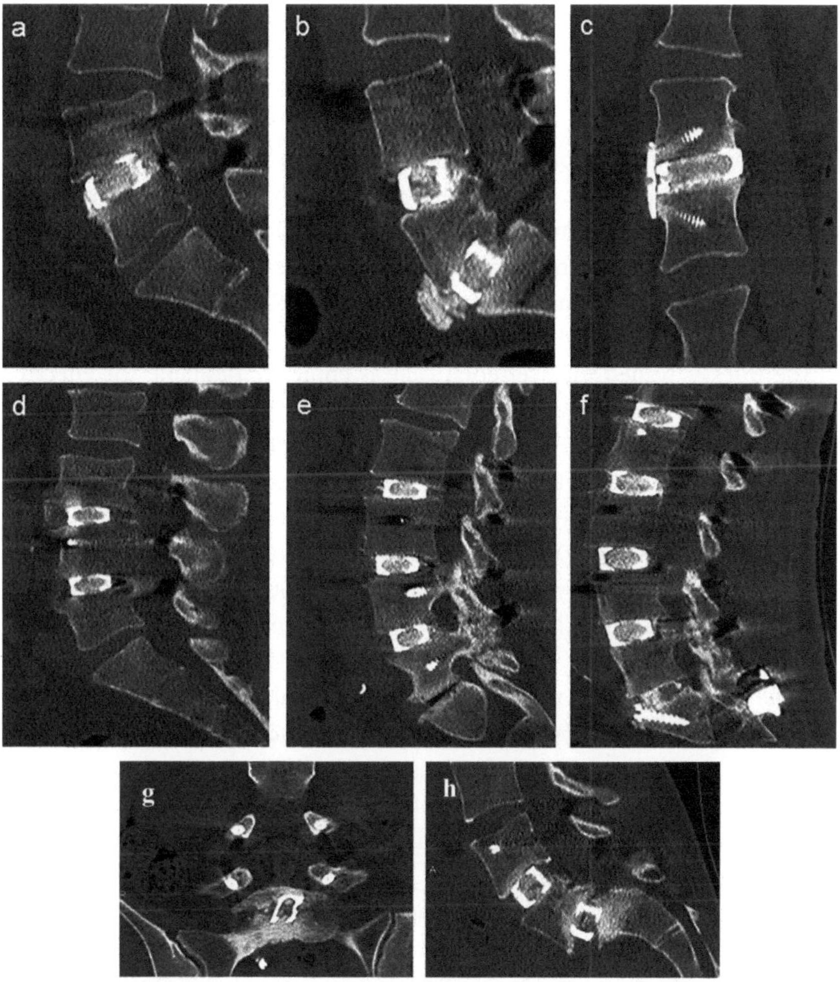

Figure 6.13 Radiographic illustration of lumbar disc injury treatment using three-dimensional printed lamellar titanium lateral lumbar intervertebral fusion (LLIF) and transforaminal lumbar intervertebral fusion (TLIF) cages (A—H). This cage was fabricated by additive manufacturing (selective high-energy laser beam melting) to form a cage in both solid and porous structures and facilitate bone on-growth, in-growth, and through-growth in interaction with the cage. The stiffness of the cage was close to that of conventional an LLIF and TLIF PEEK cage, which reduced the risk for vertebra end plate and pseudoarthrosis. Bony fusion was observed at the high rate of greater than 98% after the 1-year follow-up, as reported in the study by Mokawam et al. [17] (-F). However, the risk for nonfusion at L5/S1 was highlighted, as observed in a patient with grade II spondylolisthesis at this level, which might be due to poor stabilization of the bone—implant construct at L5/S1 (G, H). *Through the courtesy of M. Mokawem, G. Katzouraki, C.H.-J. of C., undefined 2019, Lumbar Interbody Fusion Rates with 3D-Printed Lamellar Titanium Cages Using a Silicate-Substituted Calcium Phosphate Bone Graft, Elsevier, n.d. https://www.sciencedirect.com/science/article/pii/S0967586819300244. (Accessed 3 January 2021) and permission from Elsevier.*

hydrophilic and bioactivity characteristics of the PEEK polymer at its surface, the activity of osteogenic cells and the formation of bone matrix might not be achieved at the vertebral–cage interface and fibrous tissue might be formed instead. The lack of biological integration between PEEK and bone tissue causes migration or dislocation of the cage owing to the rotational movement of the spinal column (there is nomechanical integrity between the cage and vertebra; thus, the cage might be gradually slide anteriorly or posteriorly, from which bony fusion would not be achieved and soft tissues at anterior or spinal cord at the posterior aspects of the cage might be irritated and cause pain). On the other hand, the low volume of the bone graft space in the thickly walled intervertebral cage is clinically desirable for effective bony fusion between vertebrae. Biomechanical investigations showed that the wall thickness of PEEK cages could be reduced with effective and sufficient mechanical strength. Attempts were made to enhance the osteoconductivity or bioactivity of the PEEK cage surface in contact with bone tissue, as shown in Table 6.1. (Osteoconductive surface enhancement of the PEEK polymer was reviewed by the lead author in Chapter 7 of *Osseointegration of Orthopedic Implants* and Chapters 3 and 6 of *Osseoconductive Surface Engineering for Orthopedic Implants*).

The second generation of PEEK intervertebral fusion cages was developed with an anatomic design, thinner wall, enhanced surface osteoconductivity, optimum bone grafting space, and material composition enhancement (design enhancement of intervertebral cages was discussed in Section 6.4.1 for titanium cages, which had concepts similar to those of PEEK cages). The composites of the PEEK polymer with fillers of carbon fiber and hydroxyapatite were developed as carbon fiber-reinforced PEEK (CFRPEEK) and HAPEEK. Spinal intervertebral cages are manufactured from these two composites for better biomechanical and biological performance. CFRPEEK with 30% weight carbon fiber has mechanical properties close to those of cortical bone. Thus, load transference between the bone and cage is enhanced, whereas the wall thickness is reduced compared with a cage manufactured from pure PEEK. The mechanical modulus of CFRPEEK is about 15 GPa vs. 4 GPa for pure PEEK. HAPEEK with 6%–10% weight hydroxyapatite particles has enhanced bioactivity at its surface, where the biological integrity of superior and inferior surfaces of the cage with the vertebral end plate is enhanced. Bone on-growth on the internal surface of the cage wall is facilitated, from which biological attachment of created bone tissue to the cage internal space is achieved.

Table 6.1 Different surface processing methods for osteoconductivity enhancement of polyetheretherketone cages in direct contact with vertebral bodies.

Type of osteoconductivity enhancement	Material engineering process
Porous structuring of surface	Sodium chloride microcrystal deposition
Highly thick titanium coating	Plasma spraying process
Low-thickness titanium coating	Physical vapor deposition process
Nano-enhanced the surface energy	Accelerated neutral atom beam
Placement of three-dimensional titanium porous structured scaffold through the polyetheretherketone cages	Selective laser melting and injection molding

Fig. 6.14 shows fusion treatment of the degenerated disc in the cervical subcolumn using a PEEK intervertebral fusion cage.

6.4.3 Interactions of vertebral bone with stand-alone intervertebral fusion cage and possible failures

Stand-alone fusion cages are stabilized without a conventional stabilization fixator such as screw—rod or plating systems. The cage is equipped with a fixator element such as a blade, plate, or screws through adjacent vertebrae. Stand-alone cages are beneficial for bony fusion of one degenerated intervertebral disc with enhanced mechanical integrity of the cage and adjacent vertebrae. Such a cage fixation system facilitates the achievement of a less rigid bone—implant construct compared with screw—rod and plating fixation systems. The risk for cage migration in this intervertebral implantation technique is significantly reduced owing to direct stabilization of the cage with the fixator element. Stand-alone cages have disadvantages that limit their use in bony fusion. Because of the involvement of the fixator element inside the cage, the bone grafting space might be lower than fusion cages without a fixator element. This reduces the volume of the formed bone between vertebrae. However, because fusion cages are permanently implanted inside the body, this reduced volume is not a significant disadvantage. Stand-alone cages with screws are implanted by purchasing two to four screws through the main body of the cage. During purchase of the screws through the vertebral bodies, adjacent vertebrae are compressed to the superior and anterior surfaces of the cage, and thus a preload force is generated to the cage

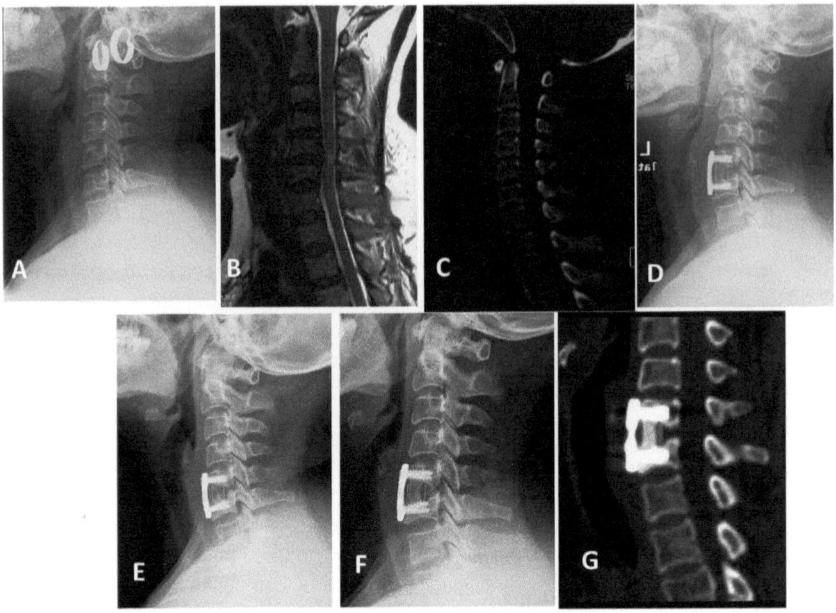

Figure 6.14 Illustration of cervical disc injury treatment with polyetheretherketone (PEEK) intervertebral fusion cage filled with local bone graft (A–G). PEEK cervical cages are normally implanted by supportive stabilization fixation using a plating system. Progression of the fusion: (E) 3 months (F, G) 12 monthsoperation. In the computed tomography scan 12 months operation, fusion of two vertebrae is seen where the bony fusion column is formed at the anterior zone of the interspace. *Through the courtesy of J. Liu, X. Xiong, A. Peng, M. Xu, ... X.C.-C.N. and, undefined 2017, A Comparison of Local Bone Graft with PEEK Cage Versus Iliac Bone Graft Used in Anterior Cervical Discectomy and Fusion, Elsevier, n.d. https://www.sciencedirect.com/science/article/pii/S0303846717300446. (Accessed 3 January 2021) and permission from Elsevier.*

bone graft and enhances bony fusion in a shorter period after the operation. Multiple implantation of the stand-alone cages in multiple adjacent degenerated intervertebral discs is challenging owing to the limited volume of vertebral bodies for implanting several fixator elements, particularly in the cervical subcolumn. This issue could be resolved by the effective design of the fixator element for insertion without impinging on other fixator elements of the stand-alone cage in adjacent vertebrae, occupying less of the vertebral body to reduce the risk for bone resorption or vertebral fracture. Clinical studies reported early results of stand-alone cages and highlighted satisfactory clinical outcomes compared with conventional cages in the treatment of degenerative disc disease [19–24]. Thus, their failures have not been reported to provide more discussion regarding various types of failure and

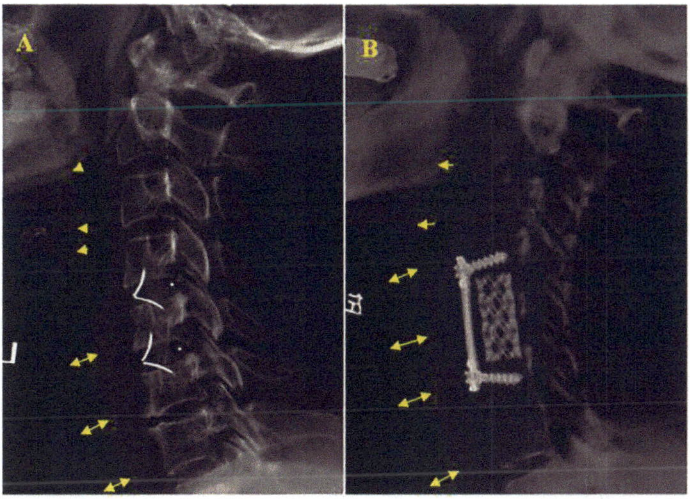

Figure 6.15 Radiographic illustration of stand-alone cage in treatment of cervical disc degenerative diseases (A). This type of stand-alone cage is designed with no projection from the vertebrae; it is known as a zero-profile cage. The fixator element has the shape of a blade, which is replaced after the placement of the polyetheretherketone cage from the anterior aspect of the cage through the vertebral bodies without interfering with the bone grafting space. This allows enhanced and increased internal space for bone graft placement inside the cage compared with other types of stand-alone cages that are fixed by screws. In a cervical subcolumn, such an advantage reduces of anterior narrowing and the risk for dysphagia in anterior cervical discectomy and fusion compared with conventional implantation methods that use anterior placement of the plate on the vertebrae and irritate the trachea (B). *Arrows* indicate the distance between the bone–implant construct and trachea. *Through the courtesy of G. Zhang, Z. Chen, Q. Liu, Z. Ji, C. Lan, H.L.-J. of Surgery, undefined 2017, Effect of swallowing function after ROI-C anterior cervical interbody fusion, Article. Josurgery. Org, n.d. http://article.josurgery.org/html/10.11648.j.js.20160406.14.html. (Accessed 8 February 2021) and Science Publishing Group.*

possible reasons. The use stand-alone cages is illustrated in Fig. 6.15. (In the next edition of this book, this part will discussed in more depth based on reported failures for this type of intervertebral fusion cage.)

6.5 Interactions of vertebral bone with corpectomy cages and possible failures

Corpectomy cages are replaced at the spinal column to treat the severe failure of vertebral bodies; thus, they are also known as vertebral body replacement [25–28]. The level of bone loss or fracture pattern in the vertebral body may necessitate removing the whole vertebra or a significant segment of it for replacement of a cage with adequate geometry, size, and bone grafting space to facilitate bony fusion between adjacent vertebrae to the affected vertebral bodies. Most developed corpectomy cages were developed with titanium alloy, with a large bone grafting space to be expandable or static. The static type is known as a mesh cage. It is designed as a tube with many peripheral holes. The mesh cage has extensive use in the cervical subcolumn as a vertebral replacement implant. The expandable corpectomy cage can expand after replacement at the affected zone to adjust the distance between two adjacent vertebrae at the desired value with adequate physiologic conditions. The bone graft space in an expandable cage is lower than that in a mesh cage. Thus, the volume of bone formation between vertebrae may not be effectively achieved. Expandable cages are normally used in the lumbar subcolumn. Various designs of expandable cages with different mechanisms have been developed and commercialized. In some models, PEEK or PEEK composite materials are used in the manufacture of expandable corpectomy cages. The mechanical mechanism for expansion in this corpectomy cage is a major challenge for effective long-term performance against physiologic loading conditions, particularly at the lumbar zone. Mechanical failure of the engaged components when it is expanded occur as a result of fatigue, which cause a significant reduction in cage height and loosening of the implant. A controversial issue in using expandable cages is their excessive compressive force at end plate zones of the involved vertebrae. When expanding of the cage, an amount of compressive force is applied to vertebral bodies at the end plate portions, which introduces a large amount of preload or constant load to bone tissue after implantation. As highlighted in Section 6.3, overloading conditions to the bone matrix at the interface of the bone and implant, while the implant has been purchased to the bone, increases the activity of osteoclast cells and forms fibrous tissue at the bone—implant interface. (The top and bottom sections of expandable cages are equipped with end plate components with adequate angular positioning featuring sharp teeth for primary subsidence or to engage the cage through the vertebra [25,28—30].) Bone resorption and the formation of fibrous tissue cause pseudarthrosis and bone loss at the interface of the bone and implant. Figs. 6.16—6.19 illustrate possible failures of corpectomy cages in the published biomechanical or clinical studies [25,28,31,32].

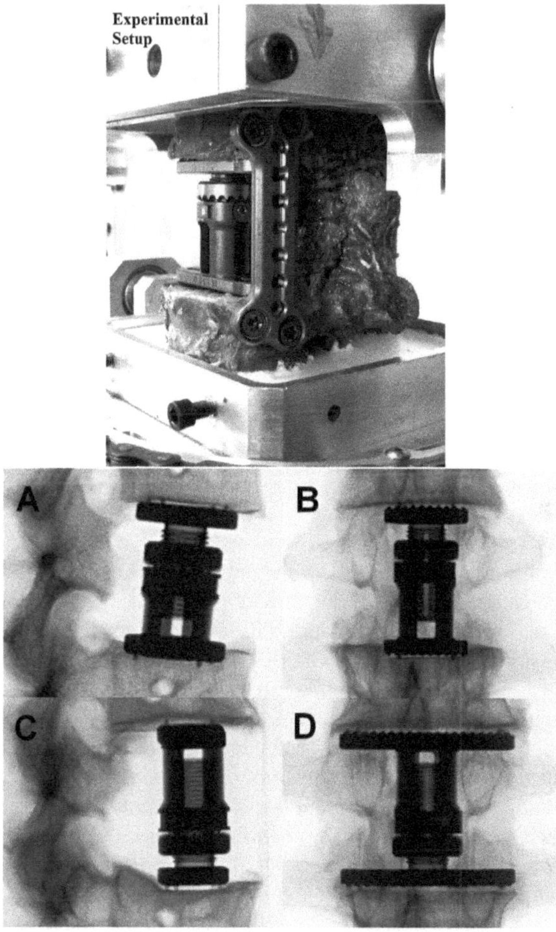

Figure 6.16 Illustration of expandable corpectomy cage subsidence in a biomechanical study using (A, B) round and (C, D) rectangular endcaps replaced in T12–L2 cadaveric spinal segments with L1 corpectomy (E). The biomechanical setup was loaded with a cyclic compressive force. The result revealed the greater resistance of corpectomy with the rectangular endcap to subsidence. The rectangular endcap had better peripheral contact surface with the end plate of the vertebrae to resist loading conditions. However, the involvement of physiologic rotational motion, particularly flexion-extension, was required to provide comprehensive biomechanical judgment regarding this achievement. Ideally, a round endcap could be designed anatomically to have a convex profile at the contact surface with the end plate. Subsidence at the interface of the metallic corpectomy and vertebrae was accompanied by end plate fracture, which is highly catastrophic. *Through the courtesy of A. Deukmedjian, J. Manwaring, ... T.L.-J. of C., undefined 2014, Corpectomy Cage Subsidence with Rectangular versus Round Endcaps, Elsevier, n.d. https://www.sciencedirect.com/science/article/pii/S0967586814001593. (Accessed 7 January 2021) and permission from Elsevier.*

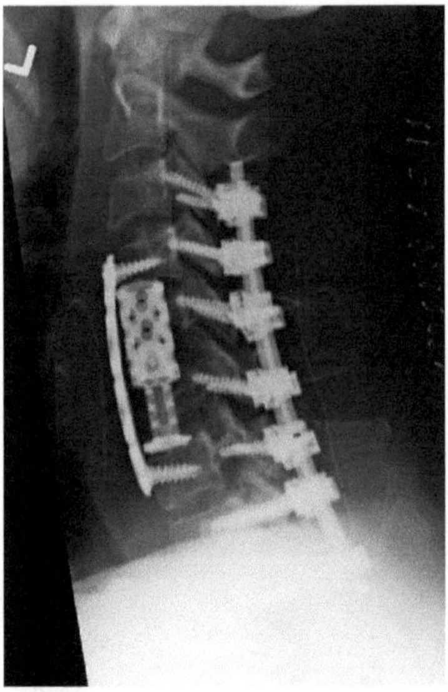

Figure 6.17 Illustration of expandable corpectomy cage for decompression of the spinal cord at the cervical zone with corpectomy of C5 and C6. Anterior plating was performed to stabilize adjacent vertebrae during fusion. Posterior screw—rod stabilization from C3 to T1 was managed to reduce the risk for cage subsidence to adjacent vertebrae caused by the excessive concentration of stress at the cage-bone interface because of the tendency of the cervical column to be rotated in three axes of physiologic motions. Poor stability of the corpectomy reconstruction causes the loss of alignment of the cervical column. *Through the courtesy of A. Awad, D. Stidd, K. Alkhalili, I. Eli, A. Baaj, Vertebral body reconstruction using expandable titanium cages after anterior decompression for cervical spondylotic myelopathy: a review, Cureus 6 (3) (2014). http://citeseerx.ist.psu.edu/viewdoc/download?doi=10.1.1.957.1153&rep=rep1&type=pdf. (Accessed 3 January 2021) and Cureus.*

6.6 Interactions of vertebral bone with expandable cages and possible failures

Intervertebral fusion cages are developed with expanding features [34,35]. Expandable intervertebral cages have multiple components using titanium alloy and PEEK polymer materials. There is a central component manufactured from titanium alloy to have sufficient mechanical strength; the other two components are manufactured from titanium alloy or PEEK polymer as end plate elements to be expanded, and which are joined with the central component. Expandable intervertebral cages have several advantages compared with conventional static cages. First, this category of fusion cages has better fit interfaces with the end plates of the vertebrae.

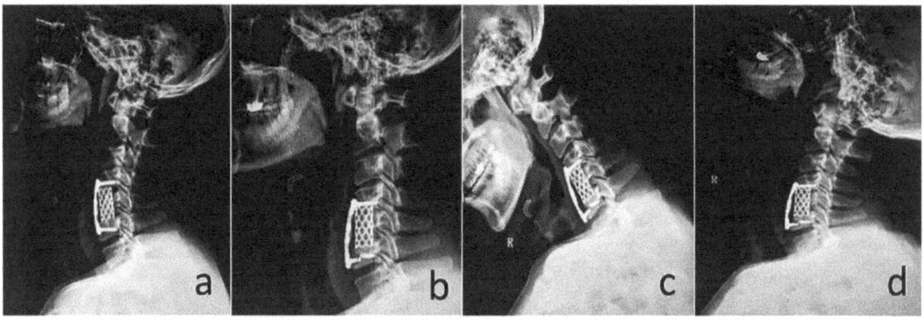

Figure 6.18 Illustration of decompression of the spinal cord at the cervical zone using a static corpectomy cage (mesh cage) in a 43-years-old man. Early screw fracture at the distal zone of fixation was diagnosed in X-ray image at 1 month after the operation. No serious subsidence was reported, and no evidence of callus formation was detected at 1 month after the operation. In general, the screws at the distal zone of the cervical plating fixation bear greater loading conditions compared with the proximal zone. Because of the concave profile of the cervical column at the anterior aspect (particularly in C5−C7), transferred effective compressive and bending stresses to the distal zone of the plate have a high tendency to move anteriorly and inferiorly, introducing a high level of compressive and shear stresses to distal screws. The compressive stress causes the screw to fracture (as occurred in this case), and shear stress causes the screw to loosen from the vertebrae, as highlighted in clinical studies [31,33]. Therefore, fixing an anterior plating system for more than two levels (in either disc fusion or vertebral corpectomy fusion) is biomechanically risky. Supportive posterior stabilization (Fig. 6.17) reduces transferred stresses to the distal section of the anterior plating fixation in cases with a high risk for failure at the distal zone of the plating fixation. *Through the courtesy of C. Ji, S. Yu, N. Yan, J. Wang, F. Hou, T. Hou, W. Cai, Risk factors for subsidence of titanium mesh cage following single-level anterior cervical corpectomy and fusion. BMC Musculoskeletal Disorders 21 (2020). https://doi.org/10.1186/s12891-019-3036-8 and BioMed Central Ltd.*

The static cage is best designed with a low profile at the frontal zone for easier insertion through the interspace of the two vertebrae, with less damage to bone tissue at the end plate portion of the vertebral bodies during impaction instrumentation. However, the insertion of expandable cages is not be managed by a large amount of impaction force, and the front tip of the cage does not need to have a low profile. Thus, the greater contact areas of the cage with vertebral end plates is effectively achievable. Likewise, the risk for cage migration in the expandable type is less than for static types. In fact, trialing or sizing of the best size for insertion of the static cage is managed with minor to significant deviations, in which the cage might be loosened and begin to migrate at the interspace. In expandable cages, sizing of the cage height is managed by expanding the cage. Thus, the surgeon ensures the presence of an adequate amount of preload on the superior and

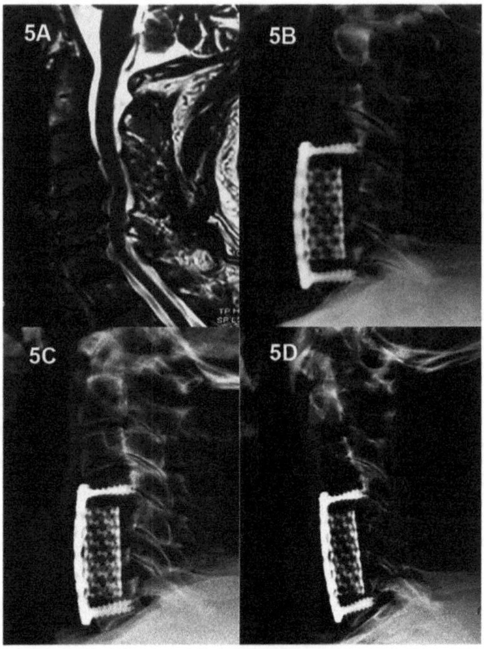

Figure 6.19 Illustration of subsidence failure in decompression treatment of the spinal cord at the cervical zone using static corpectomy cage (mesh cage) in a 46-years-old man (5A, 5B). The good stabilization of plating fixation at the distal zone causes an increased level of stress at the interface of the cage and vertebral end plate at the inferior vertebra. Because an endcap was not used when replacing the mesh or static corpectomy cage, stress concentration at the cage wall accelerated its subsidence through the vertebra at 1 year after the operation (5C). Inadequate plating fixation at the distal zone caused screw fracture or loosening, and even dissociation of the plate and screw at the screw hole, as illustrated in Fig. 6.18 (the plate introduced compressive stress to the screw head and caused significant deformation of the screw head or plate screw hole to such an extent that the plate lost engagement with the screw head and moved anteriorly). *Through the courtesy of Y. Zhang, Z. Quan, Z. Zhao, X. Luo, K. Tang, J. Li, X. Zhou, D. Jiang, Evaluation of anterior cervical reconstruction with titanium mesh cages versus nano-hydroxyapatite/polyamide66 cages after 1- or 2-level corpectomy for multilevel cervical spondylotic myelopathy: a retrospective study of 117 patients, PLoS One 9 (2014). https://doi.org/10.1371/journal.pone.0096265 and PLoS One.*

inferior aspects of the cage before vertebrae stabilization by a screw—rod or plating fixation system. Another benefit of the expandable cage is the lower risk for bone subsidence through the cage compared with the static cage. The use of expandable cages to treat disc degenerative diseases is illustrated in Figs. 6.20 and 6.21.

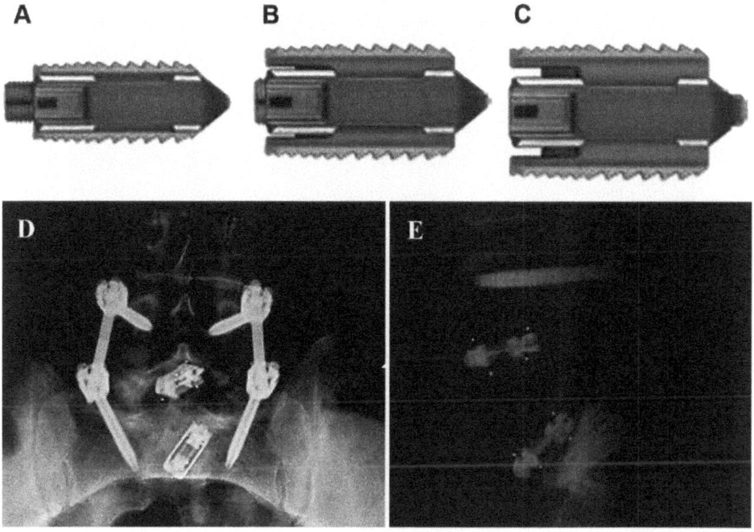

Figure 6.20 Illustration of expandable cage to treat disc fusion treatment in lumbar column (A–C). This type of fusion cage is expanded during surgery by its initial replacement at the intervertebral space in an unexpanded state (A) and then is expanded to the desired height (D, E). *Through the courtesy of C. Kim, T. Doerr, I. Luna, G. Joshua, S.S.-W. neurosurgery, undefined 2016, Minimally Invasive Transforaminal Lumbar Interbody Fusion Using Expandable Technology: A Clinical and Radiographic Analysis of 50 Patients, Elsevier, n.d. https://www.sciencedirect.com/science/article/pii/S187887501600320X. (Accessed 8 January 2021) and get permission from Elsevier.*

6.7 Interactions of vertebral bone with surface-engineered intervertebral cages and possible failures

Intervertebral fusion cages are best used on their surfaces in direct contact with bone tissue. Osteoconductive surface engineering of cage interfaces (either PEEK or titanium cages) with vertebral end plates increases the hydrophilicity characteristics of the surface and the flow of the blood, promoting osteogenic cell activity. Even the surfaces of the internal bone grafting space are advantageous for use by osteoconductive materials to enhance the biological integration of the cage and bone graft. The goal of PEEK cage surface processing is to reduce the risk for cage migration or dislocation at the intervertebral space. Two categories of surface engineering are used. One is coating with titanium powder on the surface of the fusion cages. Another second is porous structuring of PEEK or titanium cages at their interfaces with the bone. Titanium powder is normally coated by plasma

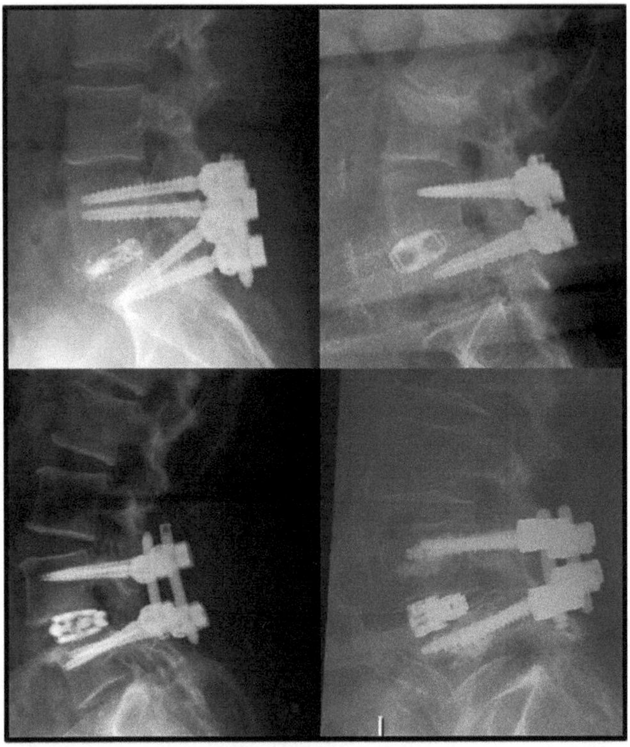

Figure 6.21 Illustration of the intraoperative subsidence of various types of TLIF expandable cages for of disc fusion treatment in the lumbar column (A–D). Although expandable fusion cages reduce the operation time, facilitate easier insertion, and improve lordosis alignment, the progression of primary subsidence during expansion occurs because of the large amount of compressive stress subjected to the end plates of adjacent vertebrae. The needed rotational torque of the surgical instrument to expand the expandable cage is transmitted to compressive force and distracts the vertebrae. The mechanical mechanism of this load transmission is designed to provide easy rotation of the surgical instrument for the surgeon with a low range of torque, while applying a high level of compressive force between vertebrae. Thus, care must be taken to perform the expansion slowly. Another effective parameter is to select the adequate expandable cage with an adequate height to reduce the extent of vertebrae distraction by the expansion of the cage, and thus the minimal amount of expansion length to reduce the risk for subsidence. (It might be beneficial to perform distraction by the distractor instrument in the surgical set until the adequate level and then manage the final expansion after inserting the expandable cage.) *Through the courtesy of C. Stickley, T. Philipp, E. Wang, J. Zhong, E.B.-T.S. Journal, undefined 2020, Expandable Cages Increase the Risk of Intraoperative Subsidence But Do Not Improve Perioperative Outcomes in Single Level Transforaminal Lumbar Interbody Fusion, Elsevier, n.d. https://www.sciencedirect.com/science/article/pii/S1529943020310779. (Accessed 3 January 2021) and permission from Elsevier.*

spraying in a vacuum environment, whereas physical vapor deposition was introduced an alternative to titanium coating on PEEK cages. The main challenge of plasma spray coating is the detachment of the coated particles when the cage impacts the intervertebral space or as a result of cyclic dynamic shear stress at the interface of the cage and end plates of vertebrae [36,37]. This failure is important in implanting static fusion cages compared with expandable cages. In expandable cages, much less impaction force and shear stress are generated at the bone–cage interface. Porous structuring of PEEK or titanium cages is another surface engineering strategy to enhance biological integration between the cage and vertebra. This process is managed on the surface of the premanufactured PEEK cage by depositing sodium chloride microcrystals through the surface of the cage and then dissolving the crystals in adequate liquid to create pores on the cage surface. (This process was completely discussed in Chapter 6 of *Osseointegration of Orthopedic Implants* by the lead author.) In titanium cages, the whole cage might be manufactured using additive manufacturing methods (e.g., selective laser melting) to form the main body of the cage as the solid structure and the surface as the porous structure with the desired geometry, pore size, and thickness. Failure of the porous structured surface during or after implantation is the main risk of this type of intervertebral fusion cage. The porous structure might crack as a result of cyclic shear stress conditions during rotational motion of the spinal column in various axes. A portion of the porous layer might detach and be released at the bone–cage interface, causing osteolysis at the interface. The advantage of porous structuring at the surface of the PEEK cage compared with titanium plasma spraying coating is greater mechanical strength. Scratch-fit implantation or replacement of the porous structured fusion cage through the intervertebral space causes initial penetration or interdigitating of the end plate layer of vertebral bodies to the pores of the cage porous layer. After replacement, primary subsidence of the vertebrae enhances the initial integrity of the cage and vertebral bodies. By further bone remodeling of the interdigitated bone tissue through the porous layer, the integrity of the cage and vertebrae is enhanced, and thus greater mechanical strength against shear stress is achieved. For an integrated layer composed of both bone matrix and titanium materials, a composite Ti-bone layer forms at the interface with higher mechanical strength compared with that of bone matrix, and thus with enhanced compressive strength. The enhancement of PEEK and titanium cages at their interfaces with vertebrae has been progressing, and sufficient clinical data are unavailable for study and have not been reported. In the next edition of this book, this part will be

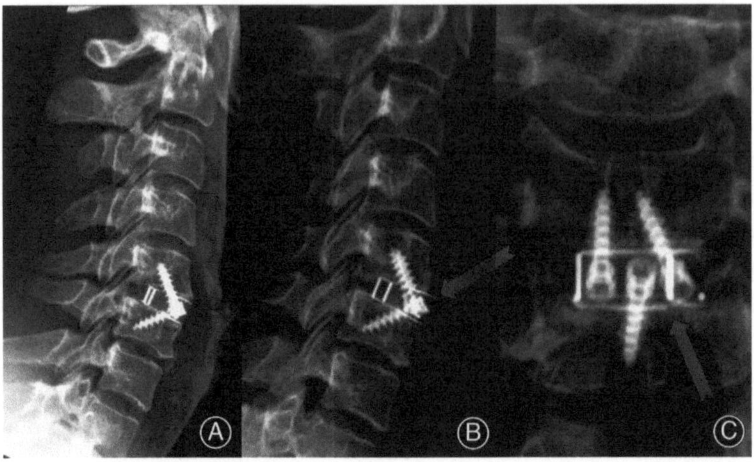

Figure 6.22 Radiographic illustration of polyetheretherketone stand-alone cage with screw fixator and plasma-spayed Ti-coating layer (A). Cage migration was observed at 9 months post-operation (B) (*red arrow*) with of titanium-coated wear and the risk for titanium particle release or delamination (C) (*red arrow*). Relative alignment of screws at the upper vertebra changed, as seen in the x-ray image at 9 months after the operation (B). The cage system rotated over the axial axis of the spinal column at the implanted zone. The X-ray does not reveal the reason for cage loosening, but it might be bone resorption at the bone–cage interface, where titanium wear progressed. *Through the courtesy of K. Phan, M.H. Pelletier, P.J. Rao, W.J. Choy, W.R. Walsh, R.J. Mobbs, Integral fixation titanium/polyetheretherketone cages for cervical arthrodesis: evolution of cage design and early radiological outcomes and fusion rates. Orthopaedic Surgery 11 (2019) 52–59. https://doi.org/10.1111/os.12413 and Chinese Orthopedic Association and John Wiley & Sons Australia, Ltd.*

more completely elaborated upon regarding clinical outcomes for surface-engineered intervertebral fusion cages. Fig. 6.22 illustrates the failure of a Ti-coated layer on the surface of a PEEK stand-alone cage fixed with titanium screws.

6.8 Interactions of vertebral bone with total disc replacement and possible failures

TDR is categorized as the need for nonfusion spinal implants. To facilitate motion of the spinal column at the affected zone, this type of implant was developed. Total discs act as mechanical joints between vertebrae to maintain the functionality of the spinal column at the implanted zone. Therefore, no fusion of the vertebrae is targeted, and no supportive stabilization system (e.g., screw–rod or plating system) is used. A TDR system is composed of at least three components: two end plate components and one medial insert. A total disc is replaced at the intervertebral space as an implantation treatment device for a severely degenerated intervertebral disc in the lumbar or cervical subcolumn. End plates are normally fabricated from cobalt-chromium, and the insert is manufactured from ultrahigh molecular weight polyethylene polymer (the traditional or first generation of the total disc was developed in only two end plate components made of titanium alloy without a median insert). Rotational motion of the vertebrae causes relative movement of end plate components to the insert component. The nonrigid structure of the total disc prostheses in interaction with the vertebrae prevents the concentration of loading conditions at the adjacent intervertebral disc. The main challenge of the total disc system is its effective integrity with vertebral bodies after implantation. Owing to the development of the end plate component from metallic materials, overloading or stress shielding occurs at the interface of the vertebrae and end plate metallic components. End plate components cover all or most of the vertebral end plates for increased buttressing during the motion of the spinal column. This prevents stress from concentrating at the edges of end plate components, which occurs when small end plate components are used. Overloading condition occurs as a result of reaction loading or energy conditions from metallic end plates, strengthening the bone matrix at the vertebral end plates. Stress shielding condition is observed at the anterior aspect of end plate components, as illustrated in Fig. 6.23, for cervical TDR [38]. The progression of stress shielding causes bone resorption and bone loss at the interface of the bone and end plate components. Stress shielding is observed in the interfaces where the effective action and reaction load are not parallel with the contact surface normal axis or very low loading conditions are transferred to part of the bone tissue. The ineffective function of TDR in treating cases with severe compression of the spinal cord was reported [39–42]; it necessitates the careful selection of patients with mild spinal injury.

6.9 Summary and discussion

Study of the complications that result after surgical implantation of spinal implants could be discussed from various facets. The spinal column might be affected at the implanted zone or the nonimplanted zone. The mechanical connectivity of all vertebrae and their involvement in the physiologic motion of the spinal column cause failures in the vertebrae with no implantation. This is a specific functional failure mechanism that occurs

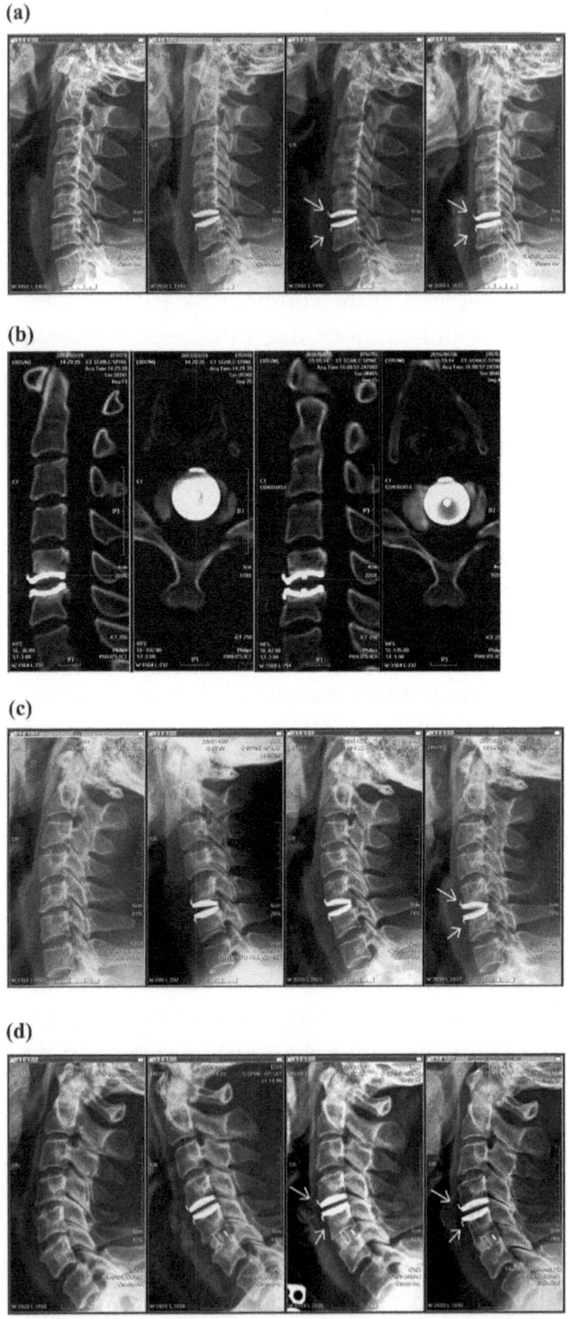

Figure 6.23 Illustration of anterior radiolucent zone and bone resorption in a specific type of total disc replacement system with anterior flange for improved stability in

compared with other treatments of bone injuries in trauma fracture fixation or joint arthroplasty. In accordance with the scope of this book, a focus on failures at the implanted zone and other possible failures are not contemplated; however, they are pointed out for a better expression of the adjacent effects on the nonimplanted vertebrae. In this chapter, the interactions of vertebrae bone with a spinal implant and possible failures were expressed separately for the screw–rod stabilization system, intervertebral fusion cages, corpectomy cages, and artificial disc prostheses. In the group of the screw–rod stabilization system and bone construct, loosening and fracture of the screw and fracture of the rod were highlighted as failures that occur to implants. Likewise, the loss of reduction or stabilization alignment as well as adjacent disc degeneration were observed to result from bone–implant construct failure. The progression of bone resorption at the interface of the bone and screw causes the screw to loosen and its dislocation to occur posteriorly, which irritates posterior soft tissues. Bone resorption at the interface of the bone and screw occurs as the result of overloading conditions, particularly in lumbar vertebrae, from which the activity of osteoclast cells

anteroposterior axis in different cases (A–D). The total disc system provides limited rotation of the spinal column at the implanted zone. Such motion alters the distribution of effective loading conditions at the interface of the vertebrae and disc replacement system. The convex profile of the artificial disc shell or end plate components at the interfaces with vertebral end plates is anatomically advantageous for load sharing at the central zone. However, it may cause less distribution of loading conditions at the peripheral areas. In addition, the processed titanium porous coating layer on end plate components enhances their biological integrity with the bone at the central zone. Furthermore, in general, the extent of effective loading conditions at the posterior aspect of the disc replacement system is greater compared with the anterior aspect. This might be because of the lordosis shaping of the cervical column, particularly at C4/7 (normally, the disc replacement system is used at the C4/C5, C5/C6, C6/C7 intervertebral spaces). Thus, the position of the cervical column is stabilized in the lordosis state in natural and extension motion and is almost straight in anterior motion. For greater stress distribution at the central and posterior aspects of the vertebrae–disc replacement construct, the risk for a low level of load transferring conditions results at the anterior zone and leads to the activity of bone resorption biological factors at the anterior zone. The progression of bone resorption was stopped at the central zone, where good osseointegration of the implant begins to be achieved. *Through the courtesy of T. Chen, W. Chen, C. Tzeng, C. Huang, C.Y.-T.S. Journal, undefined 2020, Anterior Bone Loss after Cervical Bryan Disc Arthroplasty: Insight into the Biomechanics Following Total Disc Replacement, Elsevier, n.d. https://www.sciencedirect.com/science/article/pii/S1529943020301674. (Accessed 4 January 2021) and permission from Elsevier.*

and the formation of fibrous tissue at the interface increase. Owing to the higher localization of stress at the lowest screw rows, rod or screw fracture occurs in long-term implantation. In addition, the loss of rigidity of the vertebrae and screw—rod construct occurs owing to the loss of distraction stability in the fixation of the polyaxial screws. In the group of intervertebral fusion cages, failures are observed as increased subsidence and migration of the cage at the implanted intervertebral space. The design parameters and features of the cage in conjunction with the cage materials influence the stiffness of the cage in mechanical interaction with adjacent vertebrae. Optimized stiffness of the cage is crucial to eliminate the risk for cage subsidence through the lower vertebrae or subsidence of the upper vertebrae through the cage. The bone grafting space is best maximized to have a thicker bone column between fused vertebrae for enhanced rigidity against physiologic loading and motional conditions at the fused zone. Migration of the cage is another issue that occurs as a result of the inadequate selection of cage size or pseudoarthrosis at the interface of the cage and vertebral end plate. Improper design of cage contact surfaces with the vertebrae and release or detachment of titanium-coated particles at the cage—bone interface are effective parameters that accelerate pseudoarthrosis and loosen and dislocate the cage. In this group, similar failures occur for stand-alone fusion cages, which are stabilized by a connective fixator element to the cage. In the group of corpectomy cages, the subsidence and fracture of the fixation system were highlighted as prominent failures of the bone—implant construct. The use of corpectomy cages is used more in the thoracic and cervical zones. In the cervical column, both static and expandable corpectomy cages are routinely used. Normally, corpectomy cages are used for a severe injury of the vertebral body caused by fracture or osteolysis. However, in the cervical column, it might be used for severe disc injuries that cause compression of the spinal cord at multiple levels. Subsidence of the static cage through the inferior vertebra is expected at a higher rate because there are no endcaps at the interface of the cage and vertebral end plates. The concentration of stress at the small area of the cage wall accelerates subsidence in the static cage compared with expandable cages with superior and inferior endcaps. Transferred loading conditions are increased in lower screws of plating fixation (which is normally used for the anterior stabilization of corpectomy cages), which loosens the screw from the vertebra, fractures the screw, or even dissociates the plate from the screw. In the group of artificial discs, stress shielding at the anterior zone, prosthesis subsidence, and prosthesis loosening are possible failures in patients treated with this type of spinal implant. The main point of using the artificial disc is that this implant might be inadequate for severe spinal injuries such as severe spondylosis, radiculopathy, and myelopathy. In such injuries, the spinal cord is under a high level of compression, and thus effective decompression is essentially required. Because of the motion of vertebrae at the implanted zone, particularly at the posterior zone of the lordosis cervical columns at the C5–C7, the disc height is altered, and at the minimal range, decompression is reduced. Another issue is observed at the interface of the artificial disc and vertebral end plates. Owing to the dynamic shear and compression forces generated because of the motion of vertebrae at the interface, the stability of bone and implant integrity is affected, causing subsidence, loosening, or migration of the artificial disc as a result of bone osteolysis at the interface. This chapter attempted to review, express, and discuss the main aspects of interactions of the bone with spinal implants in treating spinal injuries. However, more specific types of failures are observed in clinical practice, which were not completely included in this chapter, particularly for the latest and most recently developed spinal implants (e.g., titanium-coated intervertebral cages). In the next edition of the book, the incomplete parts as well as reported failures and interactions will be included.

6.10 Remind and learn

In this section, some questions are provided for effective reminders of the reviewed and discussed contents of this chapter. Answering these questions enhances learning about the interactions of the bone with spinal implants and possible failures.

Q 6.1: Discuss the anatomy of the spinal column. What types of tissues structure the spinal column? What are the physiologic loading conditions at this musculoskeletal structure?

Q 6.2: Discuss the structure of the vertebral and intervertebral disc. What types of loading conditions are borne by the annulus and nucleus tissues at the intervertebral disc?

Q 6.3: What type of tissue injuries might occur in the spinal column? Discuss why spinal canal narrowing happens. What is the effect of obesity on loading conditions at the spinal column? What is the effect of intervertebral disc degeneration on physiologic loading patterns between vertebrae? Refer to Fig. 6.3: discuss the reason for vertebral discoloration as the result of listhesis.

Q 6.4: What are the two main clinical strategies for spinal injury treatment using internal implantation of artificial devices or implants? What preoperative conditions of spinal injury are best studied by the surgeon to enhance surgical treatment?

Q 6.5: What types of spinal implants were developed? What are the main intended uses of these implants in treating spinal injuries?

Q 6.6: Discuss the screw—rod stabilization system and its components. What is overall loading distribution in bone—implant construct using this system? Refer to Fig. 6.4: What are the intended uses or functions of monoaxial, polyaxial, and uniaxial pedicle screws in spinal stabilization?

Q 6.7: What is the loading mechanism from the vertebrae to the screw—rod construct in two-level stabilization of the spinal column? What is the effect of such loading conditions of the adjacent intervertebral discs?

Q 6.8: What is the loading mechanism at the bone—screw—rod construct in this two-level stabilization of the spinal column? What is the effect of such a loading condition of adjacent intervertebral discs compared with a two-level construct? (Discuss the extent of reaction force from the bone—screw—rod construct to the adjacent discs.)

Q 6.9: Discuss why the extent of loading would increase from the upper level to the bottom level in a bone—screw—rod construct with more than two levels of vertebrae. What is the possible failure caused by this increased loading condition? How might such a failure be reduced in a bone—screw—rod construct with polyaxial screws?

Q 6.10: Discuss the function of the other components (e.g., cross-link, lateral offset connector, domino) of the screw—rod system.

Q 6.11: Refer to Figs. 6.7 and 6.10: Discuss why bone resorption occurs at the interface of the vertebral bone and pedicle screw. What are the advantages and disadvantages of screw placement through the posterior aspect with a greater interface with cortical bone compared with conventional purchasing of the screw from the posterolateral aspect?

Q 6.12: Refer to Fig. 6.8: What is the advantage of mono-poly-mono construct compared to polyaxial construct? What failure might occur in a polyaxial construct?

Q 6.13: Refer to Fig. 6.9: What testing method is used for a biomechanical evaluation of the bone—screw—rod stabilization construct? Discuss the effect of an unsupported length of pedicle screw on the fatigue strength of the construct.

Q 6.14: Discuss the overall characteristics of intervertebral fusion cages. What failures occur between the cage and vertebral bodies?
Q 6.15: What is the initial design concept of using titanium or titanium alloys to develop intervertebral fusion cages? What are the disadvantages of a titanium cage in the fusion treatment of degenerative disc diseases? What design concept are considered to improve the effectiveness of the titanium cage?
Q 6.16: How could the intervertebral cage be designed anatomically? Discuss the effect of anatomically designing the cage to enhance interactions between the cage and vertebral bodies and reduce failures such as cage subsidence, migration, or even the occurrence of a lack of fusion between adjacent vertebrae.
Q 6.17: What are the benefits of slots in cage walls in titanium fusion cages? What manufacturing processes are used to fabricate titanium cages? What are the advantages of additive manufacturing to develop titanium fusion cages compared with conventional processes?
Q 6.18: Refer to Fig. 6.12: What is a possible failure in a titanium cage with thin walls and lateral slots to treat disc injuries at a cervical subcolumn? Discuss the possible reasons.
Q 6.19: Refer to Fig. 6.13: Discuss the 3D printed lamellar titanium cage. What benefits might be achieved using this type of fusion cage?
Q 6.20: Discuss why the first generation of PEEK intervertebral cages was developed. What are the overall advantages and disadvantages of PEEK cages compared with titanium cages? What types of PEEK composites were used to develop intervertebral fusion cages? Discuss their advantages compared with pure PEEK cages.
Q 6.21: Discuss stand-alone fusion cages and their implantation mechanism through the vertebrae and intervertebral space. What are the advantages and disadvantages of this type of cage compared with conventional cages?
Q 6.22: Discuss the overall characteristics of corpectomy cages. What are the advantages and disadvantages of expandable and static corpectomy cages to replace vertebral bodies? Discuss why expandable cages may cause pseudarthrosis to progress at the interface of the cage and vertebral bodies.
Q 6.23: Refer to Fig. 6.16: Discuss the interactions and failures of the bone with rectangular and round endcaps under physiologic loading conditions at the implanted site.

Q 6.24: Refer to Fig. 6.17: What is the main benefit of using posterior stabilization in corpectomy treatment in the cervical subcolumn? What failure might occur when corpectomy treatment is managed by anterior plating stabilization only, particularly involving more than three levels of vertebrae?

Q 6.25: Refer to Fig. 6.18: Why might anterior plating fixation fail in corpectomy treatment in a cervical subcolumn in view of an increased level of loading conditions? What zone of plating fixation might failed, and what types of failures would result?

Q 6.26: Refer to Fig. 6.19: Discuss the reasons for static corpectomy cage subsidence through the vertebral bodies. Compare the risk for such a failure in expandable and static corpectomy cages.

Q 6.27: Discuss the overall structure and performance of expandable intervertebral fusion cages. What are the advantages of expandable cages compared with static cages in treating disc degenerative diseases? (Consider the contact surfaces with the vertebrae, the risk for loosening or migration, and the risk for subsidence in your comparison.)

Q 6.28: Refer to Fig. 6.21: What is the main failure of the expandable intervertebral fusion cage? How might such an issue be resolved during implantation of the expandable cage?

Q 6.29: What are the benefits of the biological integration of intervertebral fusion cages with vertebral bodies? What surface engineering methods might be used to enhance the biological integrity of PEEK and titanium cages? What are the possible failures of surface-engineered layer on the surface of fusion cages?

Q 6.30: Discuss the overall structure and performance of the TDR prostheses. What are the benefits of TDR in treating the disc degenerative diseases compared with static or expandable fusion cages?

Q 6.31: What is the main challenge or possible failure in replacing the total disc? Refer to Fig. 6.23: Discuss why the end plate component of the TDR might lose integration with the bone at the anterior aspect. Refer to Fig. 6.24: What type and severity of spinal injury might be intended for use of the TDR as a surgical treatment?

Q 6.32: Discuss why the concentration of loading conditions is concentrated in the posterior aspect of the cervical column in C4—C7 vertebrae. What is the effect of such increased loading conditions on the decompression function of the TDR in the posterior column?

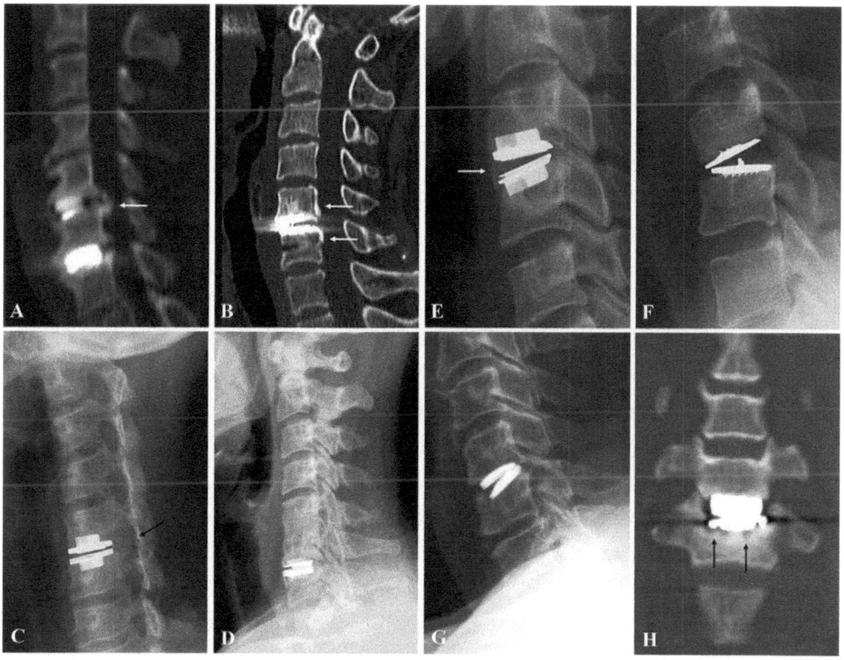

Figure 6.24 Radiographic illustration of various types of failure after implantation of total disc replacement system: severe spondylosis (*white arrow*) (A), ossification of posterior longitudinal ligament (*white arrows*) (B), foraminal stenosis (*black arrow*) (C), severe spondylosis adjacent to previous fusion (D), hyperlordotic positioning with heterotopic ossification (*white arrow*) (E), kyphotic positioning (F), subsidence (G), and osteolysis (*black arrows*) (H). Most failures of the artificial disc result from ineffective patient selection. Indications for the adequate use of the artificial disc in the cervical column are mild disc herniation and spondylosis. For patients with severe spondylosis in whom a significant extent of decompression is required, artificial disc replacement is not a good solution. Managing and maintaining the disc height at the posterior aspect would not be achieved effectively. Thus, the advantage of decompression of the TDR would be affected owing to presence of physiologic motion after replacement of the TDR in the cervical column, and therefore revision surgery will be necessitated [39–42]. *Through the courtesy of J.-B. Park, H. Chang, J.S. Yeom, K.-S. Suk, D.-H. Lee, J.-C. Lee, Revision surgeries following artificial disc replacement of cervical spine: analysis of 21 cases, Global Spine Journal 6 (2016) s-0036-1582805-s-0036-1582805. https://doi.org/10.1055/s-0036-1582805 and permission from Elsevier.*

6.11 Think and challenge

In this section, some questions are provided for further thought about the reviewed and discussed contents of this chapter. The aim is to challenge the current facets of interactions of the bone with spinal implants and possible failures by answering these questions, which may enhance new design concepts for the future development of prostheses.

Q 6.33: As highlighted in this chapter, the rigid stabilization construction of spinal vertebrae may affect adjacent intervertebral discs after a few years and other surgical interventions for the patient would be needed. Semirigid or dynamic stabilization systems were introduced as a possible solution, but semirigid systems may fracture or fail in the long term and dynamic solutions may reduce the distraction of the vertebrae and decompression of the spinal cord. Discuss what new concepts might be hypothesized to enhance spinal stabilization with a high level of functionality and satisfaction.

Q 6.34: As highlighted in this chapter, the use of 3D porous structured fusion cages from tantalum or titanium materials was mentioned as effective for enhancing bony fusion in the treatment of disc degenerative diseases. Discuss what failures might be expected for such cages in the future.

Q 6.35: As highlighted in this chapter, the risk for screw loosening may be expected in bone–screw–rod stabilization constructs. Discuss what new design concepts might be hypothesized to reduce bone resorption at the interface of the screw and vertebra.

References

[1] K.H. Yang, Modeling the spine and upper and lower extremities for impact scenarios, in: *Basic Finite Element Method as Applied to Injury Biomechanics*, Academic Press, 2018, pp. 609–654.

[2] S. Kurtz, A. Edidin, Spine Technology Handbook, Elsevier, Boston, MA, 2006.

[3] Y. Feng, B. Egan, J.W.-G.& diseases, undefined 2016, Genetic Factors in Intervertebral Disc Degeneration, Elsevier, n.d. https://www.sciencedirect.com/science/article/pii/S2352304216300149. (Accessed 5 January 2021).

[4] A. Younus, A.K.-I. Neurosurgery, undefined 2020, Lumbar Unifacet Dislocation Without an Associated Fracture-A Case Report and Literature Review, Elsevier., n.d. https://www.sciencedirect.com/science/article/pii/S221475192030534X. (Accessed 5 January 2021).

[5] P. Liu, P. Lai, C.L.-C. Biomechanics, undefined 2019, A Biomechanical Investigation of the Retentive Force of Pedicle Screw Structures for Different Screw Tulip Designs, Elsevier, n.d. https://www.sciencedirect.com/science/article/pii/S026800331831009X. (Accessed 6 January 2021).

[6] S. Irandoust, S.M.-S. Reports, undefined 2020, The interplay between bone healing and remodeling around dental implants, Nat. Commun. 10 (n.d.). https://www.nature.com/articles/s41598-020-60735-7. (Accessed 30 December 2020).

[7] C.A. Sansur, N.M. Caffes, D.M. Ibrahimi, N.L. Pratt, E.M. Lewis, A.A. Murgatroyd, B.W. Cunningham, Biomechanical fixation properties of cortical versus transpedicular screws in the osteoporotic lumbar spine: an in vitro human cadaveric model, J. Neurosur. Spine 25 (4) (2016) 467–476.

[8] S. Wray, R. Mimran, S. Vadapalli, S.S. Shetye, K. Mcgilvray, C. Puttlitz, Pedicle screw placement in the lumbar spine: effect of trajectory and screw design on acute biomechanical purchase, J. Neurosurg. Spine 22 (2015), https://doi.org/10.3171/2014.10.SPINE14205.
[9] S. Tanioka, F. Ishida, K. Kuraishi, K.T.-W. Neurosurgery, undefined 2019, A Novel Radiologic Assessment of Screw Loosening Focusing on Spatial Position Change of Screws Using an Iterative Closest Point Algorithm with Stereolithography, Elsevier, n.d. https://www.sciencedirect.com/science/article/pii/S1878875019300889. (Accessed 3 January 2021).
[10] X. Wang, C.E. Aubin, J. Coleman, J. Rawlinson, Correction capability in the 3 anatomic planes of different pedicle screw designs in scoliosis instrumentation, Clin. Spine Surg. 30 (4) (2017) E323–E330.
[11] W. Qin, K. Chen, H. Chen, P. Yang, H. Yang, H.M.-W. Neurosurgery, undefined 2020, Comparison of Poly-Axial or Poly-/Mono-Axial Mixed Screw Fixation for Treatment of Thoracolumbar Fractures with O-Arm Navigation: A Case Control Study, Elsevier, n.d. https://www.sciencedirect.com/science/article/pii/S1878875020301418. (Accessed 3 January 2021).
[12] Y.L. Chu, C.H. Chen, F.Y. Tsuang, C.J. Chiang, Y. Wu, Y.J. Kuo, Incomplete insertion of pedicle screws in a standard construct reduces the fatigue life: a biomechanical analysis, PLoS One 14 (2019), https://doi.org/10.1371/journal.pone.0224699.
[13] H. Guo, Y. Tang, D. Guo, Y. Ma, K. Yuan, Y. Li, J.P.-W. Neurosurgery, undefined 2020, Pedicle Screw Fixation in Single-Level, Double-Level or Multilevel Posterior Lumbar Fusion for Osteoporotic Spine: A Retrospective Study with a Minimum 2-Year Follow-Up, Elsevier, n.d. https://www.sciencedirect.com/science/article/pii/S1878875020309086. (Accessed 3 January 2021).
[14] J. Hyup Lee, C.-B. Kong, J. Jun Yang, H.-J. Shim, K.-H. Koo, J. Kim, C.-K. Lee, B.-S. Chang, J. Shim, Comparison of fusion rate and clinical results between CaO-SiO 2-P 2 O 5-B 2 O 3 bioactive glass ceramics spacer with titanium cages in posterior lumbar interbody fusion, Spine J. 16 (2016) 1367–1376, https://doi.org/10.1016/j.spinee.2016.07.531.
[15] A. Frost, E. Bagouri, M. Brown, V. Jasani, Osteolysis following resorbable poly-L-lactide-co-D, L-lactide PLIF cage use: a review of cases, Eur. Spine J. 21 (2012) 449–454, https://doi.org/10.1007/s00586-011-2002-4.
[16] E. Gercek, V. Arlet, J. Delisle, D. Marchesi, Subsidence of stand-alone cervical cages in anterior interbody fusion: warning, Eur. Spine J. 12 (2003) 513–516, https://doi.org/10.1007/s00586-003-0539-6.
[17] M. Mokawem, G. Katzouraki, ... C.H.-J. of C., undefined 2019, Lumbar Interbody Fusion Rates with 3D-Printed Lamellar Titanium Cages Using a Silicate-Substituted Calcium Phosphate Bone Graft, Elsevier, n.d. https://www.sciencedirect.com/science/article/pii/S0967586819300244. (Accessed 3 January 2021).
[18] J. Liu, X. Xiong, A. Peng, M. Xu, ... X.C.-C.N. and, undefined 2017, A Comparison of Local Bone Graft with PEEK Cage Versus Iliac Bone Graft Used in Anterior Cervical Discectomy and Fusion, Elsevier, n.d. https://www.sciencedirect.com/science/article/pii/S0303846717300446. (Accessed 3 January 2021).
[19] Y. Kim, J. Park, B. Moon, S. Kim, J.L.-J. of Clinical, undefined 2018, Is Stand Alone PEEK Cage the Gold Standard in Multilevel Anterior Cervical Discectomy and Fusion (ACDF)? Results of a Minimum 1-Year Follow up, Elsevier, n.d. https://www.sciencedirect.com/science/article/pii/S0967586817316934. (Accessed 8 February 2021).

[20] G. Zhang, Z. Chen, Q. Liu, Z. Ji, C. Lan, H. Lin, Effect of swallowing function after ROI-C anterior cervical interbody fusion, J. Surg. 4 (6) (2016) 141–145.
[21] J. Zhang, H. Liu, E. Bou, W. Jiang, F. Zhou, F.H.-W. Neurosurgery, undefined 2019, Comparative Study Between Anterior Cervical Discectomy and Fusion with ROI-C Cage and Laminoplasty for Multilevel Cervical Spondylotic Myelopathy Without Spinal, Elsevier, n.d. https://www.sciencedirect.com/science/article/pii/S187887501832309X. (Accessed 8 February 2021).
[22] T. Buranakarl, K. Jaisanuk, Preliminary clinical and radiographic outcomes of anterior lumbar interbody fusion (ALIF) with stand alone PEEK cage and anterior plate construct at the Bangkok spine academy, Bangkok Med. J. 6 (2013) 1–11. https://he02.tci-thaijo.org/index.php/bkkmedj/article/view/218113 (Accessed 8 February 2021).
[23] Y. Rong, Y. Luo, W. Liu, F. Gong, P. Tang, W. Cai, Clinical effects of the bridge-type ROI-C interbody fusion cage system in the treatment of cervical spondylosis with osteoporosis, Clin. Interv. Aging 13 (2018) 2543.
[24] Z.D. Wang, R.F. Zhu, H.L. Yang, M.J. Shen, G. Wang, K. Chen, M. Gan, M. Li, Zero-profile implant (Zero-p) versus plate cage benezech implant (PCB) in the treatment of single-level cervical spondylotic myelopathy, BMC Muscoskel. Disord. 16 (2015), https://doi.org/10.1186/s12891-015-0746-4.
[25] A. Awad, D. Stidd, K. Alkhalili, I. Eli, A. Baaj, Vertebral body reconstruction using expandable titanium cages after anterior decompression for cervical spondylotic myelopathy: a review, Cureus 6 (3) (2014). http://citeseerx.ist.psu.edu/viewdoc/download?doi=10.1.1.957.1153&rep=rep1&type=pdf (Accessed 3 January 2021).
[26] Mundis G.M., Eastlack R.K., Moazzaz P., Turner A.W., Cornwall G.B. Contribution of round vs. rectangular expandable cage endcaps to spinal stability in a cadaveric corpectomy model. Int. J. Spine Surg. 2015 Jan 1;9.
[27] S. Keshavarzi, C. Newman, … J.C.-, U. 2011, Expandable Titanium Cages for Thoracolumbar Vertebral Body Replacement: Initial Clinical Experience and Review of the Literature, 2011. www.amjorthopedics.com (Accessed 7 January 2021).
[28] A.R Deukmedjian, J. Manwaring, T.V. Le, A.W. Turner, J.S. Uribe, Corpectomy cage subsidence with rectangular versus round endcaps, J. Clin. Neurosci. 21 (9) (2014) 1632–1636.
[29] M. Reinhold, W. Schmoelz, F. Canto, D. Krappinger, M. Blauth, C. Knop, A new distractable implant for vertebral body replacement: Biomechanical testing of four implants for the thoracolumbar spine, Arch. Orthop. Trauma Surg. 129 (10) (2009 Oct) 1375–1382.
[30] M. Pekmezci, J.A. Tang, L. Cheng, A. Modak, R.T. McClellan, J.M. Buckley, C.P. Ames, Comparison of expandable and fixed interbody cages in a human cadaver corpectomy model, part I: endplate force characteristics, J. Neurosurg. Spine. 17 (4) (2012) 321–326.
[31] C. Ji, S. Yu, N. Yan, J. Wang, F. Hou, T. Hou, W. Cai, Risk factors for subsidence of titanium mesh cage following single-level anterior cervical corpectomy and fusion, BMC Muscoskel. Disord. 21 (2020), https://doi.org/10.1186/s12891-019-3036-8.
[32] Y. Zhang, Z. Quan, Z. Zhao, X. Luo, K. Tang, J. Li, X. Zhou, D. Jiang, Evaluation of anterior cervical reconstruction with titanium mesh cages versus nano-hydroxyapatite/polyamide66 cages after 1- or 2-level corpectomy for multilevel cervical spondylotic myelopathy: a retrospective study of 117 patients, PLoS One 9 (2014), https://doi.org/10.1371/journal.pone.0096265.
[33] M.D. Daubs, Early failures following cervical corpectomy reconstruction with titanium mesh cages and anterior plating, Spine 30 (12) (2005) 1402–1406.
[34] C. Stickley, T. Philipp, E. Wang, J. Zhong, E.B.-T.S. Journal, undefined 2020, Expandable Cages Increase the Risk of Intraoperative Subsidence But Do Not Improve Perioperative Outcomes in Single Level Transforaminal Lumbar Interbody

Fusion, Elsevier, n.d. https://www.sciencedirect.com/science/article/pii/S1529943020310779. (Accessed 3 January 2021).
[35] C. Kim, T. Doerr, I. Luna, G. Joshua, S.S.-W. Neurosurgery, undefined 2016, Minimally Invasive Transforaminal Lumbar Interbody Fusion Using Expandable Technology: A Clinical and Radiographic Analysis of 50 Patients, Elsevier, n.d. https://www.sciencedirect.com/science/article/pii/S187887501600320X. (Accessed 8 January 2021).
[36] K. Phan, M.H. Pelletier, P.J. Rao, W.J. Choy, W.R. Walsh, R.J. Mobbs, Integral fixation titanium/polyetheretherketone cages for cervical arthrodesis: evolution of cage design and early radiological outcomes and fusion rates, Orthop. Surg. 11 (2019) 52–59, https://doi.org/10.1111/os.12413.
[37] A. Kienle, N. Graf, H.W.-T.S. Journal, undefined 2016, Does Impaction of Titanium-Coated Interbody Fusion Cages into the Disc Space Cause Wear Debris or Delamination?, Elsevier, n.d. https://www.sciencedirect.com/science/article/pii/S1529943015014485. (Accessed 9 January 2021).
[38] T. Chen, W. Chen, C. Tzeng, C. Huang, C.Y.-T.S. Journal, undefined 2020, Anterior Bone Loss After Cervical Bryan Disc Arthroplasty: Insight into the Biomechanics Following Total Disc Replacement, Elsevier, n.d. https://www.sciencedirect.com/science/article/pii/S1529943020301674. (Accessed 4 January 2021).
[39] B. Skovrlj, D.H. Lee, J.M. Caridi, S.K.W. Cho, Reoperations following cervical disc replacement, Asian Spine J. 9 (2015) 471–482, https://doi.org/10.4184/asj.2015.9.3.471.
[40] K. Singh, F.M. Phillips, D.K. Park, M.A. Pelton, H.S. An, E.J. Goldberg, Factors affecting reoperations after anterior cervical discectomy and fusion within and outside of a Federal Drug Administration investigational device exemption cervical disc replacement trial, Spine J. 12 (2012) 372–378, https://doi.org/10.1016/j.spinee.2012.02.005.
[41] S.V. Nandyala, A. Marquez-Lara, S.J. Fineberg, K. Singh, Comparison of revision surgeries for one- to two-level cervical TDR and ACDF from 2002 to 2011, Spine J. 14 (2014) 2841–2846, https://doi.org/10.1016/j.spinee.2014.03.037.
[42] J.-B. Park, H. Chang, J.S. Yeom, K.-S. Suk, D.-H. Lee, J.-C. Lee, Revision surgeries following artificial disc replacement of cervical spine: analysis of 21 cases, Global Spine J. 6 (2016), https://doi.org/10.1055/s-0036-1582805 s-0036-1582805-s-0036-1582805.

CHAPTER SEVEN

Effectiveness of current orthopedic implants in interaction with bone and future perspectives

7.1 Introduction

Interaction of the bone with orthopedic implants and possible failures have been reviewed and discussed in general and detailed aspects in Chapters 1–6. Main concepts of mechanical and biological interactions of the bone and implant were explored in Chapter 1. Detailed interactions and failures of orthopedic implants in contact with bone tissue were characterized in Chapters 2–6 for joint replacement, trauma implants, and spinal implants, and possible reasons for failure were comprehensively discussed. Normally, reported interactions and failures in clinical studies are not discussed, and authors explore the success and complication rates of surgical implantation in over the short term or long term after implantation. The aim of this book was to provide a reference of comprehensive complications resulting from ineffective interactions of bone with orthopedic implants while discussing reasons not found in published clinical investigations. There was an attempt to avoid relying on reports from reviewed papers and instead to express technical judgments in accordance with the fundamentals and reliable scientific aspects. In clinical studies, the interactions of bone and orthopedic implants were not effectively used to express or discuss bone–implant construct failure directly, however, some concepts such as stress-shielding were addressed as possible technical reasons that might be occur between the bone and implant. This book attempted to define the interactions of bone and orthopedic implants in the orthopedic implant industry and then establish a systematic literature on various types of interactions and failures from the mechanical and biological viewpoints. In this chapter, the effectiveness of strategies for the design and development of orthopedic implants is discussed based on data provided in Chapters 2–6. In the final section, future perspectives and ideal concepts on this topic are expressed.

Interactions of Bone with Orthopedic Implants and Possible Failures
ISBN: 978-0-323-95410-5
https://doi.org/10.1016/B978-0-323-95410-5.00006-2

7.2 Concept of effectiveness

The concept of effectiveness is defined as the extent of successful and satisfactory outcomes of orthopedic implants in treating bone injuries in view of the interactions of bone and orthopedic implants. The aim is to discuss how design concepts can provide reliable and long-term efficacy in secure interactions of the bone with implant as an integrated construct. To make technical judgments on this topic, various factors need to be contemplated for an effective discussion. Technical factors and limitations in engineering and scientific sectors would accelerate or limit development strategies for enhancing the effectiveness of orthopedic implants. The ideal interaction between the bone and implant is briefly described in Chapter 1. As highlighted, the mechanical and biological activity of the implant in contact with bone might be a revolutionary strategy for the design and development of an orthopedic implant through which current failures might be significantly reduced. However, the achievement of such an implant needs significant involvement of all engineering and scientific aspects of orthopedic implants from the various disciplines of biomechanics, biomaterials, biomanufacturing, biochemistry, surface engineering, and so forth. In terms of biomechanics, reliable biomechanical evaluation methods need to be developed to allow the reliable design of orthopedic implants in materials, geometric, and structural terms. Relying on available standards (e.g., International Organization for Standardization (ISO) or American Society for Testing and Materials (ASTM)) is insufficient for an effective mechanical evaluation of the implant in long-term interaction with bone tissue. There are not reliable or effective testing methods for evaluating the failure of a bone—implant construct involving the soft tissues and the influence of bone remodeling. In terms of the material, the mechanically or biologically active materials close to bone tissue have not been developed that could remodel in mechanical and biological interaction with bone tissue. In manufacturing processes, additive manufacturing has provided possibilities for developing a three-dimensional (3D) implant geometry and structure, but the processing of mechanically and biologically active implants has not yet been explored. The science of bone biochemistry has had many interesting discoveries, but the controlled influence of the chemical process of bone remodeling on the implant has not been implemented in developing current implants. Although tissue engineering scaffolds would be a well-established engineering scientific field of study, most joint replacements, trauma implants, and spinal implants cannot be substituted with cell-cultured tissue scaffolds to treat severe bone injuries. All of these technical limitations along with market demand may not accelerate development strategies to achieve the ideal implant with active mechanical and biological performance in interaction with bone tissue. Therefore, the effectiveness of orthopedic implants in interaction with bone is discussed in the next sections.

7.3 Effectiveness of joint replacement prostheses in interaction with bone

Interactions of bone with joint replacement prostheses and possible failures were reviewed in Chapters 2–4 for knee, ankle, shoulder, and hip joints. The overall conditions of the knee and ankle joints might be close to each other, whereas bone–implant interactions in the hip and shoulder joints may have similar patterns in physiologic loading conditions. In the knee and ankle joints, body weight loading conditions are highly prominent compared with the range of motion, which is maintained in a limited range. In the hip joint, both body weight loading and a high range of motion greatly influence the functionality of the joint. In the shoulder or glenohumeral joint, the range of motion is prominent compared with the weight of the upper extremities. These physiologic conditions are consistent for all people and is associated with their lifestyle. With this introduction to the overall aspects of the knee, ankle, shoulder, and hip joints, effectiveness in treating bone injuries at these joints is discussed in the next subsections.

7.3.1 Effectiveness of knee joint replacement prostheses in interaction with bone

Total knee replacement is growing in the world, and many people are accepting this surgical intervention to relieve severe pain in the knee joint as effective salvage. Overall, the level of satisfaction is great and most patients return to a normal life to perform daily activities and job responsibilities reliably. Knee joint replacement reduces physiologic conditions compared with a healthy knee joint. The patient is unable to perform some specific actions of the body such as stair climbing, running, climbing, and many sports activities such as volleyball, football, and basketball as well as with a healthy knee joint. Because of the sacrifice of the anterior and/or posterior cruciate ligaments, stability and motion of the knee joint are not achieved as smoothly as in a healthy joint. Because of the use of high mechanical modulus metallic materials to manufacture femoral knee and tibial tray prostheses, the extent of action or reaction force to bone tissue that is applied from the metallic prostheses is high and may pain at the bone—implant interface. Thus, movement of the replaced knee joint need to be done carefully and activities or incidents accompanied by an impact that which would localize the extent of force to bone tissue at the interface and incur a high level of pain to the patient need to be avoided. A reduction in body weight is another factor that enhances the success rate of the total knee arthroplasty. This prevents overloading conditions at the bone—implant interface and the occurrence of bone resorption and the formation of fibrous tissue at the bone—implant interface. Overall, postoperative care is needed to enhance the effectiveness of the total knee arthroplasty in the long term. However, there are inevitable factors in the design and development of knee joint prostheses that challenge their effectiveness in long-term implantation or replacement at the knee joint. These factors are discussed in terms of interactions of the bone and implant. To reduce the effects of failure that is associated with the material, design, and structure of knee joint prostheses, two main strategies of product developers and orthopedic surgeons involve design and process improvement and an increase in type of prostheses. Design and process improvements cannot all be implemented in one model; hence, various types of prostheses might be beneficial for the end user, with specific features and benefits to enhance the medical treatment of knee joint osteoarthritis. A femoral knee prosthesis is developed made from the two materials of cobalt-chromium (CoCr) and zirconium (Zr) alloy in different design concepts of multiple curves, single curve, medial pivot, and so on, at the articulation surface. The other knee prosthesis, a tibial insert, is fabricated from ultrahigh molecular weight polyethylene (UHMWPE) as a soft material for articulation with the femoral knee to be stabilized by the tibial tray prosthesis. The design of the tibial insert at the articulation surface is matched with the design of the femoral knee. The design profile of two prostheses at the articulation might not fully coincide and might have some differences, but it provides motion and contact conditions as evaluated during the design and development process. The efficacy of femoral knee and tibial insert prosthesis articulation is associated with the sacrifice or retention of the posterior cruciate ligament (PCL) function or deficiency, collateral ligament function or deficiency, body weight, and demanded range of motion at the knee joint. These conditions are considered when selecting femoral knee and tibial insert prostheses to achieve optimal conditions at the articulation surface. Normally, in older patients with soft tissue deficiency, articulation with a high level of stability such as a medial pivot or ultracongruency with high anterior height, is appropriate to constrain mediolateral translation, internal-external rotation, and varus-valgus rotation. Conversely, in a normal condition or in younger patients with the need for a high range of motion at the knee joint, articulation with a lower extent of stability and constraints might be used. The necessary range of motion and body weight loading conditions might alter over the time, and thus the selected prostheses at the time of surgery might not be appropriate over time. This is a major limitation of orthopedic implants that cannot remodel themselves over time. Thus, failure occurs and revision of the prosthesis is the only solution to resolve the issue. Effective articulation of the knee prostheses in physiologic motion influences the interaction of the femoral knee prosthesis with the distal zone of the femur. Distal femoral condyles on either the lateral or medial sides are resected to replace the femoral knee prosthesis. Resection of the condyles is created on five cutting surfaces that completely coincide with the internal section of the femoral knee prosthesis. The bone—implant interface for this prosthesis is under compressive and shear stresses generated as a result of body weight, axial or internal-external torsion, mediolateral translation, and anteroposterior translation. All these force, moment, or motional conditions create loading conditions with the tendency to dissociate the bone and implant at the interface of the femoral knee prosthesis and distal femoral condyles. The femoral knee prosthesis has another articulation surface with the patella prosthesis or bone. Therefore, the loading conditions created at the prosthesis and femoral condyles are created at two articulation surfaces. The posterior and middle zone of the femoral knee mechanically interacts with the tibial insert, whereas its anterior zone interacts with the patella prosthesis or bone. Distribution of loading conditions on the whole surface of the femoral knee is not effective for preventing stress shielding at the prosthesis-bone interface. As reviewed in Chapter 2, radiolucent zones were reported as the result of a low range of loading condition at the anterior and posterior sections of femoral knee prostheses. These affected sections are not loaded in knee flexion at 0—60°. A higher flexion angle engages the posterior portion of the femoral knee prosthesis with the tibial insert prosthesis and creates contact force, generating a loading condition at the bone—implant interface at this zone. The high flexion of the knee joint increases engagement of the patella zone with the femoral knee prosthesis and motion of the patella through the anterior groove, in which contact loading at this zone is generated at the bone—prosthesis interface. The biological integrity of the femoral knee prosthesis with distal condyles of the femur is another challenge of bone–implant interaction at this interface. Processing of effective bone remodeling through the surface of the femoral knee prosthesis is beneficial to prevent early and long-term dissociation of the prosthesis from bone. This advantage ensures early stability of the femoral knee prosthesis, particularly in the cementless replacement method. The internal surface of the femoral knee is enhanced by bead sintering or binding of the 3D porous structure to facilitate bone ingrowth through the cementless prosthesis surface or to enhance bone cement interdigitating in a cemented prosthesis. These surface engineering processes permanently exist on the implant surface, but in excessive shear loading conditions, it might be detached from the prosthesis. Embedding or coating bioactive materials through the pores of the porous structured layer accelerates bone remodeling and is advantageous for early integration of the bone and prosthesis. Fabrication of the femoral knee prosthesis by additive manufacturing (e.g., high-energy selective laser melting) has been used to develop this prosthesis in the form of a solid-porous structure in one process. With advanced process parameter optimization to achieve the appropriate mechanical modulus and stiffness, this process could be further explored to manufacture femoral knee prostheses. The overall aspects of the femoral knee were reviewed in this section; further discussion is provided in Chapter 2.

Another prosthesis of the total knee replacement (TKR) system is the tibial tray prosthesis, which has a large area of interfaces with the proximal tibia. The tibial tray is developed in various designs and structures. Basically, this prosthesis has two parts: the tray and stem. The tray portion has a design similar to that of the inferior section of the tibial insert prosthesis, and the stem dedicates the anatomy of the proximal tibia at the metaphyseal zone. It is best to design the tray portion with a maximized contact surface with the resected section of the tibia created by a cutting instrument. It is also advantageous to design the stem with supported keels to occupy the volume of the metaphyseal zone in the mediolateral axis and posterior portion (the stem portion of the tibial tray normally occupies the central portion of the metaphyseal zone toward the anterior). The effective design of the tibial tray enhances load distribution and transference at its interface with bone. As illustrated in Chapter 2, the traditional or simply designed tibial tray in the stem portion causes subsidence of the prosthesis through the tibia or even its fracture in patients with a high value of body mass index (BMI). It may even cause stress or load concentration at the prosthesis–bone interface as well as overloading and the progression of bone resorption underneath the tray portion, loosening the prosthesis. Tibial tray surfaces at the interfaces with bone are processed for enhanced biological integrity of the prosthesis with bone tissue. Conventionally, plasma spray coating of hydroxyapatite and titanium powders is used for the conductivity of bone formation at the prosthesis surface. Such processes have an early effect on increasing osseointegration of the prosthesis, but this effectiveness in the long term is controversial. Porous structuring and the use of additive manufacturing are development techniques used by product developers to establish an osteoconductive layer with longer effectiveness in conducting the bone ingrowth at the bone—prosthesis interfaces, particularly at sections beneath the tray portion.

Failures of the total knee prostheses have been reviewed in Chapter 2. Failures of the bone in interaction with this type of orthopedic implant occur under specific circumstances, but they reveal the potential failures of developed knee joint prostheses. Thus, a rigorous pr-operative study is needed on the patient's condition, the severity of the injury, and technical features and benefits of current TKR systems on the market. Even the post-operative follow-up and careful monitoring of the patient's condition (e.g., body weight, soft tissue functionality, bone—implant interface radiographic monitoring) are important to ensure the success rate of the TKR in short-term and long-term implantation greater than 10 or 15 years. The potential occurrence of the stress shielding and overloading conditions at the bone—implant interface in replacing knee prostheses necessitates an enhanced design of the implant, implantation techniques, the incorporation of low mechanical modulus prostheses, increased bone ingrowth thickness through the internal zone of the prosthesis, and enhanced load distribution and transference between the bone and implant.

7.3.2 Effectiveness of ankle joint replacement prostheses in interaction with bone

Interactions of the bone with total ankle replacement (TAR) prostheses and possible failures were reviewed and discussed in the second section of Chapter 2. Similar to the TKR system, ankle prostheses have an articulation interface and bone—implant interfaces. The articulation interface is between the talar and insert components, whereas the tibial and talar components have interfaces with the tibia and talus, respectively. The tibia and talar components are manufactured from metallic materials (e.g., cobalt-chromium or titanium alloy), whereas the insert component is made of UHMWPE polymer. The range of motion at the TAR articulation surface is more constrained in dorsiflexion-plantarflexion compared with flexion-extension motion in the TKR, whereas protonation and supination are greatly limited. The influence of body weight load on the TAR is significant. TAR has developed over multiple generations to treat osteoarthritis at the ankle joint. The design of all prostheses of the TAR have significantly improved from the initial design to the newest ones. Stress shielding at the bone—implant interface of the stem portion of the tibia component and subsidence of the talar component are the main failures occur in the replacement of ankle prostheses. Bone stock in both distal tibia and talus is small; thus, engagement of the stem portions of both components with the metaphyseal bone is small. This limitation accelerations subsidence of the prosthesis through the metaphyseal bone. Therefore, the newest technique of talar component replacements attempts to prevent straight cutting of the talus and to ream the superior section, dedicating the curved profile of talus surface at the tibiotalar joint. Through this technique, greater bone stock is preserved for more effective engagement of the talar component through the talus. Likewise, the talar component is designed with a maximized contact surface with the cortical shell of the talus after reaming to enhance the load bearing strength of the talus against compressive loading conditions generated at the ankle joint. Furthermore, the convex profile of the talar component enables enhanced dorsiflexion-plantarflexion of the insert component through a greater contact surface of the talar component, and thus better distribution and transference of the load to a larger area of bone—implant interface. Concentrated stress or load at the central zone of the talar component is reduced and the risk for prosthesis subsidence is reduced. In the interface of the tibial component with bone tissue, both stress shielding and overloading occur. The tibial component acts like a knee tibial tray prosthesis as the tray component to stabilize the polymeric insert component. In the old design of ankle tibial tray, stress shielding was observed in the stem portion, whereas overloading was detected underneath the interface of the tray portion with the distal tibia. The stem design was modified to be an oblique peg. The oblique design of the peg transfers compressive loading to the periimplant bone. In the traditional design, the straight or vertical design of the peg or stem is not loaded owing to the perpendicular direction of compressive force with the normal axis of the contact surface of the vertical peg and bone. (The concept of stress shielding in such a condition is illustrated in Chapter 1 in more detail.) Overloading under the interface of the tray is observed in obese or overweight patients compared with what is observed in the interface of the tray portion of a knee tibial tray prosthesis with bone. This failure is inevitable, and a reduction in body weight before surgery is a solution to reduce the risk for overloading at this interface. Manufacturing the ankle tibial tray from a nonmetallic material with high compressive strength (e.g., carbon fiber—reinforced polyetheretherketone (PEEK)) is another solution to reduce the reaction force underneath the interfaces of the tray portion. However, carbon fiber reinforced polyetheretherketone (CFRPEEK) prosthesis fracture occurs in obese or overweight patients, necessitating a greater thickness of the CFRPEEK prosthesis compared with metallic prostheses. Similar to knee replacement prostheses, ankle prostheses are processed on their surfaces with various types of osteoconductive layer to facilitate biological integration of bone and prosthesis. tibial tray and talar components are plasma-sprayed with titanium powder to provide a mild porous surface with enhanced bone on-growth and ingrowth. Other types of surface engineering processes, such as bead sintering and 3D porous structuring of the surface, have not been extensively used to develop ankle prostheses, but incorporating such processes will be necessary because of complications reported for the new generation of the TAR. The risk for implant loosening and subsidence of tibial and talar components exists with the use of metallic components in contact with bone. These failures are accelerated owing to the presence of excessive compressive loading at the small volume of bone stock at the ankle joint. As highlighted, CFRPEEK material is a candidate for substitution for titanium or cobalt-chromium alloys; however, its compressive strength is not high enough to withstand body weight loading conditions at the ankle joint. The introduction of new materials with rigorous biomechanical testing and analysis is crucial for future generation of TAR tibial tray and talar prostheses. The material needs to have a mechanical modulus close to that of cortical bone with an optimal design to achieve adequate stiffness with sufficient compressive strength to ensure the mechanical safety of the prosthesis under body weight loading conditions.

7.3.3 Effectiveness of shoulder joint replacement prostheses in interaction with bone

The failures and interactions of bone with total shoulder replacement (TSR) prostheses were reviewed and discussed in Chapter 3. The TSR system is composed of glenoid and humeral prostheses. The glenoid prosthesis consists of a glenoid tray and glenoid insert, whereas the humeral prostheses includes cup-shaped and stem components assembled with other connective components. The function of the shoulder joint depends on the rotator cuff and deltoid muscles as well as ligaments at the glenohumeral joint. Loading generated at the shoulder muscles provides stability to the upper limbs at a high range of physiologic motion at the shoulder joint. The weight of the upper limbs (hand, forearm, and arm) and centrifugal force as a result of upper limb motion are completely borne by muscles at the glenohumeral joint. Motion of the upper limbs is in a high range in the abduction-adduction and flexion-extension axes. (As illustrated in Chapter 3, the attachment of rotator cuff muscles to the humeral head is interesting in light of dynamic science; it provides extensive motion at the shoulder joint.) The function of TSR prostheses is influenced by that of muscles in contraction and stretching. Likewise, the articulation of humeral and glenoid prostheses is crucial to preserve physiologic warping of the muscles, particularly the deltoid muscle, for the desirable function of the shoulder joint in daily or specific activities. TSR was developed to be anatomic and reverse. The concave and convex profile of humeral and glenoid prostheses at the articulation surface is converse in these two types. In the anatomic design, the articulation profile has a concave profile of the glenoid cavity and a convex profile of the humeral head. The anatomic TSR (aTSR) is intended for osteoarthritis in the shoulder joint with the function of all muscles at the shoulder joint and no osteolysis at the humeral head and glenoid cavity. On the other hand, reverse TSR (rTSR) is used for revision surgeries treated by aTSR or fracture fixation implants as the primary bone injury implantation method. In addition, rTSR is beneficial for patients with a deficiency in muscle function (e.g., deltoid muscle) and bone loss at the humeral head. Therefore, rTSR is considered to be salvage for severe bone and soft tissue injuries at the shoulder joint with osteoarthritis injury. The glenoid component is developed as one component made of UHMWPE polymer. In this glenoid prosthesis, the stem section occupies a large volume of the glenoid aspect of the scapula, which reduces the overall strength of the trabecular bone at the implanted zone, with prosthesis loosening. Stress shielding underneath the interface of the UHMWPE glenoid with the glenoid cavity is another possible failure, which may increase the risk for a rocking horse effect and loosen the prosthesis. The glenoid prosthesis is developed in two parts: an insert and a tray. The insert portion is made of UHMWPE and the tray portion is made of metallic materials (e.g., titanium alloy) for implantation through the glenoid aspect of the scapula to stabilize the insert at the shoulder joint (this type of glenoid prosthesis is known as a metal-backed prosthesis). The stem portion of the metallic glenoid tray component is designed to be small to occupy a smaller volume of the glenoid zone; however, because of stress shielding or overloading at the interface of the stem and trabecular bone, the risk for the rocking horse effect progresses at the interface of the glenoid tray and glenoid cavity, which increases glenoid prosthesis loosening. The metal-backed glenoid prosthesis also developed from a porous tantalum (Ta) structure to prevent glenoid tray loosening by facilitating of bone ingrowth to the internal zone of the stem portion. However, fracture of the Ta porous structure at the connection with the UHMWPE glenoid insert is a failure that occurs as a result of ineffective interaction of the glenoid prosthesis and bone. Other types of metal-backed glenoid prostheses have been developed, as reviewed and illustrated in Chapter 3. Failure is also observed in humeral prostheses in interaction with the humerus. Stress shielding effect is a major issue at the calcar, the greater tuberosity, and the lateral side of the humeral stem, as illustrated in Chapter 3. Such an issue is inevitable in long-stem prostheses such as the humeral stem and femoral stem. The internal replacement of stem designed prosthesis through the intramedullary canal of the long bones would case cause fitting of the stem with the cortical shell of the diaphyseal bone with large amount of preload conditions. This engagement results in two main failures. One is hypertrophy of the cortical shell, and the other is a low level of load transference at the proximal zone of stem prostheses at interfaces with bone, either the cortical or trabecular sections. High preload engagement of the long stem prosthesis at the distal zone decreases load distribution or sharing at the proximal zone with a much lower preload engagement. Such an issue is reduced by inward tapering of the long stem from the proximal through the distal zones in various planes.

TSR prostheses are effective salvage for treating osteoarthritis in the shoulder joint; however, the complication rate is still considerable compared with knee and ankle replacements. Significant biomechanical involvement of the muscles affects the function and performance of TSR prostheses in long after implantation, and a deficiency in their function accelerates failure at the bone—prosthesis interface, particularly in the glenoid zone. The stability of the articulation surface is a factor that deteriorates in long-term implantation owing to a deficiency in the function of muscles. This causes ineffective load transference and distribution at the articulation interface glenoid prosthesis and various types of failure to occur, as reviewed briefly in this section and comprehensively in Chapter 3. rTSR has lower independency for muscle function; however, the shoulder joint may not have effective physiologic motion, and thus bone osteolysis or resorption may occur and ultimately loosen the prostheses at the implanted zone. Further development of the TSR will be beneficial to enhance osteoarthritis treatment at the shoulder joint. Another issue in replacing the TSR is the limited or small volume of the glenoid zone of the scapula for replacing the glenoid prosthesis, which necessitates developments in design concepts. Because most progressed osteoarthritis injuries at the shoulder joint resulted from the failure of fracture fixation of the humeral head or after soft tissue deficiency in the glenohumeral joint, enhanced shoulder fracture fixation and the timely reconstruction of soft tissue deficiencies could be reliable strategies to prevent osteoarthritis at this joint.

7.3.4 Effectiveness of hip joint replacement prostheses in interaction with bone

Interactions of bone with total hip replacement (THR) prostheses and possible failures were reviewed and discussed in Chapter 4. The THR system is composed of two main parts: acetabular and femoral. The acetabular prosthesis has a acetabular cup and acetabular shell, whereas the femoral prosthesis has a femoral stem and femoral head. The hip joint is a ball-in-socket joint in which the femoral head is in articulation with the acetabulum cavity of the pelvis. The design concept of the THR uses the ball-in-socket profile, from which a high range of rotational motion in the sagittal and frontal planes is facilitated. Compared with other reviewed joints (knee, ankle, and shoulder), a high range of motion and body weight loading conditions are present at the hip joint as prominent mechanical factors. Likewise, the effective function of muscles (particularly the abduction rotator muscles) at this joint influence the efficacy of the THR prostheses in long-term implantation. The function and performance of THR prostheses in the long term are challenging because of the presence of full body weight loading conditions, the high range of physiologic motion, and the influence of the muscles. The anatomy of the proximal femur at the neck zone magnifies the complexity of the loading equilibrium at the hip joint and THR. THR prostheses are developed in different types, as reviewed in Chapter 4. The acetabular prosthesis is a as normal cup, press-fit cup, dual-mobility cup, or cup-cage; the acetabular shell have or lack screw holes. This component of the acetabular prosthesis is made of stainless steel or titanium alloy. A stainless-steel acetabular shell enables a thinner shell, whereas a titanium-made shell is thicker for better stabilization of the screw head at the shell and for sintering and coating with titanium beads or powder. Femoral prostheses have varieties in the femoral stem. This component is designed with a long, short, and ultrashort length for contact with metaphyseal and/or diaphyseal portions of the proximal third of the femur. The femoral stem is designed with one to three tapered surfaces inward from the proximal to distal zones. The proximal zone of the femoral stem might be rarely or highly dedicated to the anatomic profile of the trochanteric zone. Failures in both types of THR occur in the form of loosening, subsidence, dissociation, dislocation, migration, and bone resorption. Articulation conditions at the interface of the femoral head and acetabular liner cavity have a significant influence on the stability and integrity of femoral and acetabular prostheses in the interface the bone. Mild or severe migration and dislocation of the acetabular shell were reported in patients with a high BMI and in patients with ineffective integrity of the shell at the acetabulum cavity. Obese or overweight patients have increased contact force at the articulation surface that may cause the femoral head to jump or the shell component to dislocate. The increased contact force accelerates overloading at the interface of the bone and prosthesis. Thus, bone resorption and gradual loosening of the prosthesis occur, with catastrophic dislocation or migration, as radiographically illustrated in Chapter 4. A dual-mobility (DM) acetabular cup is advantageous in patients with an expected risk for increased contact force at the articulation surface or in those with inadequate function or deficiency of muscles at the hip joint. This type of acetabular cup facilitates the mobility of the acetabular liner inside the shell component, where increased contact force at the articulation of the femoral head and acetabular liner is converted into motion energy. Thus, its effect on the interface of the shell and acetabulum cavity is reduced. A DM cup reduces the stiffness of THR prostheses, and the patient feels smoother physiologic motion. Because of the increased laxity of the THR with a DM cup, ineffective motion of the liner through the shell occurs, in which the liner is unable to return to its initial position with excessive rotation of the liner inside the shell. Severe bone loss or the failure of fracture fixation at the acetabular wall is treated with a cup-cage prosthesis in conjunction with screw purchase through the pelvis using a multihole acetabular shell. The optimal number of screws with an effective length has highlighted for the enhanced stability of such an implantation method. Failures might occur at the interfaces of the femoral stem and bone. As highlighted in Section 7.3.3, the long stem design of such prostheses causes inevitable stress shielding at the proximal zone. As illustrated in Chapter 4, inward tapering of the stem reduces the risk for progressive bone resorption at the proximal zone of the stem with a reduction in cortical hypertrophy at the distal zone. Subsidence of the short stem is serious in some cases. During replacement of the short stem, it is challenging to stabilize the prosthesis at the optimal position with respect to the proximal third of the femur. The design of the short stem at the distal zone is challenging. The fitness of the distal zone is effectively achieved with the femoral canal in some patients, and the prosthesis might be inadequately buried through the femoral canal. To stabilize the short stem anatomically at the desired position with less fitness with the intramedullary femoral canal, it might be buried in the canal under a body weight load, with resulting prosthesis subsidence. Therefore, the biological integrity of the short stem is a key factor for the enhanced stability and integrity of the short stem with the metaphyseal section of the proximal femur in preventing subsidence. Such a scenario is considered for an ultrashort stem with greater osseointegration. The femoral stem is mainly fabricated from titanium alloy; thus, the combined porous structure and bioactively coated osteoconductive surface engineering methods are used to enhance bone ingrowth at the interface of the stem and metaphyseal section. The effectiveness of TSR prostheses in interaction with bone tissue is influenced by many factors that are inevitable. However, it is important to select the appropriate type of THR in accordance with patients' conditions at the hip joint and the type and severity of injury. New design concepts are needed for THR to enhance arthroplasty at the hip joint.

7.4 Effectiveness of trauma implants in interaction with bone

Interactions of the bone with trauma implants and possible failures were reviewed and discussed in Chapter 5. The fixation of bone fractures is a bone injury treatment method extensively used around the world. The bone might be fractured as a result of severe loading conditions in the form of an impact caused by an accident or fall. In young patients with good bone quality, the bone is fractured in an accident or a fall from a height, whereas in older patients with poor bone quality, fractures occur owing to a fall from stairs or while walking. Bone fracture fixation is managed by trauma implants, which involve trauma plating systems, intramedullary nailing systems, external fixator systems, and cannulated screw systems. This type of orthopedic implant is manufactured from stainless steel and titanium alloy materials; CFRPEEK was introduced as an alternative material for low-level load-bearing implants (e.g., distal radius plating system). Most bone fractures are fixed and healed by plating systems composed of plates and screws to reduce bone fragments in anatomic position. The advantage of plating systems is the possible placement of the plate in many aspects or the position of bone tissue from small to long bones, from metaphyseal to diaphyseal sections, and from foot fingers to skull bone. The plate is fixed by various types of screws to bone, which are known as locking screws, compression screws, lag screws, and cancellous screws. Fracture fixation by plating systems captures bone fragments by screws and establishes a rigid bone—implant construct to facilitate healing of the bone fracture at fracture gaps. The design of a plate dedicate the anatomic shape of the bone at the areas with no attachment of muscles, tendons, or ligaments. The plate has various types of screw holes in accordance with the intended screw type. This, along the whole area of the plate, provides the feasibility of screw insertion in whole sections of the plate. The plate is designed in the two forms of straight or anatomic shaping. The straight design is fixed to the midshaft of small or long bones to fix fractures at the diaphyseal zone with a thick shell of cortical layer. Using an optimal number of locking screws with an effective working length is beneficial to avoid unnecessary rigidity or flexibility of the bone—implant construct at this zone. Likewise, it is advantageous to use interfragmentary compression screws to maintain fragments at the fracture site during healing in close contact with each other. The anatomic plate is designed in two parts: straight and curved/angled. The straight portion is positioned at the diaphyseal zone of the proximal or distal third, whereas the curved/angled portion is placed on the metaphyseal zone with a thin layer of cortical bone and a large volume of trabecular bone tissue near the articular surface of the joint. It is necessary to use locking, compression, or lag screws in this zone in accordance with fragmentary conditions of the bone fracture. Bone fragment buttressing is a factor of the plate that enhances the stability of the bone—implant construct, in which the plate prevents the sliding or transverse dislocation of bone fragments while extending screw holes to larger area for more effective capture of bone fragments. Another trauma implant is the intramedullary nailing system, which is composed of a nail and screws. The nail is designed in the shape of a long tube with screw holes at the two ends; it is highly adequate for fixing bone fractures at the diaphyseal portion, particularly in long bones (e.g., femur, tibia, humerus, and ulna). The nail provides greater stability of the bone—implant construct at the midshaft of the long bones compared with plating fixation, which might fail, as illustrated in Chapter 5. Bone fracture at the midshaft is fixed from the lateral aspect using a plating system, which has sufficient stiffness against compressive and bending stresses in the frontal plane and insufficient stiffness against bending stress in the sagittal plane. Bone fractures in small bones in the foot and hand are fixed with the cannulated screw system, in which small bone fragments are reduced by inserting a guided pin and inserting screws with internal holes as cannulated screws. Based on the size of the bone and the diameter of bone fragments, one or multiple cannulated screws are used to manage fractures of the small bones (for instance, multiple cannulated screws might be needed for fracture fixation at the calcaneus). Another type of trauma implant is the external fixator system, which is used for an open fracture with severe soft tissue injury, individually or combined with internal fixation systems (e.g., plating, nailing, cannulated). Simple fractures are managed with an external fixator for a minimally invasive operation.

The bone–implant construct may fail as the result of various ineffective interactions between bone and trauma implants. Failure occurs at the interface of the screw and bone in the form of screw loosening, bone suppression, screw cutout, and screw fracture. Screw loosening is detected as a radiolucent line or zone around the screw shaft that is fully or partially threaded. This type of failure progresses as a result of stress shielding or overloading at the bone–screw interface. If the load direction is parallel to the longitudinal axis of the screw, compressive force is not transferred to the screw. In such a condition, stress shielding occurs at the interface and peri-implant bone resorption results. Normally, the bone–screw interface is loaded by compressive, shear, or torsion stress. In some cases, an extra number of screws might be purchased to bone with no load-bearing role, which may cause stress shielding. In excessive loading, during compressive, shear, or torsion stress, the reaction load from the metallic screw increases the risk for overloading and osteoclast cell activity and the formation of fibrous tissue at the bone–screw interface. When the screw is purchased through bone, an amount of preload force is expected at the interface of the bone and screw that is the initial mechanical stimulation signal for the growth of bone remodeling to resorb the damaged bone tissue during purchase and to form new bone tissue. An effective osteoconductive layer on the screw surface with a high level of adhesion strength and mechanical properties (e.g., plasma electrolyte oxidation titanium-hydroxyapatite layer) is beneficial for enhancing bone formation at the screw–bone interface. Cyclic shear stress is a main cause of screw loosening in long-term implantation. In the plating fixation of bone fractures, transferred compressive force from the screw to the plate generates a bending moment through the plate. If the plate is highly rigid, the increased amount of bending moment is concentrated at the maximum stress zone of the plate and fractures the plate in long-term implantation. On the other hand, if the plate is too flexible, large shear stress is transferred through the screw—bone interface and tends to remove the screw from bone. In long-term implantation, such overloaded shear stress causes bone resorption at the interface and accelerates the removal or loosening of the screw from bone, and the bone—implant construct fails. When the screw is not removed or dissociated from bone (e.g., when the screw has been purchased obliquely for increased strength against compression stress and enhanced loosening resistance as a result of shear stress), it causes the internal cutting of bone tissue and a cutout effect, as reviewed and illustrated in Chapter 5 for an intertrochanteric screw in the fixation of a fracture at the proximal femur. Another failure mechanism that occurs as a result of excessive compressive force at the interface of the bone and screw suppresses the bone to the screw. In such a case, progressive bone resorption occurs as a result of excessive reaction force from the metallic screw to the bone, and it collapses the bone fragment. This condition is catastrophic in the form of severe bone loss, such as that which occurs in the fixation of fractures at the humeral head, necessitating revision surgery using reverse total shoulder arthroplasty, as illustrated in Chapter 5. The screw might be fractured at the neck section or the shaft portion. The risk for fracture is greater in the neck section, where there is a reduced diameter section from the screw head to the screw shaft. The screw shaft is purchased through bone and the screw head is positioned at the screw hole of the plate; thus, the neck portion is not constrained and transferred compressive and bending stresses between bone and the implant concentrate at this section and cause a fracture in long-term cyclic loading. As discussed in Chapter 5, excessive relative motion between bone fragments as the result of ineffective fixation causes malunion, delayed union, or nonunion of bone fractures. Such failure might be observed in the fixation of bone fractures in the pelvis, distal humerus, or proximal humerus (details of these types of failures and possible reasons are provided in Chapter 5). The attachment of a high number of muscles to bone tissue is challenging for fracture fixation using plating systems. Fracture in such bones (e.g., pelvic, clavicle) affect mechanical loading equilibrium, and plating fixation may not hold the created fragments in anatomic position during healing, when the screw or plate might fracture or loosen, as illustrated in Chapter 5.

Trauma fracture fixation using trauma implants is satisfactory as effective salvage for treating bone fractures. However, the risk for failure between the bone and implant exists, necessitating adequate implant selection, appropriate fragment reduction in anatomic position, and the proper use of screw numbers to enhance the capture of bone fragments. Fracture fixation in osteoporotic bone and multifragmentary fracture patterns might be fail under physiologic loading, necessitating careful postoperative plans. An ideal concept for the interaction of the bone and implant, as presented in Chapter 1, needs to be developed for trauma implants. If the plate could be remodeled at localized stress zones to decrease stiffness and release bioactive particles or molecules through the fractured site, fracture healing could occur in a shorter time with a reduction in delayed union or nonunion complications after a couple of months.

7.5 Effectiveness of spinal implants in interaction with bone

Interactions of the bone with spinal implants and possible failures were discussed in Chapter 6. Spinal injuries compress the spinal cord and irritate the neural system, causing a high level of pain in the spinal column and other upper and lower extremities. The treatment strategy of spinal injuries is to remove the reason for spinal cord compression, which might be degeneration of the intervertebral disc, narrowing of the spinal cord canal, vertebral fracture, or vertebral dislocation. In some cases, nonsurgical methods might be useful at an early stage of spinal cord compression. However, under severe conditions, surgical intervention is inevitable to decompress the spinal cord and central neural system. Spinal implants need to be internally implanted into the spinal column for severe disc degeneration, spinal deformity, and vertebral fracture or dislocation. This type of orthopedic implant includes the categories of the screw—rod stabilization system, intervertebral fusion caging system, corpectomy caging system, and total artificial disc system. The screw—rod stabilization system has components such as the pedicle screw, rod, crosslink, lateral offset connector, and domino. The system provides rigid stability to the vertebrae at the affected zone of the spinal column from the cervical to the lumbar subcolumns. The size of components is different to stabilize vertebrae in the cervical, thoracic, and lumbar subcolumns. Screw—rod stabilization is mainly developed from titanium alloy and partially from cobalt-chromium alloy. Intervertebral fusion cages are implanted at the interspace of vertebrae to maintain disc height while facilitating of bone fusion between two vertebrae. This type of implant is manufactured from titanium alloy and PEEK. The anatomic design of the intervertebral fusion cage is important for load distribution and transference between the vertebrae and cage. Another type of spinal implant is corpectomy cages, which replace severe bone loss at the vertebral body.

This kind of orthopedic implant uses expandable and static cages manufactured from titanium alloy. Plating fixation of the vertebrae requires a specifically designed plate and screws to rigidly stabilize the cervical vertebrae from the anterior aspect and lumbar vertebrae from the anterior and lateral aspects. To maintain the functionality of the affected vertebrae close to that of physiologic conditions, total disc replacement or artificial discs replace the intervertebral space to facilitate rotation of the implanted site in physiologic motion of the spinal column in flexion-extension, axial rotation, and lateral bending. Spinal implants are increasing in design models with various of features and benefits. However, their interactions with vertebral bones might not be effective and various failures occur. The similar types of failures of screws in the fixation of bone fractures occur in the fixation of pedicle screws in vertebrae. As illustrated in Chapter 6, bone resorption at the bone—screw interface loosens the pedicle screw from its initial position. Subsidence of intervertebral and corpectomy cages through the inferior and superior adjacent vertebrae is another issue that is detected for spinal implants. A concentration of compressive stress at the cage wall accelerates overloading at the bone—cage interface. Thus, bone resorption and remodeling of the bone matrix occur at the end plate section of the vertebrae. Migration of intervertebral and corpectomy cages at the intervertebral space is another failure that results from ineffective sizing of the cage during stabilization of the vertebrae. Plating fixation can fail under physiologic loading transferred to the operated site. Plating fixation is normally used for anterior stabilization of affected vertebrae in the cervical subcolumn. Owing to the tendency of cervical vertebrae to rotate, an excessive bending moment might be transferred to the plating structure. As a result, the plating system might fail in the form of plastic deformation, screw fracture, or screw loosening. The physiologic range of motion in the cervical subcolumn may cause bone resorption at the anterior aspect and excessive compression of the posterior aspect in total disc replacement. An artificial disc is not a reliable and effective solution for treating spinal injuries with severe compression of the spinal cord. Other types of interactions and failures were described in Chapter 6 for these types of spinal implants.

Spinal implants are considered the final salvage for treating severe spinal injuries with severe and chronic pain in the spinal column and other affected limbs or organs. The treatment of spinal injuries using spinal implants is normally postponed by surgeons because of the low level of efficacy of solutions in long-term implantation. Although spinal implants relieve pain in the short term (a few years), the rigid stabilization and vertebrae fusion accelerates injures in adjacent intervertebral discs. Thus, the patient needs revision surgery to extend the levels of stabilization and/or fusion levels, or even to replace the implant through more surgical intervention and difficulties. Spinal injuries are increasing owing to the modern lifestyle. Thus, there is a good market for manufacturers in this field of orthopedic implants. Although no-rigid implantation methods were introduced to the market, their efficacy in long-term implantation is controversial and there are few intended uses for such solutions.

7.6 Future perspectives on effectiveness of orthopedic implants in interaction with bone

This book comprehensively discusses the interactions and failures of using orthopedic implants in contact with bone. Summaries of the main concepts for each category of implant are provided in this chapter. Orthopedic implants are used as the final salvage to treat severe bone injuries to relieve pain as the first goal, and then to enhance functional outcomes at the affected site. Various preoperative considerations are necessary for the effective selection of patients and implants to ensure the success rate of the implantation in the early, short, and long periods after the operation. In other words, current orthopedic implants might not be a reliable solution; however, the market demand for using implants is large and there is not a better solution for most cases. Most failures are inevitable owing to the interaction of inert components of the implant with a highly active structure such as bone tissue. Because bone can remodel itself biologically against mechanical factors, maintaining the bone structure and matrix from early to long after implantation under constant conditions is not achieved. Conversely, implants remain constant with no remodeling ability in their physical and chemical characteristics. This inhomogeneous and incompatible function and performance between the bone and implant challenges the effective integrity of the bone—implant construct as one integrated system with long-term efficacy against physiologic loading conditions or unloading conditions. Although satisfactory clinical outcomes are reported as good or excellent in many clinical studies in the use of various types of orthopedic implants, however, the design and development of an implant able to remodel would be greatly beneficial. Moreover, investigations and development in the science and engineering aspects of the biomaterial, biomechanics, and bioprocesses are crucial in this regard. Another topic for the further development of orthopedic implants is the revolutionary design modeling or geometry of the implants. There are several limitations and constraints for implanting the artificial component through bone tissue, but the design geometry and type of implantation, insertion, fixation, or replacement of the implant needs to coincide with the physiologic loading pattern and mechanisms to prevent of stress shielding and overloading zones from occurring in interaction with bone tissue. The third development concept for consideration in future is the biological release of the bioactive materials through the intended zones or areas such as the fracture gap and areas with a risk for stress shielding. Various osteoconductive surface engineering methods have been developed to enhance the bone–implant interface for better biological integration; however, the enhancement of biological activity could be further investigated and developed for the area far from the bone–implant interface. This book attempted to review and discuss interactions and failures of the traditional, conventional, and newest generations of orthopedic implants in the three main categories of total joint replacement, trauma implants, and spinal implants. Many enhancements and improvements were observed in this attempt. This shows the progress is continuing in other related science and engineering fields in the industry of orthopedic implants.

Glossary

Acetabular cup prosthesis The prosthesis for total hip replacement, which is inserted through the acetabulum aspect of the pelvis.

Acetabular liner The inner portion of the acetabular cup prosthesis, which is inserted through the acetabular shell.

Acetabular shell The outer portion of the acetabular cup prosthesis, which is inserted through the acetabulum aspect of the pelvis.

Acetabulum The semispherical cavity of the pelvis at the hip joint.

Anatomic design Design modeling of the implant in accordance with the geometry of the intended bone under healthy conditions.

Annulus The external or shell part of the intervertebral disc, which is structured from fibrous tissue with flexible tension in various axes of translation and rotation.

AP shear force Shear force generated in the anteroposterior axis at the interface of the bone and implant.

Articulation interface The contact surface at the musculoskeletal joints, where articulated bones are moving with respect to each other.

aTSR Anatomic total shoulder replacement.

Axial compressive force Shear force generated in the mediolateral axis at the interface of the bone and implant.

Axial stress Stress subjected or introduced to the craniocaudal axis.

Bead sintering A porous surface engineering process in which beads are sintered on the surface of the implant to form a fine layer to enhance the osteoconductivity of the implant.

Bioactive surface The deposition of bioactive materials (e.g., calcium phosphate, collagen, growth factors, proteins, chitosan) on the surface of orthopedic implants is known as bioactive surface engineering.

Biomimetic coating Processes A surface engineering electrochemical or chemical process in an adequate solution involving the respective ions of molecules meant to coat or form on the surface of the implant or substrate.

BMD Bone mineral density characteristic used to examine bone strengthening or weakening at the intended zone.

BMI Body mass index characteristic used to examine weight over height in meters squared. The healthy range for BMI is $18.5-25\ \mathrm{kg/m^2}$

Bone cysts The volume at which bone tissue has been weakened.

Bone suppression Excessive compression of bone tissue in interaction with hard metallic materials, implants, or prostheses.

Buttressing effect Mechanical support of bone fragments by a trauma plating system. This prevents dislocation of bone fragments buttressed by the plate.

Cannulated screw A type of screw with internal holes used to fix bone fractures in trauma injuries.

Cementless prosthesis A type of prosthesis that is replaced without bone cement. The surface of such prostheses is porous or coated with bioactive materials for bone–implant integration.

Conformity An increased contact surface at articulated prostheses in joint replacements enhances the conformity of the prostheses in which contact stress and wear are eliminated (or significantly reduced) while promoting physiologic motions at the articulated surface.

Contact constraints Constraints or limited conditions that are available at musculoskeletal joints through the concave–convex profile of bone ends and involved muscles, tendons, and ligaments, establishing the stability of the motion of bone at the joint.

Contact wear Generated wear at the articulation surface of joint prostheses as the result of increased contact stress and low material wear resistance.

Corpectomy cages A category of spinal implants designed in the shape of a cage with an internal empty space to be filled by the bone graft. These cages are used as a vertebral body replacement to treat degenerative or fractured vertebral bodies of the spinal column in one or multiple levels.

Cortical hypertrophy Increased bone formation at the cortical zone caused by excessive loading conditions.

3D printing A manufacturing process in which the implant is processed layer by layer in accordance with a predesigned model to form an implant with a three-dimensional shape.

Design conception The determination of design principles in accordance with anatomy, mechanical loading and boundary conditions, the type of bone, the patient's BMI, the patient's overall conditions, the patient's BMD, the pathologic disorder, and other physical, mechanical, and biological characteristics of the part of the body skeletal system involved with the operation with orthopedic implants. This is the main idea or strategy for designing and developing orthopedic implants for the intended performance and functionality of the implant to treat the bone injury.

Design feature Technical characteristics of the implant created or developed in the design stage based on the determined design concepts.

Diaphysis portion The midshaft of the long bone with a thick cortical layer and intramedullary hole.

Dual mobility cup An acetabular cup in which two articulation interfaces are considered to reduce increased contact stress conditions.

Energy shielding The extent of energy absorbance or fracture toughness of the implant. Greater energy shielding of the implant, particularly in metallic implants, causes the lower transference or reaction of energy or load through its interface with periprosthetic bone.

Femoral stem prosthesis A total hip replacement prostheses inserted through the femur through the proximal zone.

Fracture fixation Restoration of bone fragments in anatomic position under fixed conditions to allow the bone fracture to heal.

Gamma irradiation Irradiation used to sterilize orthopedic implants and biomaterials. This process is employed to cross-link free radicals in polymeric materials (e.g., ultra-high molecular weight polyethylene (UHMWPE)) for enhanced surface molecular integrity and higher wear resistance.

Glenoid The concave cavity of the scapula at the shoulder joint.

Horse-Rock effect This effect would be occurred when the articulation contact would tend to be eccentric. This term is normally used as the failure in glenohumeral or shoulder joint.

Injection overmolding An engineering process used to inject a polymer over the harder part with different melting temperatures in a mold.

Interconnective pores When the pores of a porous structure connect internally, in which bone growth initiates from the surface of the structure through the internal core.

Internal–external torsional torque The torque moment generated in internal–external rotation motion that influences the interface of the periprosthetic bone and implant in some cases.

Intervertebral fusion cage A category of spinal implants designed in the shape of a cage with an internal empty space to be filled with the bone graft. These cages are used to fuse two vertebral bodies while maintaining the disc height during fusion.

Load distribution The spread of load over the area or volume, which needs to be extended to a greater area of the bone and implant interface.

Load transferring Applied load from the bone to the implant and vice versa, which maintains close to physiologic interaction between them under healthy conditions.

Long-term study Evaluation of clinical outcomes after surgery after more than 5 years and for up to 15 years for joint replacement prosthesis.

Loosening Deterioration of integrity between the implant and bone, in which periprosthetic bone is weakened or has degenerated.

Metal-backed glenoid prosthesis A glenoid prosthesis for total shoulder replacement, which is developed in two polymer and metal parts. The metal part is inserted into the glenoid aspect of the scapula.

Metaphyseal zone The geometric transverse extended zone at two ends of the long bones in the lower and upper limbs, which is near the joints of the skeletal system.

Microelastic strain Strain required to induce osteogenic cells to form into mature bone matrix.

Micromotion Motion at the micrometer scale that is created at the interface of the bone and implant. An appropriate range of micromotion at the interface enhances bone formation and better integrity of the bone and implant.

Midterm study Evaluation of clinical outcomes after surgery after more than 2–5 years for the joint replacement prosthesis.

Migration Translation or rotation of the implant or prosthesis from its initial position during surgery.

Mediolateral shear force Shear force generated in the mediolateral axis at the interface of the bone and implant.

Nucleus The center of the intervertebral disc, which is structured as compressible tissue to absorb relative compressive energy between vertebrae.

Osteoconductivity Conduction of bone formation through the deep layer of the implant.

Osteoconductive surface An implant surface with enhanced ability to conduct bone formation through the deep layer of the implant. The surface might be porous structured, coated with bioactive materials, or both.

Osteoarthritis Degeneration of cartilage tissue at the joints.

Periprosthetic bone The first layer of the bone, which is in contact with the bone and has a key role in bone–implant integration. Periprosthetic bone can grow through the porous structured surface of the implant.

Physiologic functionality A term used to highlight the benefits of physiologic functioning of the injured site after surgical treatment. Bone at the injured site might not be treated with full function as well as physiologic conditions before the injury.

Physiologic loading conditions Loading conditions close to those of healthy conditions. Such conditions are used to evaluate the biomechanical behavior of the implant after implantation to bone.

Plasma electrolyte oxidation A surface engineering process for some metallic materials (e.g., titanium, tantalum, magnesium) in which the implant or substrate surface has pores created through microplasma conditions caused by a dielectric discharge.

Plasma spraying A surface engineering process in which the coating material is sprayed onto the surface of the implant or substrate by creating plasma in a spraying gun.

Porous surface The structure of an implant surface with deep pores.

Posttraumatic arthritis Degeneration of cartilage at the joint after bone fracture fixation near the joint, normally in intraarticular fractures.

Prosthesis An long-term or permanent implant that replaces severely degenerated bone or cartilage tissue.

Radiolucent area The area at bone tissue that has deteriorated in bone mineral density and cellular matrix intensity, which is lucent upon radiograph imaging.

Radiolucent line The narrow area near the implant, which is lucent to radiographic waves. This is apparent in the form of a line showing bone resorption at the interface of the periprosthetic bone and implant.

Range of motion The limit of motion at the musculoskeletal joints, which is a clinical criterion to measure clinical outcomes after the treatment of bone injuries.

Revision surgery Reoperation of the patient after primary surgery owing to the failure of bone–prosthesis integrity.

Rocking horse effect This effect occurs when articulation contact tends to be eccentric. This term normally indicates failure at the glenohumeral or shoulder joint.

rTSR Reverse total shoulder replacement

Scapular notch Severe bone resorption at the neck of the scapula and inferior aspect of the glenoid aspect of the scapula.

Screw–rod stabilization system A category of spinal implant consisting of various components such as a rod, pedicle screw, nut, cross-link, hook, lateral offset, or domino, used to stabilize vertebral bodies in anatomic condition.

Selective laser melting Rapid prototyping used for 3D printing or to form a metallic powder (mostly titanium powder) into a solid or porous structured 3D implant.

Spinal implants A category of orthopedic implant used to treat disc and vertebral injuries at various parts of the spinal column.

Stress shielding The bearing of a higher amount of mechanical stress by a harder material in a composite composed of a harder or softer material. In a bone–implant construct, bone is softer and the metallic implant is the harder material. This affects the mechanobiological reaction of the bone matrix, which decomposes and weakens.

Subsidence The gradual caving of the implant through bone tissue as a result of severe bone resorption at the bone–implant interface. This incidence is normally observed in the subsidence of a metallic implant through trabecular bone.

TAA Total ankle arthroplasty

TAR Total ankle replacement

THR Total hip replacement

TKA Total knee arthroplasty

TKR Total knee replacement

Total disc replacements A category of spinal implant used as a disc replacement that provides relative motion between two vertebral bodies in nonfusion treatment of a degenerative intervertebral disc.

Trabecular bone Spongy bone with a high level of porosity that structures the metaphyseal portion of long bones near the joints. Trabecular bone structures the whole internal volume of some bones including pelvic, vertebral, clavicle, scapula, and mandibular.

Transition layer A layer at the interface of the bone and implant that is normally porous to allow deep bone growth to create a composite layer composed of bone matrix and host material in a porous form. Such a composite layer has a mechanical transition effect and energy absorbance to prevent overloading the periprosthetic bone.

Trauma implants A category of orthopedic implants used to treat bone fractures.

Trauma plating systems A category of trauma implants in the form of a plate and screws for fixing bone fragments in anatomic position during fracture healing.

TSA Total shoulder arthroplasty

Tuberosity The prominence zone of bone in which muscles or ligaments are attached to bone.

Vapor deposition A surface engineering process in which the coating layer is formed by atom bombarding the coating material to the implant or substrate.

Index

Note: 'Page numbers followed by "f" indicate figures and "t" indicate tables.'

G

H

I

J

K

Kolibri 360 GmbH
Ferdinand-Jühlke-Straße 7
99095 Erfurt, Deutschland
produktsicherheit@kolibri360.de